Environmental History

Volume 2

Series Editor

Mauro Agnoletti

For further volumes:
http://www.springer.com/series/10168

Ian D. Rotherham
Editor

Cultural Severance and the Environment

The Ending of Traditional and Customary Practice on Commons and Landscapes Managed in Common

IUFRO
Landscape Conservation Forum
Sheffield Hallam University
SHARPENS YOUR THINKING

ENGLISH HERITAGE
the FOREST HISTORY SOCIETY
Springer

Editor
Ian D. Rotherham
Faculty of Development and Society
Sheffield Hallam University
Sheffield
UK

ISSN 2211-9019 ISSN 2211-9027 (electronic)
ISBN 978-94-007-6158-2 ISBN 978-94-007-6159-9 (eBook)
DOI 10.1007/978-94-007-6159-9
Springer Dordrecht Heidelberg New York London

Library of Congress Control Number: 2013934538

Printed on acid-free paper

Springer is part of Springer Science+Business Media (www.springer.com)

Preface

This major book explores the issue of commons, lands and rights of usage in common, traditional and customary practices, and the cultural nature of 'landscapes'. Importantly, the volume addresses the now critical matters of 'cultural severance', the presently, largely unrecognized impacts on biodiversity and human societies, and the implications for conservation and sustainability for local economies. The book takes major case studies and perspectives from around the world, and from a historical and ecological perspective addresses contemporary issues and challenges.

A standpoint of man in the contributions is that it is important or even vital to understand the past, our history, if we are to address effectively future environmental challenges. Often, this is not the case, since the environment and nature are treated as 'natural' rather than eco-cultural. Issues of common ownership and rights to natural resources present major challenges in the contemporary global world and the market forces of capital-driven economics. Yet, the long-term consequences, of the separation or severance of people from nature, are tangible and potentially disastrous at many levels. However, most contemporary actions toward conservation and sustainability fail to address this fundamental relationship between communities and local environments. This reflects perhaps, the ethos of Hardin's 1960s 'Tragedy of the commons' and from this perspective the chapters in this volume challenge such precepts and assumptions and through this, raise new and critical paradigms.

The book has developed from major international conferences and collaborations over a period of around 15 years, which culminated in a conference on '*The End of Tradition*?' held in Sheffield, UK, in 2010. However, the chapters are from individuals who are both academic researchers and practitioners. As explained in Mauro Agnoletti's chapter, the ideas and studies are now influencing bodies such as the EU, UNESCO, and FAO, with at last, some recognition by major organizations and stakeholders, of the critical state of the environment consequent on cultural severance.

In recent years, researchers have turned their attention to issues of landscape change and the eco-cultural nature of the environment. Combined with the impacts and effects of cultural severance, the break between local people and their environmental resources, the cultural nature of landscape is now better understood. However, the implicit importance and significance for conservation of biodiversity, of heritage, and consequently for activities such as tourism, are only just receiving wider recognition. The implications of widespread landscape abandonment, rural depopulation, urbanization, and severance, are dramatic and sometimes stark, with wildfires raging, ecology often in free-fall, and local communities and their traditions displaced.

A first step with all these landscapes is to recognize both the important sites and the critical issues. Then, appropriate protection and conservation must be determined and applied. Finally, there is the potential to develop new and extended commons as part of a landscape approach to future conservation. However, the cultural past, together now with issues of cultural severance, present enormous challenges for the integration of this knowledge into visions of future sustainable landscapes. Not least of these challenges is the loss of indigenous cultural and traditional knowledge, without which, much future conservation action is jeopardized. This book is intended to raise awareness, to stimulate further discussion, debate, and research, and to then turn dialogue into action.

Ian D. Rotherham

Contents

Part I Setting the Scene on Cultural Severance and its Implications

1 **Cultural Landscapes and Problems Associated with the Loss of Tradition and Custom: An Introduction and Overview** 3
Ian D. Rotherham

2 **Cultural Severance and the End of Tradition** 11
Ian D. Rotherham

3 **Globalism and the Enclosure of the Landscape Commons** 31
Kenneth R. Olwig

4 **A Natural Origin of the Commons: Interactions of People, Animals and Invisible Biodiversity** 47
Ted Green

Part II Case Studies of Cultural Landscapes from Around the World

5 **Valorising the European Rural Landscape: The Case of the Italian National Register of Historical Rural Landscapes** 59
Mauro Agnoletti

6 **Severance of a Traditional Grazing Landscape in the Himalayas: Commons and Ecosystems in Crisis?** 87
Minoti Chakravarty-Kaul

7 **Early Wood Commons and Beyond** 107
Della Hooke

8 **What, How, and Why? Collecting Traditional Knowledge on Forest Uses in Switzerland** 123
Matthias Bürgi and Martin Stuber

9 **The History of Utilization and Management of Commons and Consequences of Current Social Change in the Alpine Region of Austria** 133
Elisabeth Johann

10 **Guided Pollards and the Basque Woodland During the Early Modern Age** 147
Alvaro Aragón Ruano

11 **The Evolution of Forest Landscapes in Spain's Central Mountain Range: Different Forests for Different Traditional Uses** 161
N. López Estébanez, G. Gomez Mediavilla, G. Madrazo García de Lomana, F. Allende Álvarez and E. Sáez Pombo

12 **Of Commoners and Kings** 177
Graham Bathe

13 **The Cultural Landscape of Royal Hunting Gardens from the Fifteenth to the Eighteenth Century in Białowieża Primeval Forest** 191
Tomasz Samojlik, Ian D. Rotherham and Bogumiła Jędrzejewska

14 **The End of Common Uses and Traditional Management in a Central European Wood** 205
Péter Szabó

Part III The History and Use of Landscape Commons

15 ***'A very fair field indeed…'*: An Archaeology of the Common Lands of English Towns** 217
Mark Bowden and Nicky Smith

16 From Pasture Woodland, via Deer Park and Common, to Cultural Severance: A Case Study of the Commons of Ashampstead, Berkshire 229
Dick Greenaway

17 Changing Cornish Commons 249
Peter Herring

18 The Commons of the Ancient Parish of Sheffield 263
David Hey

19 Traditional Uses, Destruction, Survival and Restoration of Common Land: A South Yorkshire Perspective 275
Melvyn Jones

20 Abandoned Landscapes of Former German Settlement in the Czech Republic and in Slovenia 289
Petr Mares, Robin Rasin and Primoz Pipan

21 Land Management and Biodiversity Through Time in Upper Ribblesdale, North Yorkshire, UK: Understanding the Impact of Traditional Management 311
Helen Shaw and Ian Whyte

22 Policing the Commons in the Vale of York, *c*.1550–*c*.1850 323
Brodie Waddell

23 The Parliamentary Enclosure of Upland Commons in North–West England: Economic, Social and Cultural Impacts 337
Ian Whyte

Part IV Issues and Approaches for Future Commons and Cultural Landscapes

24 Biodiversity Conservation and the Traditional Management of Common Land: The Case of the New Forest 353
Adrian C. Newton

25 Looking Back to the Future: Ancient, Working Pollards and Europe's Silvo-Pastoral Systems 371
Jill Butler

26 Promoting Stewardship of *New Commons*: Lessons from WakeNature Partnership 377
Gary B. Blank and George R. Hess

27 End of Tradition, Reworking of Custom: Re-assembling *Satoyama* Woodlands on Tokyo's Urban Fringe 387
Jay Bolthouse

28 New Commons for Old: Inspiring New Cultural Traditions 401
Duncan Mackay

29 Community Grassland Conservation on a Former Common in the Wye Valley, England 413
George Peterken

30 Upland Wood Pastures 419
Peter Quelch

Part V Conclusions and Overview: The Implications of Severance for Future Landscapes

31 Concluding Thoughts on the Implications of Cultural Severance on Landscapes, Ecology and People 433
Ian D. Rotherham

Index 443

Part I
Setting the Scene on Cultural Severance and its Implications

Chapter 1
Cultural Landscapes and Problems Associated with the Loss of Tradition and Custom: An Introduction and Overview

Ian D. Rotherham

1.1 Introduction

The end of tradition and of customary practices is a massive threat to heritage, history and biodiversity (Rotherham 2009a, b, c, 2010). Indeed, it can be argued that the end of traditional land management and the impacts then of cultural severance are as big a threat as climate change to biodiversity and ecology. This book considers the threats to biodiversity from cultural change and the abandonment of traditional management. In recent decades, we have heard much about climate change and the threats that this may pose in the future but in terms of biodiversity '*The End of Tradition*' is potentially bigger and more current (Fig. 1.1).

The threats from global cultural change and abandonment of traditional, landscape management increased in the last half of the twentieth century and ten years into the twenty-first century show no signs of slowing down (Agnoletti 2006, 2007). Their impacts on global biodiversity and on people disconnected from their traditional landscapes pose real and serious economic and social problems, and these must be addressed. The contributions to the book consider fundamental issues of whether we can conserve the biodiversity of wonderful and iconic landscapes and reconnect people to their natural environment. Moreover, if we can, how might we do so and make them relevant for the twenty-first century. We cover the lessons of archaeology, history and ecology and look at the challenges for modern-day management with examples drawn from rural and urban commons, wooded landscapes, heaths, moors, coasts and wetlands. In particular, we raise critically important issues of the loss and abandonment of tradition and

I. D. Rotherham (✉)
Faculty of Development and Society, Sheffield Hallam University,
City Campus Pond Street, Sheffield S1 1WB, UK
e-mail: i.d.rotherham@shu.ac.uk

I. D. Rotherham (ed.), *Cultural Severance and the Environment*,
Environmental History 2, DOI: 10.1007/978-94-007-6159-9_1,

customary practice in terms of future sustainability, landscape quality and biodiversity. The book is international in flavour with leading researchers from around the world contributing key chapters. This important text seeks to address issues of landscape and nature conservation informed by an understanding of historical processes. This is not always easy, since as Lowenthal (1985) pointed out, the past is largely a 'foreign country'. Indeed, this historical context provides the impetus for the book since our cultural memory of past practices is short-lived (Rotherham 2007a) and even the community-based knowledge of traditions is being lost. The scene is set in Part 1 with contributions from Olwig, Rotherham and Green. In Part 2 follows a series of major case studies from around the world. Part 3 presents major contributions on the history and uses of commons and common resources. Then, looking forwards, Part 4 addresses issues and approaches for future commons and cultural landscapes. Finally, Part 5 provides a summary and overview.

The book stems from invited contributions by participants in the major conference held in Sheffield in 2010—'*The End of Tradition? Aspects of Commons and Cultural Severance in the Landscape*' held at Sheffield Hallam University from 15th to the 17th September 2010. The event included nearly fifty lectures plus displays, poster presentations, and extended discussions throughout 3 days. There was a strong community dimension with members of local groups, students, and volunteers, and the organisers involved both academics and practitioners from around the world. There were opportunities to share and compare local, national and international experiences of the important challenges facing biodiversity in the twenty-first century. This was a landmark discussion and debate with key organisations. Participating and supporting institutions included Natural England, English Heritage, the National Trust, the Biodiversity and Landscape History Research Institute, the Woodland Trust, The Wildlife Trusts, the RSPB, BANC, OPAL, the International Association for the Study of Commons, the Ancient Tree Forum, the European Society for Environmental History, and the International Union of Forest Research Organisations, and many others.

Fig. 1.1 Abandoned coppice

1.2 Conference Topics Included:

- Conservation at the crossroads
- The impacts of changes from subsistence, often rural, communities and landscapes to technology driven agri-industry and urbanisation, and the consequences for local people
- Commons in the urban landscape and community involvement
- The historical and current uses and management of traditional 'commons'
- The 'common' uses of landscapes and environmental resources now and historically, from medieval coppice woods to deer parks, from alpine pastures to grazing meadows, from coastal flats to peat bogs and fens
- The debates around perceived 're-wilding' of natural areas or 'abandonment' and
- 'Dereliction' of cultural landscapes
- The decline of biodiversity and ecology
- Future visions and actions

1.3 Conservation at the Crossroads: Cultural Severance and the End of Tradition

We suggest that the global ecosystem is now at a tipping point for historic landscapes and that changes are occurring which are very damaging for the sustainability of the ecological resource (Agnoletti 2006, 2007; Rotherham 2005, 2006). However, the issues are not merely to do with biodiversity but the ecological declines are paralleled, (and in many ways tied to), widespread degradation of entire landscapes. These changes are closely related to declining rural economies, widespread depopulation of the countryside, and importantly in terms of future economic performance, a latent erosion of tourism potential (Doncaster et al. 2006; Rotherham 2008a, b; Rotherham and Harrison 2009). In short, this is the quiet catastrophe in the countryside, with the cutting of humanity's umbilical cord with nature and a seismic shift from rural to urban living. In many ways, this also suggests an uncertain and unsustainable future with a rift in society, economy and ecology and significant breaks in rural economic functions (Fig. 1.2).

This is a growing crisis of global proportions and it feels as if nobody has noticed. The lack of recognition—by decision-makers, by agencies, by politicians, by the media, and even by many researchers, means there is a lack of a viable future vision.

1.4 Cultural Severance and Climate Change

The impacts of cultural severance (Rotherham 2007a, b, 2009a, b, c), of human-induced climate change and natural climate change, each individually and in combination, present great challenges to environmental sustainability (Rotherham

Fig. 1.2 Ancient open-grown oak now shrouded

2010). Cultural severance and associated land-use changes have impacted on and influenced climate—through destruction of vegetation, of soils, and particularly the loss of fens, bogs, heaths, moors, and other lands with extensive organic soils. These losses have released massive quantities of carbon dioxide into the atmosphere. Moves to petrol-chemically driven technologies and urban living have released huge amounts of carbon dioxide and massive quantities of waste energy as heat into the environment. With cultural severance from traditional landscapes, and fuelled by petrochemicals, agri-industry and industrial forestry have combined with urbanisation to destroy and fragment habitats and to bring about gross transformation of entire landscapes (Rotherham 2010; Rotherham 2012a, b).

These transformed landscapes and their fragmented habitats have only limited ability to respond to climate change—to moderate impacts and to mollify adverse trends—so species cannot move or adapt and biodiversity is threatened. The landscapes no longer respond to climatic pressures or for example, to extreme weather events and both floods and droughts have become commonplace. Basic ecosystem services and functions are under stress and increasingly under threat.

1.5 As yet Unrecognised and Unspoken

There needs to be recognition of the issues, their causes, the historic context, and the scale of the consequent challenges. The historic problem and its causes are indeed rooted in the past but at the same time, over time change is inevitable. The severity of the problem faced today relates to the scale of change, the time-periods of the changes, and the global nature of the effects. The associated declines in ecology and biodiversity are massive and so no sign of abating. It is worth considering particular national case study examples.

So in Britain alone, during the twentieth century we have seen:

- Catastrophic loss of lowland heaths and commons
- Almost the entire destruction of lowland wet fens, raised bogs, marshes and wet woods
- The collapse of populations of most butterflies, many farmland birds, bats, reptiles and amphibians and more
- The extinctions of huge numbers of flowering plants and ferns in many regions
- The removal of most medieval parks and their veteran trees
- The drainage of most upland moors and bogs
- The loss of most ancient unimproved pastures and meadows
- The cessation of traditional coppice management of woods and the loss of about half the ancient woods in the last 50 years
- And creeping urbanisation or gentrification of much of the countryside—the '*greying of the green*'
- Massive spread of invasive species and especially of invasive exotic or alien species
- A comprehensive failure to address the wider issues of decline beyond the cosmetic or the desperate—and no wider evidence of any recovery by key indicator species or groups of species
- The severance of people's contact with nature to a point where many can no longer recognise or identify even commonplace species

We can see the impacts and changes through history as communities interact with nature to modify, manage, and sometimes destroy the resource. Across the globe, human impacts are deeply etched into the landscape; in many cases, over centuries, there have evolved sophisticated systems to manage the resources sustainably. However, in recent times, but varying across the world, the management have become more intensive and less sustainable. The effects are seen from the developed, industrial Western countries and now increasingly in the emerging economies too. For many ecosystems and their associated habitats and species, such as woodlands and forests, peatlands, bogs, fens and heaths, and grasslands of various sorts, impacts can be tracked through time (e.g. Rotherham 1999, 2009a, b, c, 2010, 2011; Webb 1986, 1998) (Fig. 1.3).

The same processes are now happening across the globe and at an accelerating rate. From tropical rainforests to Mediterranean grassland and maquis, traditional

Fig. 1.3 Overgrown common land SSSI and abandoned wood

uses and customary management practices are abandoned, communities leave the rural areas, and landscapes become derelict. Consequences include rapid build-up of biomass and vulnerability to wildfires (Pyne 2001), loss of unique biodiversity, and depression of the rural economy. Furthermore, these landscapes and their ecologies are contested spaces. Insidious processes of globalisation overturn and replace local, long-term traditions and customary practices, rapidly displacing indigenous communities that have long-term associations with distinctive places. Some of these complex issues are beyond the immediate scope of this book, but for the first time, we have attempted to provide a broad overview of the issues and to illustrate and illuminate this with pertinent examples and case studies from around the world.

1.6 Conclusions

In terms of policies and decision-making, a number of key steps are essential. We need to: (1) Recognise the '*Eco-Cultural*' nature of landscapes and their biodiversity and (2) Re-establish links with nature. Then in order to prevent or at least limit extinctions we must: (3) Mimic traditional management methods and their impacts in conservation sites such as nature reserves; and (4) Establish social and economic links between nature, landscape and ecology. However, this will require a political shift in thinking and planning and a paradigm shift in conservation and environmentalism.

Future visions for a sustainable environment must recognise the lessons of human history and the impacts of culture and of history. Most importantly, for future hopes of improved sustainability we need to maximise ecosystem function benefits to humanity not diminish them and as climate change happens, natural and human-induced, help nature to respond and to minimise the damage.

References

Agnoletti M (ed) (2006) The conservation of cultural landscapes. CAB International, Wallingford

Agnoletti M (ed) (2007) Guidelines for the implementation of social and cultural values in sustainable forest management. A scientific contribution to the implementation of MCPFE—Vienna resolution 3. IUFRO Occasional Paper No. 19 ISSN 1024-414X

Doncaster S, Egan D, Rotherham ID Harrison K (2006) The tourism economic argument for wetlands: a case study approach. In: Proceedings of the IALE conference, water and the landscape: the landscape ecology of freshwater ecosystems, 296–300

Lowenthal D (1985) The past is a foreign country. Cambridge University Press, Cambridge

Pyne SJ (2001) Fire—a brief history. The British Museum Press, London

Rotherham ID (1999) Peat cutters and their landscapes: fundamental change in a fragile environment. In: Peatland ecology and archaeology: management of a cultural landscape. Landscape Archaeol Ecol 4:28–51

Rotherham ID (2005) Fuel and landscape—exploitation, environment, crisis and continuum. Landscape Archaeol Ecol 5:65–81

Rotherham ID (2006) Historic landscape restoration: case studies of site recovery in post-industrial South Yorkshire, England. Agnoletti M (ed) The conservation of cultural landscapes. CABI International, Wallingford, 211–224

Rotherham ID (2007a) The implications of perceptions and cultural knowledge loss for the management of wooded landscapes: a UK case-study. For Ecol Manage 249:100–115

Rotherham ID (2007b) Cultural severance—causes and consequences. In: Cultural landscapes—changing landscapes. Delegate handbook, the international association for vegetation science: 50th annual symposium, University of Swansea, Swansea, pp 43–44, 22–27 July 2007

Rotherham ID, Harrison K (2009) South Yorkshire fens past, present and future: ecology and economics as drivers for re-wilding and restoration? In: Hall M (ed) Greening history: the presence of the past in environmental restoration. Routledge Publishing, London, pp 143–153

Rotherham ID (2008a) Tourism and recreation as economic drivers in future uplands. Aspects Appl Biol 85:93–98 (Shaping a vision for the uplands)

Rotherham ID (2008b) Lessons from the past—a case study of how upland land-use has influenced the environmental resource. Aspects Appl Biol 85:85–91 (Shaping a vision for the uplands)

Rotherham ID (2009a) The importance of cultural severance in landscape ecology research. In: Dupont A Jacobs H (eds) (2009) Landscape ecology research trends. Nova Science Publishers Inc., USA, Chapter 4 pp 1–18

Rotherham ID (2009b) Habitat fragmentation and isolation in relict urban heaths—the ecological consequences and future potential. In: Rotherham ID, Bradley J (eds) Lowland heaths: ecology, history, restoration and management. Wildtrack Publishing, Sheffield, pp 106–115

Rotherham ID (2009c) Cultural severance in landscapes and the causes and consequences for lowland heaths. In: Rotherham ID, Bradley J (eds) Lowland heaths: ecology, history, restoration and management. Wildtrack Publishing, Sheffield, pp 130–143

Rotherham ID (2010) Cultural severance and the end of tradition. Landscape Archaeol Ecol 8:178–199

Rotherham ID (2011) The Implications of landscape history and cultural severance in environmental restoration in England. In: Egan D, Hjerpe E, Abrams J (eds) Integrating nature and culture: the human dimensions of ecological restoration. Island Press, Washington, pp 277–287

Rotherham ID (ed) (2012a) Trees, man, and grazing animals—a European perspective on trees and grazed treescapes. Earthscan, London

Rotherham ID (2012b) Re-interpreting wooded landscapes, shadow woods and the impacts of grazing. Earthscan, London

Webb NR (1986) Heathlands. Collins, London

Webb NR (1998) The traditional management of European heathlands. J Appl Ecol 35:987–990

Chapter 2
Cultural Severance and the End of Tradition

Ian D. Rotherham

Imagine a countryside where '*Reedbeds are dry and clogged with brambles, heathlands have vanished as scrub begins to take over. Wetlands have dwindled and rivers and canals have become clogged by invasive plants which threaten native species. The loss of money for wildlife-friendly farming has seen farmland birds resume their slide into extinction. Bat populations are clinging on to survival in isolated pockets, facing starvation due to dwindling insect populations, while the country's flower meadows have all but vanished. England's uplands have become degraded; their wildlife is in decline and their ability to lock away carbon and to provide clean drinking water for millions sadly reduced.*' This is the ***Wildlife and Countryside Link*** vision of a future 'austerity countryside' as expressed in August 2010 (*The Guardian*, 14 August 2010). The problem is that it is already happening (Fig. 2.1).

2.1 Introduction: Cutting the Umbilical Cord

In May 2010, *The Guardian* newspaper carried a headline stating that the '*Case for saving species 'more powerful than climate change*'' (Jowit 2010). This was in response to the United Nations report on biodiversity. Nevertheless, the underlying reasons for biodiversity decline are still largely overlooked. In this essay, I attempt to highlight some key issues, the lessons of history and the challenges that arise. Nature provides the umbilical cord linking humanity to Mother Earth, but in recent decades not only has the physical cord been severed, but the emotional and

I. D. Rotherham (✉)
Faculty of Development and Society, Sheffield Hallam University,
Pond Street, Sheffield S1 1WB, UK
e-mail: i.d.rotherham@shu.ac.uk

I. D. Rotherham (ed.), *Cultural Severance and the Environment*,
Environmental History 2, DOI: 10.1007/978-94-007-6159-9_2,

Fig. 2.1 Bracken infestation

psychological one too. In western countries such as Great Britain, a new generation is growing up who for the most part cannot even recognise and identify the most commonplace animals and plants. At this point, the umbilical cord of nature linking communities to their local environment is truly cut and there are major implications of separation of people and nature. This process has been growing in intensity for centuries but in the late twentieth century, it grew around the globe at an unprecedented rate. The consequences for ecology, for biodiversity, and perhaps for people must be a on a scale which approaches that of human-induced climate change. Surprisingly this is a topic almost overlooked by both popular and scientific media in recent years.

Today's landscapes and their associated biodiversity are changing on a scale that is almost beyond human comprehension (see Rotherham 2010 for example). Currently most popular and scientific literature attributes the bulk of the change and attaches the greatest elements of environmental risk to climate change. Indeed much of this seems to isolate climate change as the sole driver and even suggests that human-induced climate change is acting almost alone—a naïve and dangerous assumption. Furthermore, the human-induced elements of climate change are reduced further to causation by gases that have the ability to generate an enhanced greenhouse effect in the earth's atmosphere. Finally, the gases responsible are attributed to modern urban and industrial activity and agri-industry in its various guises. Reading both popular and scientific media today it could easily seem that this encapsulates the entirety of the global environmental crisis.

There is no disputing the intensity of the above or the scale of the associated impacts. However, this is not the whole story in terms of environmental change and associated risks. It is certainly not the only or even major threat to biodiversity and to the vast landscapes around the world that are not natural, but semi-natural and cultural in their origins, where the dramatic scale of abandonment of traditional and sustainable utilisation is the major threat. The process of abandonment of these eco-cultural landscapes is what I call '*cultural severance*' (Rotherham

2007a, b, c, d, 2009a, b, c) and represents the widespread ending of traditional land management. This occurs at a number of levels but includes both the physical ending of subsistence usage and dependence, and the psychological distancing of individuals and communities from their environments.

Human resource use in the natural landscape is a fundamental driver (Agnoletti 2006, 2007). Indeed, the interactions between people, civilisations and nature have been widely discussed (e.g. Fowler 2002; Hayman 2003; Perlin 1989). It interacts with the ecology and other environmental factors through complex social, legal, economic, and political mechanisms, facilitating and constraining usage. Almost all the landscapes observed across Europe and many other parts of the world are '***Cultural Landscapes***' (Agnoletti 2006, 2007; Rotherham 2009a, b, c, 2010), or perhaps '***Eco-Cultural***', often managed in traditional ways for millennia. In this context, observations have been made on such traditional uses and their effects. My research (e.g. Rotherham 1999a, b, 2005, 2007a, b, c, d, 2009a, b, c) to date has documented in detail the impacts of people over time for selected environments such as:

1. Wooded and forested landscapes
2. Marsh, meadow and fen
3. Heath, bog and common
4. Cultivated landscapes—field systems etc.

Yet far too often when conservationists and ecologists discuss landscapes and nature they seem to overlook this most basic and deeply embedded aspect of the '*natural*' environment; that it is not '*natural*' but '*semi-natural*', an '*eco-cultural*' resource. Within these landscapes, there are often strong links back to a more primeval ecology (see Rotherham 2012), but also forwards through long time-periods of continuity and predictability. Importantly the human and natural processes centred on micro-disturbance rather than macro-disturbance, the effective recycling of nutrients within ecosystems, and the extraction from systems of biomass. The ecological habitats produced tend to be predictable, have a strong degree of continuity through time, and are mesotrophic or oligotrophic rather than eutrophic. Human cultural utilisation has often relied on careful management of recycled nutrients sustainably within the system. When this failed, as it sometimes did, the results were catastrophic (Rotherham 2005). Overall, human usage reduced biomass, and macronutrients, especially nitrogen, were at a premium. Severance quickly reverses these trends to generate macro-disturbance, major disruption and unpredictability, lack of continuity, and massive eutrophication especially by nitrogen. These changes favour competitive and cosmopolitan species, including some ruderals, over stress tolerators and ruderals associated with traditional regular management (Fig. 2.2).

A further and final impact of cultural severance on these traditionally managed or utilised landscapes is that they lose economic and hence local utilitarian value and are destroyed by conversion to other uses. These are often disputed territories with competing social and political actors. The ecological consequences of this severance and the ending of social and economic utility is often the complete

Fig. 2.2 Overgrown common wet heath SSSI

collapse of the system and/or its transformation to other, often agri-industrial or urban-industrial uses. Oliver Rackham made the point very strongly in his seminal book *The History of the Countryside* (1986) when he argued that an ancient wood generally survived in England as long as it had local economic value. The same applies not merely to the physicality of the woodland, but also to its traditional management. Lose either or both and the site will be lost and its land-use converted to some other function. This same line of logic is pertinent to all other traditionally managed landscapes too but I develop this argument further to include not just economic value but the social and cultural values too. So a woodland or heath may survive today and be valued highly and so protected, but for its leisure and amenity or nature conservation functions. Whilst this modern cultural attachment to a site and its functions may protect it from destruction and even provide a modicum of management, it is still separated from its traditional origins and vulnerable to a slow decline through ecological succession. Importantly too in these times of increasing austerity, rather than generating an economic or resource output, the sites then require input of public funding to maintain them.

2.2 Common Lands and Lands in Common

Traditional and especially subsistence management of the landscape and its resources were often organised as 'common' either formally or informally. Common usage was then often embedded over time within a formal structure and administration of a 'common' such as a fen, heath, bog or wood. The story of the formalisation of commons and their ultimate widespread demise in Britain is well documented. Similar patterns occur around the world but with local and regional variants on the resource, the timescale, and the speed of landscape transformation when common use declines or ends. Some of the issues of perceived mismanagement of commons were highlighted in the landmark paper in 1968 by Garrett Hardin, '*The Tragedy of the Commons*' (Hardin 1968) and that seemed to help justify the ending of common rights and the imposition of other 'modern' systems of management and exploitation. He stated that human society had an inherently destructive relationship with nature and naturally over-exploited common resources. This is too complex an issue to discuss in detail here, but it seems to me that the essay overstates the negative effects of traditional common utilisation and to a large degree. His arguments also resonate back through history to so-called 'improvers' and others who sought justification for taking the common from the commoner.

Hardin's essay can be criticised on the grounds of historical inaccuracy and often failing to distinguish between 'common property' and 'open access resources'. Later ideas published by Elinor Ostrom (Ostrom 1990) and others suggested that to use Hardin's work in arguing for the privatization of resources was to overstate of the case. His use of over-grazing for example has been strongly criticized, with the suggestion that these were indeed carefully regulated commons that avoided misuse. Nevertheless, these later authors accept the existence of genuine problems in managing common resources and that the concept of the tragedy of the commons does apply to many real situations. However, this is an idea that should not be taken too literally. The concept relates to a complex structural relationship and the consequences of the interrelationships that ensure. The phrase itself summarises a concept but does not provide a precise description. Indeed the 'tragedy' as such should not be viewed necessarily as 'tragic' in a conventional sense. Furthermore, it should not be presumed that the argument is of necessity a condemnation of the processes ascribed to it. Today researchers in agrarian studies are turning to an understanding of traditional commons management to help provide insights into how common resources might be protected and conserved but without alienation of the local people whose subsistence they provide.

It also seems that Hardin's use of the term 'commons' has been widely misunderstood or misquoted which has led Hardin himself to state that he should have titled his work '*The Tragedy of the Unregulated Commons*'. However, despite the limitations and issues which arise the idea of the tragedy of the commons is very relevant in analysis of behaviour in economics and game theory, taxation and politics, and evolutionary psychology and sociology and can relate to the outcomes of individual interactions within complex systems. The relationships between this concept and the assumed imperative to privatize commonly owned resources are especially contentious. In particular, this relates to resources traditionally managed communally by local communities and subsequently enclosed or privatized, a key thrust of this essay. The excuse presented politically and economically is usually that this act 'protects' the resources for the present and the future. However, the argument frequently and sometimes very carefully overlooks former management and resources are appropriated and indigenous, (often poor), communities are alienated. The consequence is often that private or state-managed use results in a long-term deterioration of the resource and furthermore, in what we now call 'ecosystem services'.

A key issue in Hardin's debate is the degree to which individuals always behave selfishly or self-interested individuals find ways to cooperate. The argument is that collective restraint may benefit collective and individual interests. Hardin's essay was criticized by Appell (Appell 1993) as follows: '*Hardin's claim has been embraced as a sacred text by scholars and professionals in the practice of designing futures for others and imposing their own economic and environmental rationality on other social systems of which they have incomplete understanding and knowledge.*' This approach raises important issues for conservation and ecology. Indeed, his advocacy of clearly defined property rights has often been used to justify privatization or private property per se. From these discussions and debates, the opposite state to the '*tragedy of the commons*' has been described, the '*tragedy of the anticommons*'. This is where rational individuals acting alone collectively waste or destroy a particular resource by underutilization and again is an emerging theme of the current essay.

Hardin drew on examples of latter day '*commons*', such as the oceans and rivers and their fish stocks, the atmosphere, national parks and others. The example of fish stocks led others to describe this as the '*tragedy of the fishers*'. One of his key themes is the growth of human populations in relation to the resources of the planet Earth as a general global common. He advocated potential management solutions to commons problems and these included privatization, polluter pays regulations, and management regulation. Relating to his pasture and grazing analogies Hardin described the 'enclosure' of commons and the historical progression from the use of all resources as commons with unregulated access to all, through to systems where commons are 'enclosed' and subject to various methods of 'regulated use'. Here the access to the resource is prohibited, restricted or controlled. He argued that reliance on individual conscience to police commons utilisation would not work because selfish individuals, termed 'free riders', are favoured over other who are altruistic. In relation to the need to avoid over-exploitation of common resources Hardin quoted Hegel's statement as used by Engels, which was that '*......freedom is the recognition of necessity*'. Hardin indicated that 'freedom' completed the tragedy of the commons and so in recognising resources as 'commons' and understanding that they therefore needed management, people could '*..........preserve and nurture other and more precious freedoms*'.

Attempts to articulate solutions to the perceived or real tragedy of the commons are a key thrust of contemporary political philosophy. Without enlightened self-interest, there is a need for authority of some sort to deal with collective actions. Government regulation can restrict or allocate the amount of a common good made available for use by an individual. Permit systems are applied to extractive economic activities such as livestock raising, timber extraction, mining, fishing, hunting and the like. Another approach to protecting the commons is to place limits on pollution discharges and today of carbon emissions through government or international regulation. Common resources can also be the subject of cooperative regulation by the resource users themselves for mutual benefit. Finally, some resources can be converted from 'common goods' into 'private property' and it is then assumed that any new owner will have an incentive to manage it sustainably. (In fact, it is palpably clear that this is often not the case. A private owner is often inclined to capitalize the stock—in other words to cash in the value and then cut and run). In practice, this is often manifested in the imperative to increase productivity and profitability at the expense of long-term sustainability and the delivery of wider ecosystem services and benefits. The private owner has little incentive to maintain or increase any wider communal benefit and ironically in cases such as the Britain, whilst the early stages of 'improvement' were generally privately funded, the later stages during the twentieth century, were publicly subsidised by cash or by tax breaks.

2.3 Human Utilisation and the Cultural Landscape

We now recognise that such traditional cultural utilisation, whilst not always sustainable, has generated and driven many of the landscapes we now value so highly. It is important at this stage to acknowledge and recognise that much communal use of ecological resources has not been benign. Indeed, many resources were over used to the point of extinction, and in others, the uses over time transformed landscapes dramatically. However, such overuse was generally a result of sheer human population pressure, and in the case of transformation, the resulting cultural landscapes were often created over centuries, and were sophisticated systems for producing sustainable outputs with minimal external subsidies. The cultural uses created the conservation landscapes we value today and many are or were species-rich and diverse systems (Rotherham 2009a, b, c, 2010). Furthermore and presently neglected in many debates on conservation and the environment, a massive proportion of our most highly valued ecology and biodiversity depend for their existence on these traditional or customary uses. The reasons and mechanisms for this are too complex to examine here in detail, but they range from direct environmental impacts (like lowering of nutrient levels and micro-disturbance), to indirect effects through social and economic impacts (allowing people to remain and live on the land or in a particular region). A consequence of these long-term, intimate relationships between people and nature has been the creation of complex landscapes and often rich and distinctive ecologies. Many of these have evolved through generally stable and predictable patterns of human utilisation of the natural world, and the consequent response from biodiversity to adapt and evolve. A direct result of these processes is the heritage of biodiversity and landscapes that we inherit today. This includes within it the locally and regionally distinctive landscapes and their ecological character that is most highly valued for nature conservation and increasingly now for tourism and leisure (Doncaster et al. 2006; Rotherham 2008a, b).

2.4 Landscapes and Ecology—Character and Distinction

It is very clear that the wealth of distinctive landscapes that exists around the world and the associated characteristic ecology are a consequence of the backdrop of natural conditions and the history of human traditional exploitation (Agnoletti 2006, 2007). Very few locations and ecosystems can be accurately described as purely 'natural'. Yet increasingly, discussions on restoration focus on the 'natural' and not on the 'cultural' origins on desirable conservation areas. This is leading to serious problems in terms of the likely success or otherwise, and the sustainability, of even the highest profile and most well prepared and presented projects. Unfortunately, the perceived origins are generally misunderstood and therefore the hoped-for results of restoration are unlikely to be achieved (e.g. Rotherham 2011). Importantly, most restoration or conservation projects are separated from economic or local community functions, and in most cases, there seems little chance of such fundamental long-term drivers being effectively established. Of great concern, is that in most cases, it seems these issues and the absence of the key mechanisms is unrecognised by project- or site-managers. If they are, then this omission is not publicly acknowledged (Rotherham 2006).

Many so-called re-wilding or re-naturing programmes have great merits and should form an important part of any large landscape-scale restoration programme. However, they mostly lack any recognition of the cultural origins of most conservation landscapes or the dependence of many of the biodiversity targets on human utilisation (Rotherham 2011). Even the selection of suitable target areas frequently lacks any real understanding of landscape history and human cultural origins. Many even lack realistic assessments of ecosystem carrying capacity and even suitability for the species targeted for reintroduction. This severely prejudices hopes for long-term success in achieving many of the environmental, ecological and even economic outcomes and targets. Misunderstanding and misinformation follow into the more popular media such as when Vidal (2005) in *The Guardian* newspaper ran a story on re-wilding entitled '*Wild herds may stampede across Britain under plan for huge reserves*' which was potentially exciting but in its content, sadly misleading. Suggesting that the high Pennine moors might support free-roaming herds of Hecke cattle or reindeer is ludicrous. Furthermore, that this is a 'wild and natural' landscape, when it is a cultural landscape of 5,000 years standing, is entirely misleading. In regional economics terms, the idea that current upland farmers might be replaced in the economy by 'ecotourism' shows zero understanding of basic tourism economics (Rotherham 2008a, b; Anderson 2004).

2.5 So What is 'Cultural Severance'?

I define '*cultural severance*' as the breakdown of the fundamental, often subsistence relations between human communities and their local environment as manifested in the landscape and its ecology as an eco-cultural resource. This is essentially in terms of people and communities as active exploiters of the landscape and its ecology, and so is different from passive 'users' of a landscape space (such as modern tourism, leisure or recreational actors). As noted above, this may be a social and community phenomenon as well as operating at the stratum of the individual and their perceptions. It has elements that are inherently practical in nature but is also strongly psychological at every level. This separation of people and nature has occurred at evolved at various times, in different places, and at rates that vary from dramatically quick and sudden, to relatively slow and drawn out. A key process is the break in local community 'ownership' and use of the natural resource, and the imposition of essentially individual, capital-based, value and exploitation. The separation may be inherently a locally based social one, or can involve ownership and exploitation removed to a remote stakeholder without subsistence ties to the particular locale. Whilst reality may be far more complex, this is the essential process. In recent times, a major transformation has been from an essentially working countryside, to one that is largely a leisure or tourism resource. The participants are still actors and competing for the space, but their interaction with the resource is largely passive (Table 2.1).

Over long periods, the human cultural interaction with the landscape evolves and changes and various phases may be displaced as communities develop and as nature evolves. In recent times, perhaps over the last 200 years in the developed industrial countries, this relationship moves to and beyond the point of severance from nature (Table 2.1). However, it is clear that in recent history there have been a limited number of key tipping points in this process, and these relate to four major trends which are themselves closely interrelated. These key processes are:

1. Agricultural improvement
2. Industrialisation
3. Urbanisation
4. Globalisation

Each of these represents a complex of human interactions with natural resources, of economic and cultural evolution, and of social change with competition for the control of space and other resources. The net result though is a loss of traditional management, often of established patterns of ownership, frequently dramatic decline in locally or regionally distinct landscapes, and associated with

Table 2.1 British eco-cultural landscapes now largely severed from their subsistence past include

1.	Fens and marshes
2.	Bogs
3.	Coppice woods
4.	Ancient forest, parks and chases
5.	Moors, heaths and commons
6.	Ancient meadows and grazing lands such as sheep-walk downland
7.	Ancient sand dune systems
8.	Ancient arable lands

these a loss or transference of economic process, and a catastrophic reduction in ecological conservation value (Table 2.2). Distinctive wildlife and plant species associated with particular niches maintained in these landscapes are quickly lost. Invasive and often aggressive species sometimes including exotics may then displace the native or at least long-established ecology.

2.6 The Mechanism of Impacts

One consequence of severance is a move to different land management perhaps for features managed consistently for centuries as part of social and economic systems. This might be for example, woodland or heath having lost social and economic functions converted to farmland, the original habitat destroyed. All associated species are consequently lost. Alternatively, traditional management may cease or change radically, but the site remains physically intact. For example, mediaeval coppice wood abandoned and then replanted with native or exotic hardwoods as high forest, or cleared and replanted with exotic conifers as a working plantation (Rotherham and Jones 2000).

A heath, once central to the local economy, if its function and value to local people is lost, may be grubbed up and 'improved', and so destroyed (Webb 1986, 1998; Rotherham 2009b, c). Sometimes the site might be physically intact but abandoned in terms of management and use. This severance triggers a successional change to birch wood and a gradual loss of open heathland species (Rotherham 2009b, c). A further option is that like the coppice wood, it is planted with exotic conifers. Alternatively, heath or similar commonland maintained as open grazing loses other socio-economic functions; its ecology is changed and major successional shift occurs. The traditional and customary uses that maintained the ecosystem included harvesting gorse, bracken, ling and small wood for fuel, cutting wood for construction, bracken for bedding, or turf for fuel or roofing, holly, bramble, and gorse as fodder, and grass meadows cut for hay. The site becomes nutrient enriched and the low, open vegetation is replaced by taller more rank species. Rich mosaics of ecological habitats are converted to a few distinct areas but with limited diversity and stress-tolerant and habitat specialist species are generally lost (Table 2.3).

In both these examples, the ecological consequences vary from total loss and replacement by a very limited ecology of intensive agriculture or agri-forestry, to a deflection at varying speeds into a new ecological successional process. This may reflect natural drivers of the ecological systems released by the ending of traditional management, but generally fuelled too by eutrophication (nutrient-enrichment) influences. The ecological changes are modified if new or different management is overlaid to displace that traditionally applied. In such cases, there is an interaction between the abandoned drivers and those of the newly imposed systems. A general observation is that the new approaches are usually 'owned' and managed at a distance from the resource, and they replace labour-intensive

Table 2.2 The impacts of severance on these British eco-cultural landscapes

1. **Fens:** spread mostly over period from 1600 AD until mid-1900s with almost total loss of both function and ultimately of around 99 % of the resource; examples include the Yorkshire or Northern Fens with 3,000–4,000 km^2 lost between 1650 and 1900, and the East Anglian or Southern Fens with 4,000 km^2+ lost between 1650 and 1950
2. **Bogs:** gradually abandoned, drained, or worked-out as fuel turbaries over several centuries followed by rapid drainage and removal in both upland and lowland environments from 1800 through to the late 1900s. Almost all lowland raised mires destroyed and those that remain are hugely modified
3. **Coppice woods:** from enclosed mediaeval woods, to wooded commons and early industrial coppice woods, these were a resource of huge important to communities across the country. Many were converted to high forest in the 1800s as wood charcoal for industry became less important; and almost all the others were abandoned, converted, or grubbed out for agriculture from 1850 to 1950. From 1950 to 1990 many more were lost and a sizeable proportion converted to conifer plantations. Traditional skills of *woodmanship* were almost entirely lost
4. **Ancient forest, parks and chases:** these complex economically functional landscapes give some insights into how large areas of primeval Europe may have looked and the ecologies that it may have generated. The landscapes of these areas were essentially multifunctional to provide many resources alongside hunting and meat. A very few locations have remained largely intact, though most were lost from 1600 AD onwards. This was through conversion to farmland or to ornamental landscaped parks. Sites that remain physically intact have lost most of their complexity and are separated from their social and economic functions
5. **Moors, heaths and commons:** until around 1700 AD, through until the late 1800s these were the distinctive open landscapes of all parts of the country. Many sites probably included extensive wooded commons and were managed as such. At the end of the Parliamentary enclosures they were reduced in area dramatically with a few lowland groupings of intractable heathland such as the Lizard in Cornwall or the New Forest and Dorset heaths, and extensive upland moors but now separated spatially and economically from the landscapes down in the valleys and lowlands. Even where sites remain however, their traditional functions were either abandoned or changed to specific economic uses especially sheep grazing and intensive grouse farming
6. **Ancient meadows and grazing lands such as sheep-walk downland:** like moors, heaths and other commons, until around 1700 through until the late 1800s these were the distinctive open landscapes of all parts of the country. Through the impact of Parliamentary enclosures, they too were drastically reduced in area. Twentieth century wartime improvements removed most sheep-walk areas, and a decline in mixed farming often with local subsistence and in the use of marginal lands too, meant most areas were destroyed or abandoned
7. **Ancient sand dune systems:** although severely damaged by coastal urban development and conversion to recreational or tourism uses the remaining locations are of major conservation value. Many sites and dune systems were 'stabilised' by the planting of exotic conifers during the 1800s and 1900s. However, in the pre-improvement landscapes these ecosystems were exploited as a part of the complex of heaths, grasslands and other commons
8. **Ancient arable lands:** with post-medieval improvements in farming and especially because of Parliamentary enclosures almost all open field arable land was lost. With the advent of modern herbicides and cultivation systems, the distinctive ecology of these systems was quickly driven to extinction. With mechanisation and improvement, the economic and social functions of these landscapes were transformed and communities squeezed off the land and into the emerging cities. It is also likely that the original open field landscapes were in fact far more complex, diverse and ecological rich than has previously been recognised

Table 2.3 Some examples of species declines in Britain through cultural severance

a) **The White Admiral butterfly (*Ladoga camilla*):** this strictly woodland butterfly does not cope well with the regular disturbance of coppicing but is a high forest species in woods with shady rides and some open areas for male territorial displays. Mid-twentieth century abandonment of coppicing in lowland English woods benefited this butterfly and they even colonised into young conifer plantations too. They decline in the latter as the conditions become too shady. Conversion of woodland to farmland clearly removes the insect entirely
b) **The Nightjar (*Caprimulgus europaeus*):** open heath and open areas in woodlands such as traditional coppice are favoured habitat but this bird also does well in young conifer plantations. However, this is a short-term benefit unless the habitat is opened up by clear felling, and numbers swiftly decline after 20 years or so
c) **Hen Harrier (*Circus cyaneus*):** one of our birds of prey most closely associated with moors and heaths they also do well in young conifer plantations and so breeding numbers may increase in the first 20 years after conversion. However, from that time onwards, the effect of the plantation is to have displaced the traditional habitat and the bird's population then declines
d) **Black Grouse (*Tetrao tetrix*):** this large game bird favours mixed heath, unimproved grassland and a mosaic of woodland. Again, they do well in the early years of conversion to conifer agri-forestry but decline as the trees mature

systems sustained by the local community, with technologically applied management. The latter is often delivered by a few people, sometimes living locally but increasingly over time, by small numbers of specialist contractors who have no long-term relationship to the resource.

A further consequence of these changes is that a landscape resource, which was utilized in many differing ways by specialist, locally based craftsmen, is reduced in complexity to become a much-simplified system. The inherent complexity of a subsistence ecology linked through traditional management to a locally functional subsistence economy is broken. Economic and social functions are operating at political levels and at key points in history for each landscape these are contested spaces operated upon by socially, politically, and economically polarised actors. Throughout history and over long periods there are critical tensions between the competing interests, which may balance to produce stability and continuity for this shared resource. However, when for varying reasons, the social, economic, political or ecological sustainability break down, periods of dramatic and sometimes catastrophic change ensue. Throughout history, such changes have generally led to moves towards less communal ownership, reliance, utilization and determination of functions, to ones which are more individual and centralised, and that often apply technology and chemistry to the landscape in place of skill and physical labour. Furthermore, demand for individual profit and financial return for the few resource owners, replaces the need for local sustainability. Acquisition of control of these resources is a political act as well an economic one, and it further manifests itself it terms of enhanced political power and influence as well as in sheer material wealth (Fig. 2.3).

Fig. 2.3 Overgrown commonland SSSI

In England during the eighteenth century, enclosures of commonland were swiftly followed by the imposition of grand stately houses and halls with their gardens and parks. These were clear statements of power and unfettered political and economic influence. They are further surrounded by parklands, which in turn have sturdy walls to keep the commoners out. The peasant communities, displaced from their lands became wage-slave labourers, tenant farmers, or drifted to the growing cities to power emerging industries. The consequence was the ending of traditional use but also a contestation of spaces and functions, which in this case ran for centuries. Separated from their legitimate means to subsist in their landscape these peasants and commoners became poachers. Their overlord was both judge and jury in the local legal system, and employed armed gamekeepers to ensure compliance in the new landscape. In uplands and lowlands, access to land was denied and it took 200 years before the access to countryside legislation re-established recreational access where subsistence access was once a right.

2.7 Some Examples of Ecological Consequences

The overall trend in ecology wrought by these changes at every level from local to international is a catastrophic decline in diversity and local distinction (Tables 2.4 and 2.5; Fig. 2.4) (Rotherham 2009a, b, c, 2010). There is a loss of stress-tolerant species as they are replaced by homogenous and simplified landscapes and by ecological competitors. With globalisation, this has ultimately generated a '*Disneyfication*' of ecology and landscape. However, whilst this general and widespread trend is obvious and is undeniable, the specific impacts vary and the short-term effects of some modifications and functional changes can be beneficial to certain species. Clearly, when habitats such as bog, woodland, fen and heath are simply lost by conversion to other uses then wholesale extinction of those

Fig. 2.4 Overgrown heath

associated species results. In this case, the calculation of loss is simply an assessment of the plants and animals associated with the particular landscape and its habitats factored against the areas of land converted. I give a few examples later, but essentially this is total loss. Where some element of the original landscape remains then there is the potential for a degree of recovery or for some species to adapt and to thrive. Unfortunately, this may sometimes be a transient gain and the species decline as ecological successions change the conditions to disfavour them.

2.8 Embedding Culture in Nature

The last 10 years have witnessed a remarkable debate on the origins of ancient landscapes, and centred on Europe, a re-visiting of ecological and evolutionary origins (see Rotherham 2012). This has developed to a significant degree through the seminal writings of three individuals namely Frans Vera (2000), Oliver Rackham (1981, 1986), and George Peterken (1981, 1996). Essentially the discussion has been around a vision of how Western Europe looked in prehistoric times and associated with this, how its ecology functioned and developed. Much has been said about the evidence-base for such a vision and the interpretation of that same information. In summary, it is likely that this vast primeval landscape had the following attributes (Rotherham 2009a, b, c, 2010, 2013a, b):

Wooded landscapes: woodland and forest of various types, in varying proportions depending on terrain, topography, geology and climate; from high mountain forest, to lowland floodplain forest and more isolated woodland patches within a mosaic of other habitats.

Open grassland, heath, sand dunes, and savannah: in varying amounts varying as above and again juxtaposed with other habitat types; and in upland and northern zones both Alpine and Tundra forms of these.

Wetlands: from vast valley-bottom floodlands and their great meandering rivers to expansive peat bogs and fens, and with large areas of wet woodland; scattered throughout these landscapes would be a complex and intimate matrix of pools, ponds, marshes, bogs, fens, and drier ground with considerable temporal and seasonal fluctuation.

Disturbance features: scattered across all the other communities from mountain rock-fall zones and erosion areas, to riverine erosion and deposition habitats, and wildfire areas caused by lightning strikes.

Coastal zone landscapes: flats, cliffs, dunes, and extensive marshes, fens, grasslands and heaths.

These habitats would be in some long-term balance and in proportions that can be debated and argued about. The landscapes and their habitats occurred in a context of a vast continuum of a mosaic of interacting ecologies, which fluxed seamlessly across the entirety of the continent. These were not static but changed through their internal and interactive dynamics and in response to climate change and sea level rise and fall. It is onto the basic template that the human footprint was imposed indelibly and increasingly; from the first hunter-gatherer communities to the techno-fix cultures of today. However, for several thousand years the basic process was one of gradual imposition of human influence and the assimilation of nature into locally and regionally distinct landscapes. Over time, whole areas and regions were transformed beyond recognition but much of the basic ecosystem function remained, though altered, within the now not natural but eco-cultural landscapes. Even today, we can see a degree of resonance in the contemporary landscapes of the very few remaining medieval deer parks, with aspects of the primeval ecology. We can also see evidence of human modification and utilisation in for example the great worked trees, be they pollards or coppices, which remained scattered throughout today's countryside.

For individual regions, countries and cultures we can recognise pulses of increased human impact and often periods of retraction and release. Nevertheless, gradually throughout history the impact of human usage and landscape transformation has increased. However, until the advent of petro-chemically subsidised economies, most landscape use had to be inherently self-sustaining and often at a local subsistence level. When the utilisation became unsustainable, as happened not infrequently, the society and its systems changed, declined or died out entirely. A society that lived beyond the means of its regional environment to support it, or above the level that modest trading could subsidise, could not survive (Rotherham 2005).

The consequence of these limitations was the evolution of often regionally distinctive and often complex systems of landscape resource utilisation to regulate exploitation and to ensure sustainability. A result of this long-term predictable land management was the development of distinctive cultural landscapes from the original primeval environment, and the selection and development of ecologies best able to adapt to these new scenarios. In Western Europe, these systems and their distinctive landscapes and ecologies were generally well established by the medieval period and many of the particularly landscapes and the most highly

Table 2.4 Examples of species increases with cultural severance in Britain

There has been a huge increase in range and abundance of common, catholic, competitive species and especially of invasives. The latter include large numbers of aggressive alien invasives. Some of the major increases are noted below
1. Bracken (*Pteridium aquilinum*): across moors, heaths and grasslands as a cultural artefact of changes in management
2. Birch (*Betula pendula* and *B. pubescens*): massive spread over heaths and moorland fringe as a result of abandonment of traditional management
3. Sycamore (*Acer pseudoplatanus*): huge spread and colonisation throughout many woods and grasslands too as a result of cultural changes and especially of disturbance and eutrophication of soils; the debate on whether this species is native or alien in Britain still continues
4. Grey Squirrel (*Sciurus carolinensis*): another aggressive invader but the success is largely due to cultural severance as coppice woods were re-planted or abandoned to high forest
5. Tall Fall Oat Grass (*Arrhenatherum elatius*): this is one of a number of competitive species that have spread across abandoned and eutrophicated grasslands

valued habitats and ecologies we have today, descended to us from this time. It is the traditional utilisation and management of these unique and often semi-natural and eco-cultural landscapes that is rapidly ending across Britain and now all over Europe. For heritage and nature conservation, the implications are potentially catastrophic and yet largely either unrecognised or ignored (Tables 2.4 and 2.5).

Table 2.5 Examples of species declines with cultural severance in Britain

1.	**Wetland birds:** massive declines of species such as Bittern (*Botaurus stellaris*), Spoonbill (*Platalea leucorodia*), Marsh Harrier (*Circus aeruginosus*), Night Heron (*Nycticorax nycticorax*), Crane (*Melornis grus*), Black Tern (*Childonias niger*), Black-tailed Godwit (*Limosa limosa*) in traditionally managed lowland fens and washlands, but with some very localised recovery with habitat creation
2.	**Wetland flora:** huge losses of typical spcies such as Fen Ragwort (Senecio paludosus), Fen Violet (*Viola stagnina*), Heath Dog Violet (*Viola canina* ssp *montana*), Milk Parsley (*Peucedanum palustre*), Snakeshead Fritillary (*Fritillaria melagris*) and Fen Woodrush (*Luzula pallidula*)), in traditionally managed lowland fens and washlands
3.	**Wetland invertebrates:** Large Wolf Spider (*Lycosa paludicola*), Raft Spider (*Dolomedes fimbriatus*) and Fen Raft Spider (*Dolomedes plantarius*), Large Heath (*Coenonympha tullia*), Large Copper (*Lycaena dispar*), Marsh Fritillary (*Eurodryas aurina*) and Swallowtail (*Papilio machaon*) Butterflies, in traditionally-managed lowland fens and washlands
4.	**Heathland and grassland birds:** Skylark (*Alauda arvensis*), Woodlark (*Lulluala arborea*), Nightjar (*Caprimulgus europaeus*), Cuckoo (*Cuculus canorus*), Red-backed Shrike (*Lanius collurio*), Black Grouse (*Tetrao tetrix*), Stone-curlew (*Burhinus oedicnemus*), and Great Bustard (*Otis tarda*)
5.	**Heathland and grassland invertebrates:** losses of many species such as Dark Green Fritillary (*Argynnis aglaja*), High Brown Fritillary (*Argynnis adippe*), Large Blue (*Maculinea arion*), Adonis Blue (*Lysandra bellargus*), Chalkhill Blue (*Lysandra coridon*), Glow-worm (*Lampyris noctiluca*), Orb-web spider (*Araneus quadratus*), and Silver-spotted Skipper (*Hesperia comma*), though direct habitat loss and through successional change following abandonment
6.	**Heathland and grassland flora:** specific losses such as Pasque Flower (*Pulsatilla vulgaris*) and Dodder (*Cuscuta epithymum*) but removal of entire flora from most of the lowland landscape, and desiccation and degradation of upland areas too

(continued)

Table 2.5 (continued)

7.	**Heathland and grassland herptiles:** Adder (*Vipera berus*), Smooth Snake (*Coronella austriaca*), Common Lizard (*Lacerta vivipera*), Sand Lizard (*Lacerta agilis*); massive declines of amphibians too
8.	**Woodland birds:** many species have declined but particular examples include Nightingale (*Luscinia megarhynchos*) and Woodlark (*Lulluala arborea*) in the lowlands and Capercaillie (*Tetrao urogallus*) in the north
9.	**Woodland flora:** loss of rare woodland species such as Herb Paris (*Paris quadrifolia*), and contraction in range of many other woodland 'indicator' flowers with destruction of sites e.g. Bluebell (*Endymion non-scriptus*), Wood Anemone (*Anemone nemorosa*); veteran trees largely removed
10.	**Woodland invertebrates:** Declines in particular species such as Stag Beetle (*Lucanus cervus*), Lesser Stag Beetle (*Dorcus parallipedus*), Purple Emperor butterfly (*Apatura iris*), Duke of Burgundy Fritillary (*Hamearis lucina*), Purple Hairstreak butterfly (*Quercusia quercus*), Wood White butterfly (*Leptidea sinapis*), Chequered Skipper (*Carterocephalus palaemon*), and loss of many dead wood invertebrates from most areas

2.9 Conclusions

The nature and intensity of exploitation have affected many landscapes such as those highlighted across Great Britain (e.g. Rotherham, 2005, 2007a, b, c, d, 2008a, b). This may be through the actual direct impact of extraction and processing, or of harvesting and processing, and of use. It may be that the processes have indirect impacts through landscape change, through pollution, and through the disposal of associated wastes. Use had major effects through the social need to protect or even to establish the resource, such as coppice wood, peat bog, fen, or common heath; important when otherwise these would have been removed from the landscape. Literature and records give insight into resource uses and availability, providing data for detailed interrogation on comparative values, costs, and trends. These aid interpretation of impacts, the nature, and intensity of landscape exploitation; directly through extraction or harvesting, processing, and use, or indirect landscape change perhaps caused by pollution or waste disposal.

In determining these changes, both crisis and continuum have played a part (Rotherham 2005). The transformation from traditional subsistence use to petrochemically driven industrial exploitation has been defining in many environments. Environmental conditions and resources, economic and political or social forces, and the interactions or competitions involved in these, have been crucial in determining the impacts on land-use and landscape. In many ways, the landscape provides a continuum punctuated by crises for the community and for the environment. The interaction of community, resource utilisation, and environment has been a driving force in the evolution of the eco-cultural landscape that we inherit. However, scholars, politicians and practitioners rarely appreciate the fundamental nature of this relationship and the sophistication and totality in which subsistence and other traditional communities interacted with their environment. The medieval landscape was rather like the traditional family pig, with everything used except the squeak. It is perhaps this totality of use that most eludes us today.

There are opportunities to re-build connectivity with nature, but with society increasingly remote from the natural world as a resource, this is very difficult. We now seek ways to link people to nature through education and conservation but this must extend to a genuine long-term social and economic value system too. There are attempts such as in Britain, particularly by the RSPB and the National Trust to join landscape to communities and to the economy but much more will be needed in the future. Approaches now seek to join the following in coherent local and regional projects to do this:

1. Nature conservation
2. Tourism, leisure and amenity
3. Local provenance food and drink and farming extensification
4. Economic function
5. Ecosystem function and value

The truth is that we cannot turn back the socio-economic clock and there are often good sound reasons why traditional management is no longer undertaken. However, the impacts and implications of cultural severance from past tradition and the abandonment of eco-cultural landscapes are massive. We need now to recognise what is happening and to attempt as far as is possible, to re-construct a viable and sustainable future. It will not be easy but the penalty of failure will be even harder to bear. A key challenge is to re-establish local economic function in ways that do not compromise or destroy either ecology or heritage interest. Too frequently, in initiatives such as biofuel extraction from semi-natural ancient woodlands, in England, or the imposition of industrial wind-farm landscapes across heritage-rich countryside, this is not the case. There are effective models for the use of areas as fuel lots in both Europe and in North America, and perhaps that could be a way to conserve lowland heaths and commons and to re-join their connections to local economies and to local people. However, to be effective, such initiatives need to be sustainable and non-damaging utilisation not industrial exploitation. Moreover, these ideas need to resonate with local communities and where possible, to be immersed within them; common and commoner need reuniting.

References

Agnoletti M (ed) (2006) The conservation of cultural landscapes. CAB International, Wallingford

Agnoletti M (ed) (2007) Guidelines for the implementation of social and cultural values in sustainable forest management. A scientific contribution to the implementation of MCPFE—Vienna Resolution 3. IUFRO Occasional Paper No. 19, ISSN 1024–414X

Anderson P (ed) (2004) Upland ecology, tourism and access. In: Proceedings of the 18th Conference of the Institute of Ecology and Environmental Management, IEEM, Buxton, Winchester, 25–27 November 2003

Appell GN (1993) Hardin's myth of the commons: the tragedy of conceptual confusions. Working Paper 8. In: Phillips ME (ed) Social Transformation and Adaptation Research Institute. Indiana University, USA
Doncaster S, Egan D, Rotherham ID, Harrison K (2006) The tourism economic argument for wetlands: a case study approach. In: Proceedings of the IALE Conference, water and the landscape: the landscape ecology of freshwater ecosystems, Oxford Brookes University, UK, 5–8 September 2006, pp 296–300
Fowler J (2002) Landscapes and lives. The Scottish forest through the ages. Canongate Books, Edinburgh
Hardin G (1968) The Tragedy of the Commons. Science 162 (No. 3859):1243–1248
Hayman R (2003) Trees. Woodlands and western civilization. Hambledon and London, London
Jowit J (2010) Case for saving species 'more powerful than climate change'. *The Guardian*, Saturday 22 May 2010, p 1
Ostrom E (1990) Governing the commons. Cambridge University Press, Cambridge
Perlin J (1989) A forest journey. Harvard University Press, Massachusetts
Peterken GF (1981) Woodland conservation and management. Chapman and Hall, London
Peterken GF (1996) Natural woodland—ecology and conservation in northern temperate regions. Cambridge University Press, Cambridge
Rackham O (1980) Ancient woodland: its history, vegetation and uses in England. Edward Arnold, London
Rackham O (1986) The history of the countryside. Dent, London
Rotherham ID (1999a) Peat cutters and their Landscapes: fundamental change in a fragile environment. In: Peatland ecology and archaeology: management of a cultural landscape. Landscape Archaeology and Ecology 4:28–51
Rotherham ID (ed) (1999b) Peatland ecology and archaeology: management of a cultural landscape. Landscape Archaeology and Ecology 4:99
Rotherham ID (2005) Fuel and landscape—exploitation, environment, crisis and continuum. Landscape Archaeology and Ecology 5:65–81
Rotherham ID (2006) Historic landscape restoration: case studies of site recovery in post-industrial South Yorkshire, England. In: Agnoletti M (ed) The conservation of cultural landscapes. CAB International, Wallingford, pp 211–224
Rotherham ID (2007a) Cultural severance—causes and consequences. In: The International Association for Vegetation Science: 50th Annual Symposium, University of Swansea, Swansea, 22–27 July 2007. Cultural landscapes—changing landscapes, Delegate Handbook, pp 43–44
Rotherham ID (2007b) The ecology and economics of medieval deer parks. In: Rotherham ID (ed) The history, ecology and archaeology of medieval parks and parklands. Wildtrack Publishing, Sheffield, pp 86–102
Rotherham ID (2007c) The historical ecology of medieval deer parks and the implications for conservation. In: Liddiard R (ed) The medieval deer park: new perspectives. Windgather Press, Macclesfield, pp 79–96
Rotherham ID (2007d) The implications of perceptions and cultural knowledge loss for the management of wooded landscapes: a UK case-study. For Ecol Manage 249:100–115
Rotherham ID (2008a) Lessons from the past—a case study of how upland land-use has influenced the environmental resource. Aspects of Applied Biology, shaping a vision for the uplands 85:85–91
Rotherham ID (2008b) Tourism and recreation as economic drivers in future uplands. Aspects of Applied Biology, shaping a vision for the uplands 85:93–98
Rotherham ID (2009a) Cultural severance in landscapes and the causes and consequences for lowland heaths. In: Rotherham ID, Bradley J (eds) Lowland heaths: ecology, history, restoration and management. Wildtrack Publishing, Sheffield, pp 130–143
Rotherham ID (2009b) Habitat fragmentation and isolation in relict urban heaths—the ecological consequences and future potential. In: Rotherham ID, Bradley J (eds) Lowland heaths: ecology, history, restoration and management. Wildtrack Publishing, Sheffield, pp 106–115

Rotherham ID (2009c) The importance of cultural severance in landscape ecology research. In: Dupont A, Jacobs H (eds) Landscape ecology research trends, ISBN 978-1-60456-672-7. Nova Science Publishers Inc., USA, pp 1–18

Rotherham ID (2010) Cultural severance and the end of tradition. Landscape Archaeology and Ecology 8:178–199

Rotherham ID (2011) The implications of landscape history and cultural severance in environmental restoration in England. In: Egan D, Hjerpe E, Abrams J (eds) Integrating nature and culture: the human dimensions of ecological restoration. Island Press, Washington DC, pp 277–287

Rotherham ID (2012) Traditional woodland management: the implications of cultural severance andknowledge loss. In: Rotherham ID, Jones M, Handley C (eds) Working & Walking in thefootsteps of Ghosts. vol 1: the wooded landscape. Wildtrack Publishing, Sheffield, p 223–264

Rotherham ID (ed) (2013a) Trees, man, & grazing animals—a European perspective on trees and grazed treescapes. Earthscan, London (in press)

Rotherham ID (2013b) Re-interpreting wooded landscapes, shadow woods and the impacts of grazing. In: Rotherham ID (ed) (2013) Trees, man, & grazing animals—a European perspective on trees and grazed treescapes. Earthscan, London (in press)

Rotherham ID, Jones M (2000) The impact of economic, social and political factors on the ecology of small English woodlands: a case study of the ancient woods in South Yorkshire, England. In: Agnoletti M, Anderson S (eds) Forest history: international studies in socio-economic and forest ecosystem change. CAB International, Wallingford, pp 397–410

Vera FHW (2000) Grazing ecology and forest history. CABI Publishing, Oxon, UK

Vidal J (2005) Wild herds may stampede across Britain under plan for huge reserves. *The Guardian*, Thursday October 27 2010

Webb NR (1986) Heathlands. Collins, London

Webb NR (1998) The traditional management of European heathlands. J Appl Ecol 35:987–990

Chapter 3
Globalism and the Enclosure of the Landscape Commons

Kenneth R. Olwig

3.1 Introduction

'Enclosure,' it will be argued, severs people from the landscape of not just the material commons, but the 'cultural commons,' thereby breaking the living bonds of custom that motivate sustainable use. Globalism, it will be further argued, is the contemporary manifestation of the enclosure movement, and thereby a threat to both the material and the cultural commons. Enclosure leaves behind, however, a residue of the cultural commons, as *picturesque tradition*, that easily dissolves into death by nostalgia. But before delving into enclosure, globalism and picturesque tradition, the meaning of the 'cultural commons' in relation to the material commons will be explored.

'The commons,' in this analysis, is primarily an abstracted version of the historical commons as found notably in England (Rodgers et al. 2011), but also in other European countries more generally. The commons, as understood here, is the material landscape of common lands shared by a community of commoners with customary use rights in that land. Such a commons becomes a 'cultural commons' when the commons provides a foundation for culture in the dictionary sense of 'the customs, arts, social institutions, and achievements of a particular nation, people, or other social group' (NOAD 2005 custom). As a cultural commons the commons ceases to be simply a material landscape, and gains symbolic meaning by becoming the referent for a metaphor, growing out of daily life, through which the material commons is transferred into the realm of ideas, becoming 'a thing regarded as representative or symbolic of something else, especially something abstract' (NOAD 2005 metaphor). The commons thereby becomes something one can think *with* (as a symbol and idea), as well as

K. R. Olwig (✉)
Department of Landscape Architecture, Planning and Heritage, Swedish University of Agricultural Sciences, Box 58, Sundsvägen 4–6 23053 Alnarp, Sweden
e-mail: kenneth.olwig@ltj.slu.se; kenneth.olwig@slu.se

I. D. Rotherham (ed.), *Cultural Severance and the Environment*,
Environmental History 2, DOI: 10.1007/978-94-007-6159-9_3,

about (as a material phenomenon). Thinking *with* the idea of the commons allows its extrapolation to include a broader idea of the commons that is more widely applicable to contemporary social and environmental issues.

3.2 The Commons Landscape

As common *land*, the commons is part of the physical landscape, but as metaphor and symbol it carries broader meaning of *land*scape in the sense of being both material and an expression of culture, as in landscape art. The *land* in *land*scape becomes *landscape* through the addition of the suffix *-scape*, which is a spelling variant of the suffix *-ship*, meaning *quality, character*, or *condition*, as found in words like citizen*ship*, friend*ship* or fellow*ship* (Olwig 2002, pp. 18–19). Thus, when an individual becomes part of a community of fellows this person becomes a fellow, and between these fellows, a more abstract quality or character of *fellowship* arises, in the sense of an: 'association, esp. with people who share one's interests*: they valued fun and good fellowship as the cement of the community'* (NOAD 2005 fellowship). Likewise, when the various *lands* of an area, e.g. the common lands, the arable lands, the meadowlands, the parklands, the town lands, etc., are conceived as 'land*scape*', then the sense of a more generalized and abstract character, condition or quality of these lands as a single 'land' emerges, which is something an artist might wish to capture. This is an expression of the community shaping and unifying the land. Ambrogio Lorenzetti's famous fourteenth century painting in the *Palazzo Pubblico* of Siena thus contrasts the lands of a city characterized by a harmonious commonwealth with the lands of a disharmonious polity. Similarly, the influential *pastoral* artistic genre posits an ideal '*Arcadian*' pastoralist society rooted in the sharing of common land (Putnam 1970). This pastoral ideal provides a measure against which to assess the quality or character of a land according to the degree of social harmony or disharmony, much as in Lorenzetti's painting. Not surprisingly, the pastoral has been a persistent theme in landscape art, poetry and, not the least, landscape architecture. The cultural commons, as exemplified by the heritage of the pastoral, is thus part of a larger landscape, understood not just as a material phenomenon, but also as a cultural, social and political phenomenon (Williams 1973) (Fig. 3.1).

The concept of fellowship is a good example of the way the commons functions as a master metaphor, grounding ancillary metaphors in the material and social processes that generate a sustainable commons. The word *fellow* derives from the Old Norse, *felagi*, meaning someone who is a member of an institution or association (*lag*), for the grazing of sheep or cattle (*fe*), in common (Olwig 2002, p 19). The substantive association between the commons as a form of land and its more symbolic aspects is palpable on many English village commons where the turf may still be grazed by animals of the commoners whose dwellings surround the commons, along with a church and a public house (Fig. 3.2).

Fig. 3.1 Ambrogio Lorenzetti, allegory of good government, ca. 1338–1340

Fig. 3.2 Village commons, England, note the lack of a fence, and that the animals are not tied. Photograph KRO

As a further symbol of community this commons provides a setting for a game of cricket. This symbolism is less clear on a typical New England village commons, where the grass is machine mowed. This same symbolic distancing is also found at a typical English college, where only fellows of the college are allowed to walk on the grass of the landscaped court or quad around which the buildings of the college, including both a church and refectory (instead of a pub) are scattered in an often pastoral, park like, setting. This effectively 'common land' does not graze sheep, just as the fellows no longer are likely to be future "pastors" in a church, but the grass is mowed to create a symbolic landscape that could look as if it had been tended by a commonality of fellows caring for their flock (who are now students rather than sheep). A landscape symbol of this kind thus often works today more on the level of practice than of conscious symbolism, but it is assiduously maintained along with the customary rights of access to such landscapes.

The physical commons not only supplies a material referent for the cultural commons, it is also engaged in a reciprocal relationship with the cultural commons. The use of the commons by commoners who have use rights in the commons was typically governed by custom, and if that failed, by custom as formalized in common law (Milsom 1981). The Latin for custom is *moralis*, the root of the concepts of both *mores* and *morality*. There is thus a reciprocity between the customary practice of fellows raising sheep together and the sense of morality, belonging to the broader realm of culture and community, which helps, in turn, to stimulate the moral behaviour necessary to sustain the use of a shared resource (Gudeman 2001, p. 28). It is the connection between commonplace, everyday, customary practice, a quotidian landscape, and the more elevated cultural sphere of morality and ethics, that makes the commons such a potent symbol, and a natural metaphor in which concrete practices and the abstract are intertwined. The landscapes of ancient English colleges, such as those of Cambridge University, were historically rooted in religious orders, and had the training of clerics as a central purpose. The idea of Christian fellowship was, of course, key to these orders, and the metaphorical language and paraphernalia of shepherding, from the bishop's staff to the role of the 'pastor' in tending the parish 'flock,' can be traced back to the good shepherd of the Bible, but also to other influential sources, such as the wise shepherd statesman of Plato (1961).

The commons arguably form the heart of what the historian E.P. Thompson called 'the moral economy,' by which he meant an economic alternative to the money economy, which was based upon exchange and custom (Thompson 1993). It is in the commons that the commoners learn to collectively harvest and manage common resources. Writing from the perspective of economic anthropology, Stephen Gudeman argues that: 'the commons is the material thing or knowledge a people have in common, what they share, so that what happens to a commons is not a physical incident but a social event.' It therefore follows that: 'Taking away the commons destroys community, and destroying a complex of relationships demolishes a commons' (Gudeman 2001, p. 27). The commons do not stand-alone, however, because they usually existed, historically, in tandem with more intensively cultivated individual gardens and arable land parcels used during part of the agricultural cycle by an individual farming unit. If the primary use of the arable fields is agricultural *production,* the commons is more related to agricultural, and social, *reproduction*. The commons, thus, as a grazing resource, help provide the nutrients, through the manure of the grazing animals, as well as supplementary materials (e.g. as brushwood, bedding, litter, etc.), for intensive horticultural production and agricultural production on the arable, while providing subsistence foods (berries, mushrooms, herbs, etc.), especially for the poorer labourers. They

Fig. 3.3 Hikers on a Bennachie forest, an historic commons outside Aberdeen. Photograph KRO

thereby create an important ecological and social buffer for the community of commoners. A commons is thus characterized by considerable natural diversity compared with the more uniform environment of the arable fields. Finally, the commons are also a place for sporting activities (e.g. early football and golf) and recreation (e.g. walking) (Olwig 2008). The sharing of the commons is thus important for both the physical and social sustenance of the village community. Finally, the commons, as a locus of social cooperation, play a role as the social glue binding the commoners of varying social status together as a community (Gudeman 2001, p. 28), giving them great symbolic power (Fig. 3.3).

The meaning of the landscape of the commons as a symbol of fellowship and citizenship is exemplified by the American historical geographer Donald Meinig (1979) in an article entitled 'Symbolic Landscapes: Some Idealizations of American Communities.' He begins with a quote to the effect that 'to the entire world, a steepled church, set in its frame of white wooden houses around a manicured common, remains a scene which says 'New England'', continuing (Fig .3.4).

> … drawing simply upon one's experience as an American (which is, after all, an appropriate way to judge a national symbol) it seems clear that such scenes carry connotations of continuity (of not just something important in our past, but a viable bond between past and present), of stability, quiet prosperity, cohesion and intimacy. Taken as a whole, the image of the New England village is widely assumed to symbolize for many people the best we have known of an intimate, family-centered, God-fearing, morally conscious, industrious, thrifty, democratic community (1979, p. 165).

Though Meinig's purplish prose indeed 'idealizes' the New England village, many New England townships are still working *direct* democracies. These townships still have regular town meetings in which the citizenry meets to discuss and vote on the governance of the community, just as they did at the time of their seventeenth-century founding on the basis of customs taken by the settlers from their origination in 'old' England. In old England, similar ideals are likewise still attached to the village with, as noted, its church and pub on the commons. Cricket

Fig. 3.4 New England township commons, massachusetts. Photograph KRO

is thus characteristically played on the village commons, the word 'cricket' becoming synonymous with customary morality. Thus, when something 'isn't cricket' it is 'a thing contrary to traditional standards of fairness or rectitude' (NOAD 2005: cricket).

The New England village commons, as Meinig indicates, tends today to be a carefully 'manicured' and landscaped lawn, more a public park than the rough irregular working commons of the original settlement (Wood 1997), but commons-like institutions can still exist, favoured by the structure of the community's organization. The township of South Orleans on Cape Cod, Massachusetts, thus has designated large areas of land as nature reserves for the protection of common water resources, but it is also a common area to which the citizenry has a use right for suitable forms of recreation. There is also a communal waste disposal centre where waste paper and bottles become a shared resource when sold for recycling by town charities. Hence when it is picked up at the exchange booth and put to use again, one person's 'junk' facilitates another's 'waste not want not' subsistence. Finally, this 'commons' provides a shared space, with its recycling booths, where the citizenry can chat with neighbours, perhaps about an upcoming town meeting.[1]

Townships like these are rare today, in part because enclosure has destroyed the shared use rights in land that are at the heart of the "*-scape/-ship*" that holds the polity of the town together as a community (Gudeman 2001, p. 27). Enclosure is the process by which the commons are surveyed and turned into private property. In the following the process of enclosure will be abstracted in very broad terms as a basis for a further examination of the implications of enclosure for the 'cultural commons.'

[1] This information is based upon fieldwork in South Orleans, Massachusetts.

3.3 The 'Enclosure' of the Commons

Under the field systems inherited from medieval times the area primarily used for arable cultivation would be cultivated in strips belonging to individual farms, but within larger common fields (usually two or three in number) that would periodically be put in fallow and grazed in common, just as nearby meadows would be grazed in common. Vaster commons would be located on the more marginal land, usually further from the village, but a diminutive commons, the village green, would also typically be located at the core of the village (Butlin 1982; Clayden 1985). Enclosure generally involves the bounding of the common lands within an area of square feet in Euclidean space as demarcated, for example, on a cadastral map, at which point it becomes private property. The term enclosure is used to describe both: (1) the process whereby farms with strips dispersed throughout the common arable fields could have these replaced by a contiguous space of bounded property (sometimes distant from the original farmstead), and (2) the process by which primarily pastoral commons on the vaster more marginal lands, often towards the periphery of the manor, have been turned into private property, generally under an estate-owner (Butlin 1982; Clayden 1985).

In medieval times, manors were not owned as private property, but were rather the inalienable domain (meaning they could not, in principle, be sold) of a noble family to which the farmers and others under the manor were bound by reciprocal feudal ties and inherited customary rights (Olwig 2005a, b). By the time of the great enclosure movements, stretching from the sixteenth to the nineteenth centuries, saleable private property had been introduced. However, until the present day, the rights to the commons defy modern notions of property because, though the soil of the common belongs to the Lord of the Manor or estate owner, because, though the use rights to many resources of the commons belong to the farms of the commoners. So, in principle, one can own a commons, but not have a significant right to its use. Even when the 'owner' of the commons may not profit especially from the direct use of the commons, the 'owner' may have an interest in the commons being managed sustainably because this benefits the entire economy and society of the larger area in which the commons' 'owner' may have an economic and social interest (e.g. through the rents and services from well tenanted farms and shops). Today, however, the manor owner may no longer have much of a stake in the management of the commons, and the role of the 'owner' might be taken over by an organization of commoners (Rodgers et al. 2011), or the board of a National Park. The latter would have the ecological, economic and social health of both the Park and the nation as its larger concern. The new organization playing the role of the 'owner' can through local courts, for example, help adjudicate the sustainable management and use of the commons based on custom and law. Particularly in areas that historically carried common use rights, such as beaches and open uncultivated lands (e.g. heaths and moors), common citizens often have similar use rights to lands they do not own (Edwards 1995)—which in Britain have often been formalized as National Parks.

Commons were historically defined in terms of use rights to differing resources making up a plurality of commons, not to a delimited spatial area. As differing farms and social groups held rights to differing resources a commons need not have a clearly defined boundary line in-so-far-as differing resources might be distributed unevenly over differing areas. Enclosure, however, changes this by incorporating the former diversity of the commons, within the uniform space of the cadastral map, as properties within a contiguous space with well-defined boundaries. The lines of latitude and longitude forming the map's graticule create an abstract uniform geometric space within which the lines demarcating different properties are drawn. Stakes driven into the ground are used to affix the abstract locations within the map's space to the earth's variegated terrain, which can then be *enclosed* with hedgerows, walls or fences. Before enclosure agricultural systems typically sought to distribute access to different soil qualities (e.g. moist, dry, sandy, clay) equitably amongst farms as strips or patches, thus favouring crop variety. The concentration of an individual farm owner's resources within the uniform space of a bounded property, on the other hand, encouraged farmers to transform this space into equally uniform cropland, the result being the modern landscape of uniformly cultivated, low diversity, geometric fields (Kjærgaard 1994). The once varied terrain of the landscape thereby often comes to resemble the space of the map with its quadratic graticule. The commons typically lay at the spatial and environmental margins of the agrarian village core. Depending on the quality of the soil, these might be enclosed either for cultivation in arable fields or for more specialized intensive grazing or even as a landscape garden or hunting park.

Whereas the use rights to the commons are to resources distributed unevenly within the topography of the landscape, the property right to land applies to an abstract uniform undifferentiated area of bounded space on the cadastral map. Whereas use rights are often bound to differing farmsteads according to time-out-of-mind custom, and thus are difficult to sell, the property right is to a territorial space that can be easily measured and sold according to an abstract and singular medium of monetary exchange. Likewise, whereas the resources accessible through use rights to the commons concerned primarily resources important to subsistence consumption and exchange, enclosure under the ownership of a single proprietor facilitated the use of the land for the more specialized production of a market product. The property owner, furthermore, need not be present, or contribute to the community of commoners, to profit from the land. Enclosure thus transforms the commons from the place of a community defined by use rights to material resources, into a property demarcated in an abstract space that can be managed from a distance.

3.4 The Enclosure of the 'Cultural Commons'

Whereas the variety of customary use rights developing in a specific place, which are typical of the commons, are foundational to the complexities of common law as built upon precedence, the rights to the universal space of property lend themselves to universalistic concepts of 'Natural' legal rights, such as those identified with the Enlightenment philosopher John Locke. Such rights, as formulated in many of the world's constitutions, thereby often give property a foundational status to society. The loss of customary rights to the land through enclosure, which was particularly difficult for the poor, was thus paralleled by the rise of property rights as a foundation for liberal "democratic" society (e.g. the individualized right to vote) (Gudeman 2001, p. 28; Olwig 2005a, b; Olwig 2011a).

Enclosure works toward the severance of people from their cultural commons by promoting an alternative social model that is not based upon rights relative to the sharing of common resources, but rather upon an alternative set of cultural values founded upon absolute individual rights to property associated not only with the philosophy of John Locke, but also with the economic liberalism of Adam Smith (Olwig 2002, p. 162). From the perspective of economic liberalism, the commons came to be seen as the symbolic antithesis of morality, and the commons took on especially negative connotations. This was because the assumption was that economically motivated individuals with use rights to the commons would naturally seek to maximize their own resource use at the expense of the other commoners, and the sustainability of the land. This thereby created what has come to be known as a 'tragedy of the commons'—an old argument for enclosure that was consciously re-appropriated by the biologist Garret (Hardin 1968).[2] There was nothing tragic, however, about the sharing of a commons as long the commoners did not behave as the individualistic economic man promoted by the economic liberals, but rather behaved as a member of a community exercising the social control necessary to prevent commoners from taking more than their share (Gudeman 2001, pp. 27–28).

3.5 The Enclosure of Landscape as Scenery

The top down vertical projection of the cadastral map provided the primary tool by which the material commons were enclosed as individual private properties. The horizontal view of the landscape as perspectival scenery, on the other hand, had the same effect upon the cultural commons. The perspectival technique of drawing and planning landscape as scenery developed particularly out of cartography, being largely based on a change of projection from the map's top down view to the horizontal view of the spectator (Fig. 3.5).

[2] Hardin traces the historical background for his ideas in (Hardin and Baden 1977).

Fig. 3.5 Illustration from Albrecht Dürer, *Underweyssung der Messung* (Nuremberg 1538)

It had the effect of focusing the view upon the individual eye of an individual perceiver, giving that onlooker a distanced 'commanding' view (Barrell 1972; Cosgrove 1984; Daniels 1989). Within the area of this landscape gaze the various lands of an area are united within a uniform abstract space extending to the globe's horizon (Olwig 2011b). It was, furthermore, common for estate owners to create, often from former common lands, pastoral landscape garden/parks, with hidden fences (the so-called "ha–ha") intended to create the illusion of a unbounded common, which were inspired by pastoral art (Olwig 2002, pp. 99–124). The commons were thereby not only enclosed materially, but also aesthetically, with symbolic reference to pastoral art, which had hitherto celebrated the values attached to a fellowship of commoners (Williams 1973) (Fig. 3.6).

The transformation of the commons into landscape scenery, as if on a stage, had the effect of transforming *custom* into *picturesque tradition,* performed against the background of landscape scenery. Custom thereby becomes, figuratively speaking, a form of *costume* that is donned on special performative occasions. Following the distinction between custom and tradition made by Eric Hobsbawm in the book, *The Invention of Tradition* (1983) it can be argued that.

> 1 The object and characteristic of 'traditions,' including invented ones, is invariance.... 'Custom' cannot afford to be invariant, because even in 'traditional' societies life is not so. Customary or common law still shows this combination of flexibility in substance and formal adherence to precedent. The difference between 'tradition' and 'custom' in our sense is indeed well illustrated here. 'Custom' is what judges do; 'tradition' (in this instance invented tradition) is the wig, robe and other formal paraphernalia and ritualized practices surrounding their substantial action.... Inventing traditions, it is assumed here, is essentially a process of formalization and ritualization, characterized by reference to the past, if only by imposing repetition (1983, pp. 2–3).

When the grazing landscape of custom is frozen as pastoral landscape scenery it becomes *picturesque* and severed from everyday life as an unchanging tradition, whereas custom, though rooted in precedent, is nevertheless constantly changing, as a legal principle, to suit changing times. It thus makes a difference if one is celebrating midsummer with a bonfire or a maypole as part of a yearly custom intended, in part, to make manifest one's use right to certain forms of wood in a common

Fig. 3.6 Stowe Park, with invisible Ha Ha in foreground. Photograph KRO

woodland (Bushaway 1982), or the celebration is simply the performance of a cultural tradition with no link to use rights and obligations. Customs have, as happened about a century ago in London, thus provided the basis for lawsuits seeking to prevent the expropriation of common land for private housing development, thereby opening the way for the subsequent transformation of common lands, with the commoners' approval, into public parks (Eversley 1910; Clayden 1985). Working agricultural commons on the London fringe thus evolved into recreational park landscapes that nevertheless maintain their character as a shared community resource. The performance of a tradition may help strengthen a local sense of identity (Pearson 2006), but without a connection to customary rights and obligations in a community commons, both the commons and the traditions can be reduced to a picturesque nostalgia, from which an individualized citizenry is easily severed.

The political landscape of the commons changes depending upon whether it is practiced by a community sharing use rights and obligations, or property owners using it as the background against which various social, recreational and cultural activities are performed. As noted, private property was a keystone in the liberal economic and social ideas of thinkers like Smith. By contrast the ideal of a commons can provide a powerful metaphor for the political left, ranging from grass-roots democrats to socialists and anarchists (Casarino and Negri 2008; Olwig 2005a, b) (Fig. 3.7).

Historically, however, as indicated earlier, many agricultural systems have combined an intense, specialized, production oriented horticultural and arable regime where particular land parcels are (at least periodically) under the stewardship of individual farms, with extensive grazing and subsistence regimes involving complex shared use rights and obligations in the diverse landscape of the commons. The commons, thus, can also be part of a larger system that allows for more individualized and hierarchical stewardship of land parcels suitable for intensive production. This dialectic between individualized and communal access to productive and reproductive resources fits well with the theories of the anthropologist Victor Turner (1974) concerning the character of community. He has thus argued that societies do not necessarily have a monolithic constitution as either hierarchical and

Fig. 3.7 *Left* commemorative plaque at quarry where the workers' movement triggered the Kinder Scout Trespass in 1932, Hayfield, Peak District. *Right* Annual Mayday Demonstration on the Copenhagen Commons. The sign reads: "Take the Grass Roots Seriously." Pictures KRO

individualized, or flat and communitarian, but rather can have need of both forms of social organization under differing circumstances (1974).

According to Turner (1974), the means by which societies deal with the need, in some contexts, to maintain hierarchical status relations (as between a citizen and a policeman or a judge in daily legal situations) and in other contexts to maintain a sense of equal community solidarity (as in the case of disaster) is to mediate the mental threshold between them through liminal, threshold crossing rituals, such as pilgrimages. A pilgrimage, as described by Turner (1974), thus typically moves from the urban or village centre of a society, where society is structured according to hierarchical status, and outward to a religious site located in a more marginal rural area of the sort often devoted to grazing commons. In making this geographical movement the pilgrims cross an imperceptible threshold, or limen, which allows a change of mind-set so that a more egalitarian sense of 'communitas' develops. 'Communitas' is the state or character of community that develops between people in a community, much as fellowship develops between fellows. The same sense of communitas can also occur when people cross a limen in time, as in the case of a liminal holiday (holy day) event, such as the bonfires or maypoles of spring and midsummer, where, in the spirit of comedy, class differences can be turned on end and communitas generated (Bakhtin 1984). Severance from the cultural commons, which functions as a landscape of communitas, thus can threaten the balance that holds a sustainable society together.

3.6 Globalism and the Enclosure of the Landscape Commons

A map is basically a flattened out section of a globe. There is no essential difference between the space of the globe, that of the map, and that of the perspective representation of landscape. A globe is a perfect geometrical shape with a uniform spatial surface, whereas the earth has an irregular ball-like topological shape. The space of land surveyed as property is thus fundamentally a global space in the sense of a universalized space that, in principle, can be scaled up from the smallest square unit of property to the level of the globe itself (Marston 2000). Scale requires a common denominator, be it pitch in music or the map's absolute space. Property is scalable because it can be valued according to scalable square spatial units as valued according to a uniform absolute monetary measure in a market that can factor in everything from spatial location and view to mineral deposits and soil. Enclosure, thus, was not simply a local matter, but a product of a society in which the enclosers were increasingly engaging in a global economy, and many of the estate owners who benefited from enclosure owned property in colonies in widely scattered corners of the globe (Seymour and Stephen Daniels 1998).

'Globalism' can be defined as a mode of thought deriving from the practice of 'thinking globally,' both literally and figuratively (Olwig 2011b).[3] The ideology of globalism is presently identified particularly with neoliberal economic arguments for a single world economy (Stiglitz 2003; Herod 2009). The global economy was from the start, as has been seen, facilitated by the enclosure of the historical European commons as property. The commons, however, can also be conceptualized more broadly in world terms than the historical European example has allowed. Gudeman does this when he argues that.

> The commons is a shared interest or value. It is the patrimony or legacy of a community and refers to anything that contributes to the material and social sustenance of a people with a shared identity: land, buildings, seed stock, knowledge of practices, a transportation network, an educational system, or rituals (Gudeman 2001: 27).

The world's landscapes, as a material and cultural phenomenon, might following this line of thinking, be regarded, to some degree, as a commons.[4] The notion of enclosure can thus be applied when global economic interests cause the enclosure, privatization and intensification of the industrializing world's community forests (Goldman 1998), along with the paths, ponds, etc., that, together with the historical commons, make up the common landscapes of industrialized Europe.

The macro-and micro-climate is a component of the landscape commons of the world; and it is affected when forests are destroyed, commons are ploughed under; ponds and wetlands are drained; or wind breaking hedges and fences are removed to create uniform production areas for the world economy. Ironically, the same globalist argumentation that has helped bring about the privatization, intensification and uniformity of the landscape, also has been applied to the solution to the environmental problems that have resulted. This is particularly notable with regard to global warming. Globalist thought, as noted, involves scaling from the global to the local according to singular common denominators such as space, property, or

---

[3] 'Globalism' is defined by the *Oxford English Dictionary* as 'the belief, theory, or practice of adopting or pursuing a political course, economic system, etc., based on global rather than national principles; an outlook that reflects an awareness of global scale, issues, or implications' (O.E.D 1989).

[4] This is arguably the *European Landscape Convention's* premise (Olwig 2007).

money. Global warming, from the globalist perspective, thus trumps other environmental threats because it involves factors such as heat and carbon circulation that apparently can be scaled from the global down to the local and back up to the global and which permeate the space of the globe. Globalist argumentation has thus lead to measures such the production of large scale nuclear plants, carbon credits and the planting of 'CO_2 neutral' energy crops, the effect of which can be measured in terms of the single, scalable and "monetizable" variable of carbon. Such solutions to global warming emphasize the carbon free *production* of the energy necessary to the continued expansion of the global economy that is central to the globalist worldview. But it is also, ironically, this expansion that is arguably behind the landscape destruction that brings about both climate change and the vulnerability to climate change.

Rather than tackle issues of climate change via globalist modes of thought it would arguably make more sense to fight climate change, the depletion of landscape diversity, and the impoverishment of our shared environment, by refocusing on the principles and practices that have historically produced sustainable commons. Thus, rather than focus on the production and sale of greater amounts of carbon neutral power, thereby helping to stimulate the economic growth that encourages even more resource use, we might pay attention to energy perceived in terms of the conservation and reproduction of the common resources embedded in the landscape over generations. Such resources range from sheltering hedges, with their reserves of wood fuel, to the "carbon-sinks" of bogs, to buildings that nestle in the terrain where they are not robbed of heat by the wind. The energy savings from well insulated and well positioned buildings and infrastructure save energy more cheaply than CO_2 neutral energy can be produced (Lovins 2004; Nadaï and van der Horst 2010). Even the production of energy is more efficient when it occurs as part of a community effort in which waste heat and economic benefits can be shared amongst households, and the environmental consequences are near at hand, and subject to public debate (Krauss 2010).

3.7 Conclusions

It has been argued here that the commons should not be seen as a singular phenomenon, but as a component of larger landscapes incorporating both commons, and individually controlled non-commons. This suggests that the 'cultural commons,' with its 'communitas' ideal, should not be seen as an argument against the existence of individually controlled access to bounded resources, but rather as an argument for the need to counterbalance individual acquisitiveness with community needs and control. When the commons are under threat, there is thus arguably a need to maintain the social and cultural practices that have produced them in order to preserve those dimensions of the landscape that provide, via their diversity, a reproductive environmental, social and cultural buffer against over exploitation by private interests.

In the forgoing the focus has been upon the historical commons of England in particular. This has not precluded, however, the application of the analysis to a broader understanding of the commons at a worldwide scope. A commons in this analysis is something to which people have use rights and obligations, and which, through the shared exercise of those rights and obligations, produces a community of commoners sharing cultural norms and practices. Such a commons may take many forms, ranging from a tropical rain forest, to a national or local park, to the waste disposal area of a New England township, to an academic discipline of a community of scholars studying the commons. It is important to such commons that there be a living connection between such 'material' commons and the cultural commons that help constitute social 'communitas'. Without an interaction between living customs, including time out of mind rituals and holidays, and the substantive use of commons, people will be severed from both their material commons and the community identity necessary to maintain them, and all they will be left with is the nostalgia of picturesque tradition[5].

References

Bakhtin M (1984) Rabelais and his world. Indiana University Press, Bloomington

Barrell J (1972) The idea of landscape and the sense of place, 1730–1840. Cambridge University Press, Cambridge

Bushaway B (1982) By rite: custom, ceremony and community in England 1700–1880. University Press, Junction Books, London

Butlin RA (1982) The transformation of rural England c 1580–1800. Oxford University Press, Oxford

Casarino C, Negri A (2008) In praise of the common: a conversation on philosophy and politics. University of Minnesota Press, Minneapolis

Clayden P (1985) Our common land: the law and history of commons and village greens. The Open Spaces Society, Henley-on-Thames

Cosgrove D (1984) Social formation and symbolic landscape. Croom Helm, London

Daniels S (1989) Marxism, culture and the duplicity of landscape. In: Peet R, Thrift N (eds) New models in geography II. Unwin and Hyman, London, pp 196–220

Edwards V (1995) The New Forest commons, Hampshire: horse-riders, mountain-bike riders, free-riders. Department of Land and Construction Management. University of Portsmouth, Portsmouth

Eversley L (1910) Commons, forests and footpaths. Cassell, London

Goldman M (ed) (1998) Privatizing nature: political strugles for the global commons. Pluto Press, London

Gudeman S (2001) The anthropology of economy: community, market and culture. Blackwell, Oxford

Hardin G (1968) The tragedy of the commons. Science 162:1243–1248

Hardin G, Baden J (1977) Managing the commons. W. H Freeman, San Francisco

Herod A (2009) Geographies of globalization: a critical introduction. Wiley-Blackwell, Oxford

---

[5] I thank Graham Bathe for his insightful critical reading of this text. Any remaining errors are fully my responsibility.

Hobsbawm E (1983) Introduction. In: Ranger T, Hobsbawm E (eds) The invention of tradition. Cambridge University Press, Cambridge, pp 1–14
Kjærgaard T (1994) The Danish revolution, 1500–1800: an ecohistorical interpretation. Cambridge University Press, Cambridge
Krauss W (2010) The 'Dingpolitik' of wind energy in northern German landscapes: an ethnographic case study. Landsc Res 35(2):195–208
Lovins AB (2004) Energy efficiency, taxonomic overview. Encyclopedia of energy, vol 2. Elsevier, Oxford, pp 383–401
Marston SA (2000) The social construction of scale. Prog Hum Geogr 24(2):219–242
Meinig DW (1979) Symbolic landscapes: some idealizations of American communities. In Meinig DW (ed) The interpretation of ordinary landscapes: geographical essays. Oxford University Press, New York, pp 164–192
Milsom SFC (1981) Historical foundations of the common law. Butterworths, London
Nadaï A, Van der Horst D (2010) Landscapes of energies. Landsc Res 35(2):143–155
NOAD (2005) New Oxford American dictionary. McKean E (ed) Oxford University Press, Oxford
O.E.D. (1989) Oxford English dictionary. Clarendon Press, Oxford
Olwig KR (2002) Landscape, nature and the body politic. University of Wisconsin Press, Madison
Olwig KR (2005a) The landscape of 'customary' law versus that of 'natural law. Landsc Res 30(3):299–320
Olwig KR (2005b) Representation and alienation in the political landscape. Cult Geogr 12(1):19–40
Olwig KR (2007) The practice of landscape 'conventions' and the just landscape: the case of the European Landscape Convention. Landsc Res 32(5):579–594
Olwig KR (2008) Performing on the landscape vs. doing landscape: perambulatory practice, sight and the sense of belonging. In: Ingold T, Vergunst JL (eds) Ways of walking: ethnography and practice on foot. Ashgate, Aldershot, pp 81–91
Olwig KR (2011a) The right rights to the right landscape? In: Egoz S, Pungetti G, Makhzoumi J (eds) The right to landscape: contesting landscape and human rights. Ashgate, Aldershot, pp 39–50
Olwig KR (2011b) The earth is not a globe: landscape versus the 'globalist' agenda. Landsc Res 36(4):401–415
Pearson M (2006) 'In Comes I': performance, memory and landscape. University of Exeter Press, Exeter
Plato (1961) Statesman. In: Hamilton E, Cairns H (eds) The collected dialogues of Plato. Pantheon, New York, pp 1018–1085
Putnam MCJ (1970) Virgil's pastoral art. Princeton University Press, Princeton
Rodgers CP, Straughton EA, Winchester AJL, Pieraccini M (2011) Contested common land: environmental governance past and present. Earthscan, London
Seymour S, Stephen Daniels CW (1998) Estate and empire: Sir George Cornwall's management of Mocas, Herefordshire and La Taste, Grenada, 1771–1819. J Hist Geogr 24:313–351
Stiglitz JE (2003) Globalization and its discontents. Norton, New York
Thompson EP (1993) Customs in common. Penguin, London
Turner V (1974) Dramas, fields and metaphors: symbolic action in human society. Cornell University Press, Ithica
Williams R (1973) The country and the city. Oxford University Press, New York
Wood JS (1997) The New England village. Johns Hopkins University Press, Baltimore

Chapter 4
A Natural Origin of the Commons: Interactions of People, Animals and Invisible Biodiversity

Ted Green

4.1 Introduction

Prior to the publication of Frans Vera's work *Grazing Ecology and Forest History* (Vera 2000) ancient oaks and other open grown trees, especially light demanding species, had led field ecologists to examine the early development of wooded landscapes. In the UK, scholars such as Francis Rose (*pers. comm.* 1993) as early as the 1960s, Oliver Rackham (1986, Mick Crawley (Dobson and Crawley 1994) and Keith Alexander (1999), to begin to question the long held beliefs about landscape origins. This was especially in terms of the extent of the natural closed-canopy forest of Europe and ideas were emerging that areas of temperate Europe were perhaps not covered with a continuous unbroken tree cover. Indeed, perhaps some areas never were treed (Green 1998), and Francis Rose (1993) had already come to the conclusion that some horse grazed meadows in Norfolk had always remained open (*pers. comm.* 1993). To support Vera it is suggested that, in the 'Vera landscape' of savannah, commons and wastes, parkland, wood-pasture, groves and forest, pests and diseases could create glades and large open areas in woodland or other vegetation. In some cases, these agents could also maintain them, and at times, they could play a significant role. Browsing and grazing animals including Wild Boar, *Sus scrofa*, Beaver, *Castor fiber* and in time man, could continue to maintain the open spaces until they themselves were affected by diseases and pests after stress events. These might be such as prolonged drought, leading to a sudden large drop in numbers. Thus this 'window' of small numbers of surviving animals would allow the colonisation to some degree by trees and shrubs, especially around or near the margins of isolated trees, small groves and woodland edges. The cycle of partial tree cover could then begin again—until the

T. Green (✉)
Ancient Tree Forum, 22 Reeve Road, Holyport, Maidenhead SL6 2LS, UK
e-mail: edwardgreen629@btinternet.com

I. D. Rotherham (ed.), *Cultural Severance and the Environment*,
Environmental History 2, DOI: 10.1007/978-94-007-6159-9_4,

arrival of man in sufficient numbers to have a lasting impact. Man colonised the savannah and glades, and gradually expanded and modified them for the final time, or perhaps until he too succumbs.

In supporting Vera's hypothesis on the role of grazing animals, the question remained about where, and how, large grazing animals—such as bison, *Bison bonasus*, tarpan, *Equus przewalski gmelini*, aurochs, *Bos primigenius*, elk, *Alces alces* and red deer, *Cervus elaphus*, could find sufficient food in the reduced light levels of extensive closed-canopy forest (Green 1998). Equally, in ideal circumstances, they all have different feeding strategies and therefore hard to contemplate how they could survive on the restricted range of plants available. The assumption was that forests were usually comprised of both shade tolerant and light-demanding tree species of all ages—'assumption' is used because it appears that there is very little understanding of the interactions and consequences between shade tolerant and light demanding trees when growing together.

Whilst the majority of shade tolerant trees appear to be able to grow successfully as open-grown trees in totally exposed conditions similar to light demanding species there are for example, in times of extremes of temperature and drought, these species will show adverse symptoms and react much earlier in comparison with light demanding species. In fact, Vera points out that light-demanding trees in competition with shade tolerant trees in a forest, will eventually be out-competed, and even if they manage to achieve some sort of crown in the canopy it will be severely reduced and, in consequence, reduced pollen and seed production which is especially noticeable with Oaks. Light-demanding trees, especially Oak and Scot's Pine, and perhaps several fruit bearing trees and shrubs such as Hawthorn, Hazel, Blackthorn and Crab Apple, are good examples of plants adapted to open conditions as all have difficulty producing pollen and fruit in shaded conditions. The decline of Oak in ungrazed non-intervention forest reserves is well known, e.g. in Fontainebleau Forest where the former Oak areas have become dominated by Beech—the key change in management being removal of grazing by the local commoners. From evidence of the 1987 hurricane in the UK and the hurricane in France at the end of 1999, it is clear how difficult, if not impossible, it is for grazing animals to move and exploit open areas in the chaos of fallen trees and the mass of fallen limbs. It would take many years before animals could return to these areas unless Man intervened and cleared away the impenetrable walls of decaying or 'layering' wood that would otherwise take several years to decay. The protection could perhaps result in tree regeneration that could go straight to a thicket stage outstripping the reach of animals on their return. Evidence from the 1987 hurricane in the New Forest, has shown that the only subsequently successful tree regeneration—other than Vera's normal thorn-protected trees, is in these impenetrable areas. Areas and glades that could support grazing animals are not created. Any treeless areas within continuous forest have been explained away as the result of abiotic events such as fire and hurricanes, and could include landslips, avalanches and even Beaver work. However, experts in the UK such as Oliver Rackham (1986) were questioning these explanations and commented that "*broadleaved woodland burns like wet asbestos*". Charles Edlin (1956), in charge of fire fighting in the south-east of England for the Forestry Commission from 1940 to 1945, commented that—during the hot dry summer of the 'Battle of Britain' in the skies in 1940—"*Incendiary bombs dropped either in clusters or singly and which ignited in broadleaved woodland of any kind simply refused to burn at any time*". Furthermore, the interval between the occurrences of hurricanes can often be centuries and, in the case of both wind and fire, the event could be incredibly random. The randomness would apply to the interval between occurrences, extent of occurrence, region of event and the severity of the fire or hurricane. It seems highly improbable that large herbivores evolved to rely for survival on such tenuous and unpredictable systems.

4.2 Open Areas in Forests

The explanation of the origins of glades and open areas in woodland and forests provided by Frans Vera (2000) is that open areas expand from the centre of ageing forest through the death of trees, which in most cases are the oldest trees. Stress and increased age are considered to assist the natural biological and physical agents involved in the decline of the trees. Trees in company with other large, long-lived organisms seldom die naturally from the effects of a single organism. In simple terms, physical actions together with the natural ageing process appear to allow pathogens and parasites to accelerate their host's demise. Therefore, stress brought about through, say, drought or competition can allow pathogens, pests and parasites to gain the upper hand and accelerate the demise of any surviving trees.

Unfortunately, there remain only a few very small areas of Europe where forests with expanding and receding open areas are being allowed to develop through natural processes. So examples are very limited with or without grazing animals; however there are situations where small glades may be created in high winds when the crown of a tree snaps off or a tree blows over or loses the majority of its limbs; a sudden event creating a space in the canopy for perhaps only short time. In addition, when an individual tree dies and remains standing it has usually declined over a period of years allowing the surrounding trees in both cases to gradually exploit the space and to close-over, especially with broadleaved trees.

4.3 The Role of Pathogens and Pests

While many plants, animals and other organisms classified as economic pests or diseases are well studied and researched in order to reduce their impact on the commercial world, the part that they play in the natural world is rarely considered. It is true to say that most ecologists pay scant regard to—or have little or no understanding of—the part 'pathogens' and 'pests' play in 'managing' natural ecosystems (Green 1992; Dobson and Crawley 1994). It cannot be stressed enough that the vast bulk of fungi, protozoa, bacteria, nematodes, invertebrates and the billions of other micro-organisms that occupy the planet, are essential and fundamental to the very life of the natural world in the complex web of life—they provide the essential 'ecosystem services' that are so popular in the academic world today. Although the 'pests' and 'pathogens' appear to be detrimental to their hosts, they are essential in the natural world to the continuation of the ecosystem as a whole. By accepting the role of pests and diseases in a woodland or 'tree-ed' ecosystem it becomes much clearer how open areas—usually recognised whilst they are only a small area—could be initially created and developed perhaps with the aid for example of small mammals, nematodes, molluscs, invertebrates and micro-organisms.

4.4 The Role of Fungi

However, there are examples of natural fungal growth patterns—probably in partnership with other micro-organisms—in plant communities best illustrated by looking at 'fairy rings' which can be often be seen in old meadowland or in an organic lawn, when the toadstools often appear in a perfect circle, where the vegetation is usually more vigorous and brighter in colour. At Stonehenge some of the rings in the ancient sward have been calculated to be centuries old, and one ring of honey fungus group *Armillaria ostoyae* in a North American conifer forest in Washington State is up to 50 km in diameter and covers 600 hectares is said to be the largest known organism on the Earth. Another fungus-related open area in the Malheur National Forest in Oregon is 890 hectares and approximately 2,400 years old! If, in these examples, *Armillaria ostoyae* is the major pathogen of trees and shrubs, then one could argue that—as the ring of *Armillaria* expands and progresses through the forest—it contributes to the death of the odd tree or clump and creates glades. Possibly the best illustration in the UK to date of this very natural event are two clearings in an area of 60-year-old Yew *Taxus baccata* at Kingley Vale, Sussex. The Yew colonised open chalk downland when it ceased to be grazed from the 1940s and developed an even-aged closed canopy. However, the causes of the death of the Yews remain unidentified. In today's environment, so changed and manipulated by Man, natural open areas are rarely able to form. However, in the two examples in North America in the conifer forests of both Oregon and Washington, massive areas of trees have died, with the fungus *Armillaria* claimed to be the agent. However, the term 'claimed' is used because opinions are changing in favour of not blaming one single organism for the demise of another species, but rather a series of contributing factors and suites of organisms. Chris Brasier (1999) points out that primary and recurrent *Phytophthora* root damage are likely also to predispose trees, to attack by other organisms, such as *Armillaria* or bark beetles.

4.5 Dispersal of Fungal and Fungal Like Spores by Water and Animals Including of Course Man

Another good example, in a different type of habitat, are the fungal-like pathogens the *Phytophthoras* or the heather beetle, *Lochmaea suturalis* which often cause a quite sudden and dramatic death to quite large patches of the man-made monoculture Heather deserts of *Calluna vulgaris*, so typical of the southern and upland heathlands in Britain today. The mosaic of dead and dying patches is quite a good illustration to compare with a pathogen or a pest in closed-canopy forest and woodland. This is especially the case where Man is growing light-demanding tree species and shade-tolerant trees leading to the light-demanding trees being constantly under stress from competing shade-tolerant species. The latter are much more suited to the growing conditions. Although the Heather example is a highly unnatural system, it shows how a pathogen or pest could move through a plant community, creating a mosaic of different stages of growth. In this case, Man has unintentionally helped the pathogen or pest, because Heather moor is not only a man-made habitat subject to regular burning (which removes essential minerals and micronutrients together with a whole suite of the recycling organisms, many of which can assist in a form of biological control on pathogens), it is also often severely stressed. This is because it is growing in poor ground conditions caused by centuries of sheep occupation, a land use that often causes continual compaction with the subsequent waterlogging that aids the advance of some *Phytophthora* species. There is also a whole suite of damaging veterinary products that are never tested for their environmental impact and which are secreted by these alien animals. Incidentally, one of the very successful strategies for the transportation of spores around the world, and used by some species of the *Phytophthora* group that can be pathogenic to plants, is water. With today's modern world, mass-movement of rooted plants and even Man's walking boots, spores can be widely dispersed very easily! Therefore, it is interesting to speculate that a well-used, compacted, animal or man-made track across grassland, heather, moorland or through forest can sometimes hold water for a length of time that could allow spores to migrate. Animals including Man could transport spores, too, on their feet, legs and vehicle tyres, to and from the water. The situation would be similar at regular communal watering places and wallows, which attract animals from the surrounding area. Depending on species behaviour, and under normal circumstances where individuals, groups, or herds never mix, the watering place or water-filled track could be a major point of cross-contamination not only for passing on animal diseases and parasites but also to plants via the animal vector.

4.6 Interaction Between Diseases of Grazing Animals and Tree Generation in Open Areas

There are also examples of diseases and parasites affecting populations of grazing animals by reducing their numbers, which in turn would allow the generation of trees and shrubs. In the UK when the Rabbit, *Oryctolagus cuniculus* population virtually disappeared in the mid-1950s due to the virus *Myxomatosis*, Oak became established from acorns buried by jays and wood mice in open areas in the resulting ungrazed meadows up to 75 m from the nearest seed source. Several other tree species appeared including Elm, *Ulmus procera* suckers that continued to extend annually until the return of the Rabbit. However, further tree establishment ceased in subsequent years when the meadows became too dense and rank. Similarly, in East Africa, the different age classes of the groves of *Acacia tortilis* can be linked directly to the periods of reduction in populations of grazing animals through outbreaks of disease (Dobson and Crawley 1994). There is also an account in *The Drove Roads of Scotland* (Haldane 2006) of drovers bringing cattle to markets in the south from as far afield as the Isles losing all their animals to disease before they had reached their resting destinations in Norfolk and Suffolk before sale. "*There, the cattle disease which raged that autumn throughout England attacked them, and his drove died first in scores and then in hundreds*"; "*our conditions are such that several drovers have run their beasts and left them dying in the lanes and highways and nobody to own them*". Sir Arthur Oliver, Principal of the Royal Veterinary College writes: "*A serious outbreak of cattle plague (rinderpest) occurred in Great Britain in 1745 following it's extension throughout the greater part of the continent*". The outbreak lasted until 1757 though cattle plague existed in some parts up to 1770, and there is no record of any other major outbreak of cattle plague in this country again until 1865. The mortality was exceedingly high, probably running to 90 per cent in badly affected herds. It is very unlikely that the cause of the cattle's death was specific and therefore quite possible that a large proportion of other grazing and browsing animals, whether domestic or wild, could have succumbed over a large area. Obviously, a reduction in animal numbers would have positive implications for tree and shrub colonisation for several years. Perhaps it is worth posing the question about the impact the decrease in the Human population had on tree cover after The Black Death. Therefore, there can be situations in the forest where trees die, assisted by diseases, and glades/savannah can develop. These areas are grazed by herbivores perhaps later assisted by Man but when animal numbers are reduced by diseases, often exacerbated by periods of prolonged drought, there are opportunities for the forest to regenerate to some extent and generally around the margins of surviving isolated trees, groves and edges of woodland (Figs. 4.1, 4.2).

Fig. 4.1 Animal tracks criss-crossing heathland—many may be centuries old and could be focal points for the transmission of diseases. Credit Ted Green

Fig. 4.2 Oak generation and elm suckering after the crash in the rabbit population from myxomatosis in the mid-1950s. Credit Ted Green

4.7 Oaks and Jays

Evidence for tree and shrub generation is found in the fascinating evolutionary relationship between the Oak and the Jay, *Garrulus glandarius* and other members of the crow family. Azure-winged Magpies, *Cyanopica cyanus* and the evergreen oaks, *Quercus ilex* in Spain and Portugal, the nutcracker, *Nucifraga caryocatactes* and the Arolla Pine, *Pinus cembra*, and other tree seeds in European mountain ranges, and the Rook, *Corvus frugilegus* and the Walnut, *Juglans* ssp. It is also very likely, but as yet not widely recognised, that small mammals such as Wood Mice and other rodents can play an equally significant role in tree seed dispersal (M.Crawley and Willoughby pers. comm.). In most cases, trees that get successfully established will either be of an isolated open-grown form or small groves created by a rodent cache after the demise of its gatherer and, in both cases, protected by thorns. Thrushes and several mammal species are perhaps the key agents in distributing berries and other soft fruit tree and shrub seed with excreta playing a significant role. In other words, again in support of Vera, an acorn buried in an open space by a Jay or a small mammal and left to germinate can become a tree which, when able to grow unrestricted without competition from other trees, can develop into what is termed an 'open-grown' form, optimising its light gathering power by developing a shape similar to a sphere. Thus producing copious volumes of pollen. In later years, if the browsing animals re-appear, the lower limbs within reach could be browsed and the light levels increased beneath the canopy. In 1904, William Menzies speculated about the Oaks at Windsor: "*It is interesting to conjecture how these ancient trees, both in the Park and the Forest, sprung up, and survived, unprotected as they were, against cattle and deer. It is probable that many of them must have come up in the thickets of thorns and bushes, and by the sides of roads and division fences*".

Menzies (1904) also quotes an old writer, Arthur Standish (1613), who speaks of bushes as "*the Mother and Nurse of Trees*", and that "*but for them there would be no timber in the common land*" and goes on to say "*There is an old forest proverb: the Thorn bush is the mother of the Oak*". It is remarkable to think that even by 1613 this proverb was considered old! In the UK, Ancient trees especially open grown have an incredible wealth of associated biodiversity—visible and invisible—and their great age provides biological links and continuity with past generations of trees. As Rackham has stated, 10,000 oaks of 100 years old are no substitute for one 500-year-old oak (Rackham 1986). They are also significant features in those landscapes with perhaps the closest resemblance to the savannah that graced the European landscape (Figs. 4.3, 4.4).

Fig. 4.3 Solitary oak generation—the 'thorn is the mother of the oak'. Credit Ted Green

Fig. 4.4 Scots pine establishment within the protection of juniper (smell could perhaps also act as a deterrant to browsing animals). Credit Ted Green

References

Alexander KNA (1999) The invertebrates of Britain's wood-pastures. Br Wildl 11(2):108–117

Brasier C (1999) Phytophthora pathogens of trees; their rising profile in Europe. Forestry information note

Dobson A, Crawley M (1994) Pathogens and structure of plant communities. Tree 9(10):393–398

Edlin HL (1956) Trees woods and man. New Naturalist, Collins

Green T (1992) The forgotten army. Br Wildl 4(2):85–86

Green T (1998) Fungi first. In: Proceedings of bring back the bison. Farnborough, Hants

Haldane ARB (2006) The drove roads of Scotland. Birlinn Ltd, Edinburgh

Menzies W (1904) Windsor park and forest. London

Rackham O (1986) The history of the Countryside. J. Dent and Sons, London

Rose F (1993) Ancient British woodlands and their epiphytes. Br Wildl 5:83–93

Vera FWM (2000) Grazing ecology and forest history. CABI Publishing, Wallingford

Further Readings

Green T (1996) Pollarding—origins and some practical hints. Br Wildl 18(2):100–105

Green T (2000) Growing downwards. Tree Line, UK and Ireland Chapter of ISA

Green T (2001) Should ancient trees be designated as sites of special scientific interest. Br Wildl 12:164–166

Green T (2002) The role of invisible biodiversity in pasture landscapes. In: Redecker B, Finck P, Hardtle W, Riecken U, Schroder E (eds) Pasture landscapes and nature conservation. Springer, Netherlands

Green T (2010) Natural origin of the commons: people, animals and invisible biodiversity. Landsc Archaeol Ecol 8(1):57–62

Green T (2013) Ancient trees and wood pastures. Trees, forested landscapes and grazing animals. Routledge, UK, pp 127–142

Merryweather JW (2001) Meet the glomales—the ecology of mycorrhizal. Br Wildl 13(2):86–93

Merryweather JW (2006) Secrets of the soil. Resurgence 235:26–28

Merryweather JW (2007) Planting trees or woodland? An Ecologist's perspective. Br Wildl 18(4):250–258

Part II
Case Studies of Cultural Landscapes from Around the World

Chapter 5
Valorising the European Rural Landscape: The Case of the Italian National Register of Historical Rural Landscapes

Mauro Agnoletti

5.1 The Italian Case in the European and International Context

Europe's rural areas are diverse in terms of geography, population, demography, economic and social structures as well as labour markets. It is this diversity that is part of its richness and that has also created an extraordinary diversity of landscapes, which today are playing an increasingly important role. Europe's rural landscape represents 91 % of the territory in EU 27 and about 56 % of the population live in predominantly and significantly rural areas. Rural areas generate 45 % of gross value added in EU 27 and 53 % of the employment. Nevertheless, many of Europe's rural landscapes face a common challenge as their dynamics and quality are mostly triggered by socio-economic developments affecting the rural world. The changes in technology, culture and economy at world level threaten them, the traditional landscapes and the biodiversity on which they are based, and even the structure of rural society (Agnoletti 2006). Complex landscapes where the population had to establish specific management practices to adapt to the local environment, are rapidly disappearing. Farmers are often compelled to develop innovation and to adopt unsustainable practices. In doing so, they are overexploiting resources and contributing to the genetic erosion and loss of the cultural identity of places. This interrupts the transmission of important cultural heritage from one generation to another. The accumulated knowledge and the experience in the management and in the use of local resources is a significant wealth at world level expressing the cultural identity of each ethnic group. Unfortunately, there has

M. Agnoletti (✉)
Laboratory for Landscape and Cultural Heritage, Facoltà di Agraria – Università di Firenze, Via San Bonaventura 13 50145 Florence, Italy
e-mail: mauro.agnoletti@unifi.it
URL: www.landscape.unifi.it

I. D. Rotherham (ed.), *Cultural Severance and the Environment*, Environmental History 2, DOI: 10.1007/978-94-007-6159-9_5,

not been a clear recognition of the significance of landscape resources in rural development strategies. Neither has the role of rural landscape for society been recognised in key international documents or in policies concerning sustainability. In this respect, political entities like the European Union (EU) are very interesting case studies of the effects of a common agricultural policy affecting different nations with different histories and for the influence. Despite the evident importance of the rural regions for landscape quality and socioeconomic aspects, agricultural policy in the past decades has favoured the degradation of cultural landscapes. There has been little recognition of the relationships between typical products and local landscapes, or services like agri-tourism, based on landscape, and few actions for the conservation of the cultural values of traditional landscapes, or for saving the cultural identity of the European regions. From this point of view, the case of Italy is an interesting one. Italy has not only been a rural country for most of its history, and one of the most important ones in Europe, but since the sixteenth the value of its rural landscape has been recognized by many European travellers. Today, rural tourism has become one of the most important revenues for farmers. The inclusion of landscape among the strategic objectives of the *National Strategic Plan for Rural Development* 2007–2013, developed by the Ministry of Agriculture, Food and Forestry, has finally recognized the importance of landscape resources for economic, environmental and social reasons. The Ministry has been funding regional governments to support planning and conservation according to the European Landscape Convention, and promoted a number of projects for the conservation and valorisation of rural landscapes. The Convention already took into consideration the European rural landscape not only during its preparatory works, but also in organizing a conference dedicated to the rural heritage in Sibiu (Romania) in 2007 (Council of Europe 2009). Although few initiatives have been undertaken to promote the Convention in the framework of the Common Agricultural Policy, the case of Italy shows that a landscape-based approach in rural policy, is probably one of the best ways to take integrate economic development, environmental conservation and quality of life. However, it should be made clear that the concept of landscape conservation should not necessarily overlap that of nature conservation, which may imply very different strategies and actions.

The national catalogue of historical rural landscapes is one of the projects promoted by the Ministry of Agriculture. Based on the idea of creating a catalogue, this has evolved into an open register following the decision to establish a national inventory of historical rural landscapes and traditional practices, and developing specific political initiatives (Agnoletti 2012). In the meantime, there has been the development of a planning approach to the whole rural territory, considering landscape as the most effecting paradigm for the development of the rural territory (Agnoletti 2010b). In this respect, as also occurred in urban planning where historical city centres were preserved for their value all over Europe, the first concern was for the most fragile historical landscapes. This view seems to be very useful also for the future development of the Common Agricultural Policy, which will place more attention on environment and landscape (Van Der Ploeg 2006).

Concerning the relevance to the European Landscape Convention, the project specifically refers to article 6.C.1, requiring identification and assessment. According to this each party undertakes: a. (i) to identify its own landscapes throughout its territory; (ii) to analyse their characteristics and the forces and pressures transforming; (iii) to take note of changes; b: to assess the landscapes thus identified, taking into account the particular values assigned to them by the interested parties and the population concerned. The projects also meet the aims of the FAO research programme "Globally Important Agricultural Heritage Systems", that is developing an inventory of traditional rural landscapes mainly in developing countries. According to an agreement between the Italian Ministry of Agriculture and FAO, some of the areas of the national register will be proposed for GIAHS. Both the National Register and FAO GIAHS are putting into evidence the importance of biodiversity at landscape level. In this respect, another important aspect of this research is the relationships with the Joint Programme UNESCO-CBD on biocultural diversity. This is particularly important since the biodiversity is usually related to natural habitats for wild species, while in this case it is the product of the historical relationships between man and nature and the traditional practices developed by farmers.

5.2 The Dynamics of Italian Landscape in the Last Century

To fully understand both the situation "captured" by the catalogue, and the urgency of such an investigation, we need to note the evolution of the Italian rural landscape since the country's unification. This is not so much in terms of socio-economic changes, but rather as regards land use, which gives a measure of the dramatic changes that affected the quality of the landscape in this period. It is undoubtedly a limited time-frame, considering the remote historical origins of the Italian landscape. However, as environmental historians have shown (McNeill 2002), this has been a time when the abundance and intensity of changes at the global level has occurred with a swiftness that has no precedent in the history of human civilization, and Italy is no exception. At least until the second post-war period, much of the country's rural landscape was still strongly influenced by traditional agro-silvo-pastoral models developed during the previous century. Sometimes these origins go all the way back to the Etruscan period and Greek civilization. The following decades, however, witnessed deep transformations. Due to demographic growth and the expansion of agriculture into mountain areas, the rural landscape attained the peak of its development in the decades between the late nineteenth and early twentieth centuries. The resulting landscape was one of great complexity, enhanced by the stratification of the prints left by so many civilizations on the land, and the country's complex orography and climatic variability. In the second post-war period, however, we observe a gradual simplification and homogenization of the rural landscape, that lead to situation like the one observed for Tuscany were more than 45 % of landscape diversity was lost due to abandonment and industrialization in agriculture (Agnoletti 2010a). At national level in the last century, we observe the strong reduction of farmed land, the increase of woodland and unproductive land, including also urban areas, as urban expansion partially accounts for the increase of unproductive surface in our country (Fig. 5.1).

Urban growth is often branded as the main enemy of the rural landscape, something on which there is usually a broad agreement among the public, farmers and environmentalists. While it is true that the permanence of agriculture acts as a barrier against urban expansion, it is equally true that the most significant changes in the rural sector are due to abandonment, on the one hand, and endogenous changes on the other. These are not as obvious, but much more in-depth and enduring. In this respect, many European rural landscapes, although protecting the countryside against abandonment urban sprawl, are often simply open air "factories" of wheat, corn, or wine, retaining very few elements of what once was a rich heritage of traditional farming systems.

Urban surface, according to the most up-to-date European mapping system (Corine Land Cover 2000), does not exceed 5 % of the total surface of Italy. It is true however, that scattered urbanization eludes Corine. The Italian Ministry of Agriculture, Food and Forest Policies hence resolved to establish a new category

Fig. 5.1 1:250,000 map of the Italian territory resulting from an interpretation of Corine Level 4 data. The map highlights the polarization of the rural landscape, which today appears divided between forest areas (in *green*), prevalently located in mountain areas, and agricultural areas (in *beige*). Although the adopted scale overemphasizes the phenomenon, socioeconomic dynamics have indeed undermined the historical integration between woods, pastures and agriculture, reducing the complexity of Italy's landscape mosaic and biodiversity by favouring, instead, simplification and structural homogeneity

of rural area labelled "*poli urbani*", including areas still classified as rural, but with high settlement densities. The analysis of the five first-level CLC classes in 2000 and 1990 shows that agricultural areas are the prevalent category in terms of total surface, but also the category that changed most significantly, with a 1,434 km^2 decline. In relative terms, instead, the class that evolved the most from 1990 to 2000 is that of artificial surfaces, with a 6 % increase. Extending the analysis to the second level of Corine, the land-use class that expanded the most in absolute terms is that of wooded areas (by over 800 km^2). Interestingly, over 350 km^2 of scrublands and herbaceous areas have evolved into woods. Within the class of artificial areas, although urban areas for residential purposes have expanded the most in absolute terms (over 500 km^2), in percentage terms the largest expansion was that of industrial, commercial and infrastructure areas (10.68 %). This bears witness to the strong impulse to urbanization over the last years, whose visual impact on the public is higher than that of changes in agriculture. These can only be perceived by a trained eye, capable of interpreting changes in the rural landscape mosaic. In other words, while the great majority of the public can perceive the higher aesthetic quality of a Tuscan farmhouse compared to a suburban condominium, not all can appreciate the difference between a mixed cultivation and an industrial monoculture.

Together with farmed land, Italian forest landscape can also be interpreted historically in terms of changes to the natural vegetation brought about by human beings, following a well-defined historical sequence of culturally determined landscapes. The beauty of Italian forest landscapes was celebrated as much as that of the country's rural landscapes by the "Grand Tour" travellers. Stendhal and Shelley were impressed by the splendid, dense chestnut groves extending down the slopes of the mountains around the Como Lake almost to its banks. Edward Lear describes with admiration groups of huge hollies and oaks, as well as the incredibly diverse landscapes he encountered during a journey to Calabria in 1847, which he contrasts with the "forests dense as carpets" and "monotonous expanses of greenery" found in other countries. Like its agricultural landscape, the wooded landscape of Italy today appears simpler and more homogeneous than in the past. Its diversity is presently mainly a matter of specific composition rather than spatial arrangement. This is partially a result of the presently clear-cut separation between the woods and agriculture, after many centuries of integration. The natural substrate of the Italian forest landscape was modified long before the Roman period, but the public is largely unaware of our forests' historically determined character. This is partly due to the scientific trends of recent years, which have seen a prevalence of environmental approaches in the study and management of forests, constantly looking for the protection of "natural areas". However, this quest fails to take adequate account of centuries of human influence. The truth is that the actions of human beings in historical and proto-historical times constantly modified the ecosystem. Identifying truly "natural" landscapes in Italy is thus not an easy task (Moreno 1988). The last few decades have witnessed a trend in forest studies to relegate the historical reality of wooded landscapes in the background in favour of a naturalistic interpretation. This of course has affected planning and management policies and led to conflicts with farmers and livestock breeders. Significantly, the catalogue highlights many cases of woods that are losing their historical characteristics due not only to abandonment of traditional practices, but due to management policies aimed at transforming them into formations considered 'more natural'.

One of the interesting elements highlighted by the graph in Fig. 5.2 is the relationship between forest surface and demographic trends. As we can see, from the unification of the country to ca. 1910, demographic growth went hand in hand with a shrinking of the wooded surface. This is a typical landscape trend in developing countries, where the woods give way to pastures and fields to meet the urgent food demands of a growing population. In spite of some not negligible problems in the data-recording criteria, it seems certain that from the 1920s onward there was a stable reversal in this trend, with a more than twofold increase of forest surface, although accurate statistics are not available (Agnoletti 2010b). Thus, in this period the ratio between population and woods extension changed, since the latter continued to expand independently of demographic growth; an indication that Italian society's food supply no longer depended on the availability of cultivable land. The 1920s thus marked the end of the last phase of surface reduction in the history of Italian woods, which had seen several expansion and reduction cycles from the Roman period onwards. The new expansion was the result of the gradual abandonment of mountain and high hill areas, a trend that is already apparent during the Fascist period and became unstoppable in the second post-war period. The secondary forestation process affected all of the country's regions, especially those where the abandonment of agriculture and animal

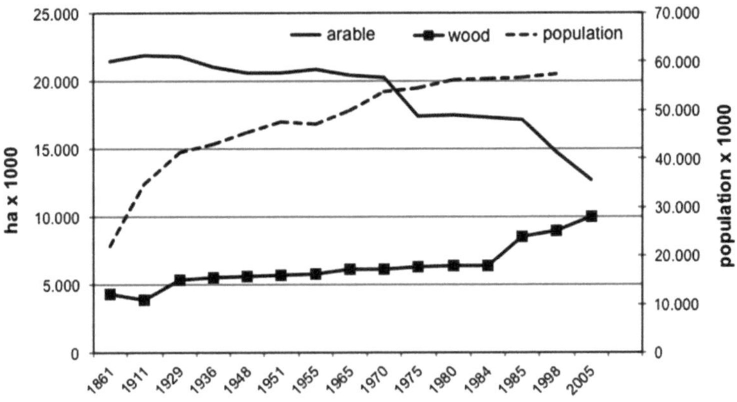

Fig. 5.2 Evolution of agricultural, wooded, and unproductive surfaces and of the Italian population from 1861 to 2007. One can observe the strong reduction of agricultural surfaces and the increase of woodlands. The increase of woodlands is often considered a positive phenomenon. However, concerning the rural landscapes surveyed, abandonment followed by vegetation dynamics and successional change are recognized as the most important threats. One side effect, and significant in terms of sustainability, is the need to import food from abroad. Similar trends are found in many industrialized countries

husbandry was more intense, even extending to lower altitudes. This led to a gradual reduction of the pre-existing landscape mosaic, a strong and often uncontrolled increase of wild fauna, and a strong decrease of cultivated land. Along with the reduction of agricultural surface, to which it is indissolubly tied, reforestation is one of the most important phenomena to affect the Italian rural landscape in the last century. The expansion of the woods from 10 % of the national territory in 1920 to the present 34 % has changed the face of whole regions. This statistic, however, also partially reflects changes in the notion of "woods". The new forest inventory of 2007, regards as "forest formations" populations of trees or shrubs meeting all three of the following requirements: a surface larger than 5,000 m^2; a foliage cover percentage higher than 10 %; and an area width higher than 20 m. Clearly, however, the above criteria also gather under the heading "woods" shrub and areas that are actually pastures or wood pastures. These would require distinctive management approaches to preserve adequately their roles in the landscape. The advance of woods contributes to reducing the biodiversity of complex rural landscape mosaics, at such a rate that in Tuscany about 70 % of biodiversity has been lost since the nineteenth century in mountain landscapes. This biodiversity, as indicated by studies of the Tuscan landscape monitoring system on some mountain areas in the region, arose from a great variety of land uses that have given way to a homogenization and 'banalization' of the landscape (Agnoletti 2010a). It is true, although not always, that the expansion of woods can increase biodiversity because of the increase of the number of tree species. Concomitantly, however, there is a decrease in herbaceous species associated with meadows and pastures, and in animal species populating cultivated habitats, as well as a reduction of diversity at the landscape scale. Almo Farina (1993) provides a significant testimony of this trend. His research indicates that the replacing of olive groves with woods has determined a reduction of avifaunal diversity.

5.3 Objectives of the Project

The catalogue is meant as a testimony, not only for the importance of the Italian landscape as one of the most representative historical expressions of the country's cultural identity, due to the important role of rural civilization in its history, but also of the universal value of the Italian landscape in the cultural heritage of humanity. This latter value sometimes seems to have been forgotten today. Nowadays, we are witnessing increasing interest in the subject at the European level, as Roberto Gambino (1994) explains, that the need to preserve the identity and meaning of places expressed by the current "demand for landscape" reflects a deeper malaise. This in turn is certainly to do with globalization processes and their effects: on the one hand, homologation and modernization; on the other, imbalances and inequalities. In this perspective, the introduction of landscape as a concern of national agricultural policies reflects a change in the conception of the role of this resource, as well as that of rural land in general. The role of landscape and its perception has indeed changed over time. Today it is no longer an elite aesthetic and cultural construct, isolated from its socioeconomic context; it has become, instead, an essential element in the definition of an adequate development model for the national rural context.

The prevalence of aesthetic considerations in past conceptions of landscape, as well as their more recent superimposition on the concept of "nature", has led to an emphasis on deterioration caused by urban dynamics. This is along with criteria for the estimation of landscape quality based on its ecological characteristics, reductively understood as its flora and fauna, or as a series of natural habitats. All this has pushed in the background both the strong human print on our country's landscape and the fact that, while urban expansion certainly played a role in this, the transformation of the rural landscape was largely endogenous, something that few have remarked. While it is evident, as Emilio Sereni explained (1961), the agrarian landscape is "the form that man, in the course and for the ends of his agricultural productive activities, impresses on the natural landscape"; it is equally evident that not all agricultures produce good landscapes. Unfortunately, as confirmed by the data gathered by the catalogue, ordinary conservation legislation based on protected area systems or landscape restrictions are ineffective as a means to preserve the rural landscape. It is this realization that persuaded all of the scholars who contributed to our catalogue of the need to draw it up. Furthermore, it is finally time for the issue to be addressed by agricultural policies. Conserving the quality of a rural landscape, which by its own nature is always evolving, can only be done by setting up socio-economic systems capable of supporting and reproducing it; hence the decisive importance of strategies and actions undertaken in the framework of agricultural policies. The new guiding principle for rural development policies, associating them with local development, is a major step forward in this direction. In the local dimension of Italian rural policies, the landscape dimension plays a paradigmatic role, as it corresponds to the transition from individual business projects to projects at territorial scale, for which a landscape-oriented approach is undoubtedly more suitable, because of the peculiar characteristics of our country, than an industrial or environmental one, even in a development perspective. Indeed, today the notion that conservation is an obstacle to development in any form has given way to the realization that conservation is the new face of innovation in contemporary society. An authentic innovation is one that adds to a store of values slowly accumulated over the ages. Conversely, there can be no authentic conservation without innovation. In this perspective, the restoration and promotion actions implemented in Italy by the recent *National Rural Development Plan* (2007–2013) have already introduced instruments by which the Italian regions can begin to modify the orientation of their Rural Development Plans to address landscape issues.

The research is not meant as an exhaustive overview of Italy's landscape heritage. Rather, it intends to set methodological guidelines for the identification and classification of landscapes of historical interest, and, at the same time, to provide a preliminary sample of the substance and state of the country's landscape heritage. Hopefully, this will be a first step in the drawing up of a true comprehensive inventory of the Italian rural landscape, on the desirability of which there appears to be a wide consensus today among both scholars and agricultural policy makers. We decided not to focus on the strictly environmental features of Italian rural landscapes—climate, geomorphology, vegetation—since these have been largely described in existing literature. We strove, instead, to take a more detailed look at the structure and organization of rural landscapes. Thus, we did not focus on ecological and naturalistic aspects, or aesthetic ones, although these are mentioned in the individual area descriptions. Rather, we adopted as our landmark Emilio Sereni's pioneering work (1961), which examined the "forms" impressed by man on the natural substrate, but left open the question of their characterization and conservation at the national scale. Our purpose was to carry forward Sereni's work by combining traditional historiographies of agriculture, forestry and, more in general, the rural landscape. The approaches highlight the material elements of landscape structure and fieldwork, as found in important studies by European scholars, especially English ones such as Oliver Rackham (1986), and also in some remarkable investigations conducted in Italy by work groups led by Diego Moreno (1990) on the agro-pastoral sector and Pietro Piussi on forests (1996).

One of the methodological problems we had to deal with in the initial stage of our research was the definition of its spatial and chronological scale. As regards the chronological scale, no limits were set. The origin of the landscapes under investigation were traced as far back as available sources allowed. As regards the spatial scale, we decided to survey areas with extensions between 500 and 2,000 ha, large enough, that is, to include management units such as the typical Italian sharecropping farm or the *latifundium*. These encompassed spatial relationships between land uses, in consideration of the importance of the spatial scale in UNESCO parameters for world heritage sites. In the area descriptions, we decided to indicate only the geographical coordinates of the centre of each area, leaving the construction of a GIS database to a later stage. The main reason for this was the difficulty of accurately determining the geographical boundaries of areas with non-contiguous cultivated zones.

The criterion of "significance" employed in our selection of areas is not wholly patterned after the criteria in the UNESCO convention, which select sites proposed for inclusion based on "exceptional universal values", indicating a broad range of elements to be taken into consideration. In this research, instead, the emphasis is on the *national* value of landscapes, although the heritage value of some of the sampled areas undoubtedly transcends the national boundaries. By "significance", we mean here the set of "values" expressed by a given landscape, and, above all, the "historical persistence" of the local agricultural fabric. The deep changes the Italian rural sector went through have enhanced the special value of historical landscape forms that have shown strong resiliency. We left the case-by-case evaluation of significance to the specific competence and individual sensibilities of our local investigators, who were prompted for further elaboration after the first draft of the area descriptions was completed. One of the first questions the scientific committee had to deal with was the meaning to be assigned to the expression "historical landscape", a key concept for the definition of the criterion of "significance" employed to select the areas. Every landscape has a history. The use of the expression "historical landscape" could hence appear semantically inappropriate or even misleading, without further specification. We therefore looked at the work conducted on rural landscapes by some international institutions, most notably the UNESCO World Heritage Convention, the FAO and the IUFRO.

As regards the UNESCO World Heritage List, the catalogued landscapes certainly belong in the "cultural landscape" category, being the result of the combined work of human beings and nature, as defined in Article 1 of the convention regarding the category "continuing landscapes" (Fowler 2003). These are defined as still vital landscapes playing an active role in society, associated with traditional lifestyles and, although continuously evolving, retaining conspicuous testimonies of their historical evolution. This is a definition that well suits the historical evolution of Italian rural landscapes. The traditional concept the UNESCO document refers to is taken up again in the FAO's project "Globally Important Ingenious Agricultural Heritage Systems" (GIAHS). This initiative's aim is to promote traditional agricultural practices and the landscapes associated with them at the global level. Actually, the notion of the association of traditional knowledge and landscapes was already contained in the "Agenda 21" document produced by the UN Environment and Development conference at Rio de Janeiro in 1992, and has hence been circulating for quite some time, notably in connection with the concept of sustainability. The FAO project specifies that traditional practices provide a fundamental contribution to the world natural and cultural heritage, as well as the biodiversity of rural areas. They reflect the evolution of humanity in its relationship with the natural world, producing landscapes of great beauty, but also a variety of services and products, food security and a good standard of living. These are agro-silvo-pastoral landscapes, typically found in densely populated areas, which are significant at the global scale and must be preserved. FAO's project, developed in the year 2000, was followed in 2003 by one from the Ministerial Conference for the Protection of Forests in Europe (MCPFE). This is slowly working at introducing these same values in the concept of "sustainable forest management". Although these values form the third pillar of Sustainable Forest Management at the European level, no policy-orienting resolution to promote them has been adopted by member countries so far, versus about twenty resolutions on ecological and economic values (IUFRO 2007). Recently, an international workgroup was charged by the Conference with drawing up scientific guidelines that again placed the accent on traditional knowledge and the landscapes they are associated with, and on landscape conservation as a fundamental aspect of sustainability.

In view of the above considerations, we decided to regard as historical landscapes those characterized by the use of traditional practices. Methodological distinctions regarding the definition of "traditional" induced us to develop a parallel research project, financed by the Ministry of University, specifically devoted to traditional landscapes. According to the guidelines of this project, landscapes are "traditional" when they have been established in a given territory for a long time, perhaps many centuries, and appear to have stabilized, or to evolved very slowly. They are generally maintained by practices and techniques requiring few external energy inputs, whether in the form of mechanization and irrigation or of chemical fertilizers and agro-drugs. Their crop fabric is characterized by long historical persistence and a strong connection with the local social and economic systems that produced them. Their stability, or slow evolution, is evidence of harmonious integration of production, the environment and culture in a given area or region (Antorp 1997). Consequently, our selection criteria were not restricted to the permanence in a landscape of traditional practices determining its historical character. We employed the concepts of "significance" and "integrity" that were also included among the UNESCO criteria, but added that of "vulnerability" to emphasize the risk factors for each of the described areas.

The data are not the result of a systematic inventory, and hence have a limited statistic value. The purpose of this first sampling was only to single out some of the most significant Italian landscapes, described their historical features, integrity and vulnerability. All the selected areas retain valuable vestiges of their historical origin (Fig. 5.3). In many areas, crop continuity goes back at least 2,000 years, as in the case of the *alberata aversana* or the centuriated areas of the Po River Plain, although the crop fabric has often changed. The catalogued landscapes frequently show strong connections with the historical events that determined the evolution of local socioeconomic conditions and the main forms of holding management. Thus, sharecropping gave rise to polycultural mosaics, *latifundia* to extensive agriculture, and land reclaiming to regular farmland grids. The persistence of landscapes of Etruscan, Greek or Roman origin bears witness to the strength of the rural print left by these civilizations, which survived not only the barbaric invasions but also the recent and possibly more violent impact of the industrialization of agriculture. From a technical point of view, the influence of ancient agriculture appears independent of the land ownership system. Ancient techniques have survived through the ages down to our day, when sharecroppers and latifundists no longer exist, but the cultivation methods they employed are still in use. Most of the

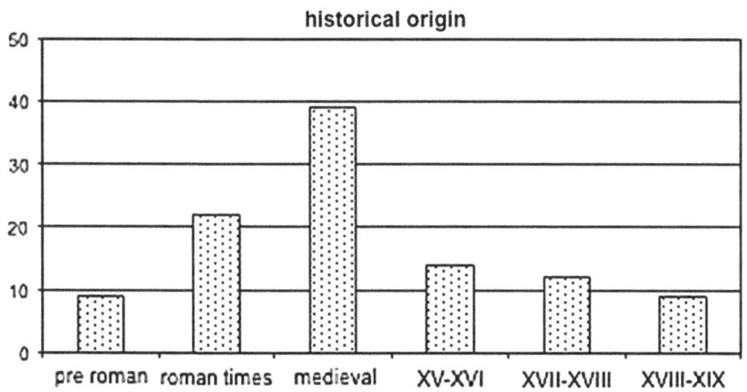

Fig. 5.3 The landscapes surveyed show remarkable historical persistence, and still feature crop management systems and cultivation practices going more than two thousand years back

recorded landscapes date from the Middle Ages. This bears witness to the fecundity of the revival of Italian agriculture in this period and its role in shaping the landscape and identity of much of the national territory. On the other hand, our catalogue includes more recent landscapes reflecting the slow development of traditional agricultural techniques. These have been forging landscapes well into the contemporary age, and thanks to their remarkable adaptation to local environmental characteristics, in part still endure.

Our study was carried out in such a way as to allow us to analyse the different landscape categories found in the selected areas, organizing the data on different hierarchical levels. As one can see (Fig. 5.4), among the areas surveyed there is a prevalence of agricultural landscapes, followed by mixed landscapes with arable land, and with woodlands, and pastures. This result partially depends on the higher degree of complexity of the historical agricultural landscape (Fig. 5.5). However, it also reflects the difficulty of finding woodland areas retaining their historical characteristics, due to the decline of woodland management in the last century and the gradual loss of their integration with the agricultural and pastoral activities that once helped to diversify them.

As to pastures, they are fewer in number due both to their significant decline and to their lower variety in comparison with agricultural and forest landscapes. The mixed landscapes characterized by polycultures recorded in the catalogue confirm that elaborate agricultural mosaics of small-scale cultivations persist in many Italian regions. This is not just on small farms, but also on separately managed holdings belonging to large farms. As regards categories of crops, in most of the selected areas vineyards are the main crop, followed by olive and grain. This is typical not just of the landscape of Italy, but of that of the Mediterranean as a whole. To quote Fernand Braudel, everywhere in the Mediterranean "can be found the same eternal trinity: wheat, olives, and vines, born of the climate and history; in other words, an identical agricultural civilization, identical ways of dominating the environment" (Braudel 1986). This statement eloquently

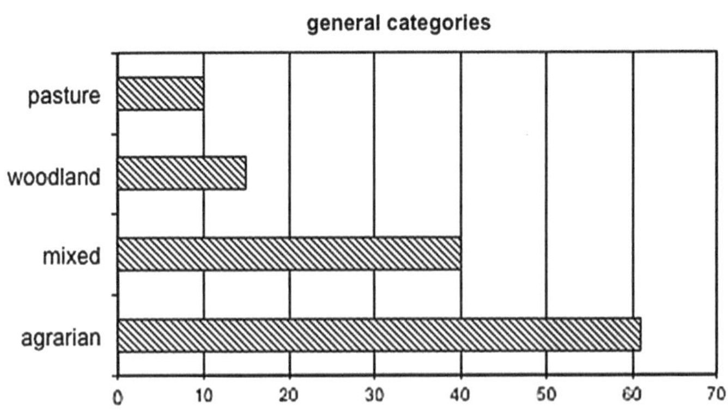

Fig. 5.4 Number of areas in each general category of rural landscape

summarizes the value of such landscapes, which in Italy assume special significance in the light of the importance of mixed cultivation, which is placed immediately after the first three in terms of frequency. This form of cultivation, typical of central Italy, is among those that most embody that beauty of the Italian rural landscape that made such an impression on many foreign scholars. Henri Desplanques, a French geographer, wrote that the agricultural landscapes of the hills of the Tuscan-Umbrian-Marchigian area were constructed as if with "no other concern than beauty" in mind (Desplanques 2006).

5.4 Mixed-Cultivation Landscapes

Among all the landscape types surveyed, mixed cultivations deserve a special attention as they represent perhaps the most important traditional form of agriculture. Mixed cultivation is the growing of different crops in the same field. It reflects the need to make the most of soil nutrients, a limited resource, to obtain a variety of products. In Italian agronomy textbooks of the 1950s, this technique was still described as a valid alternative to specialized crops. It provided the best opportunity, especially in the sharecropping system, to exploit the cultivable surface to the utmost to provide for the subsistence of peasant families. That some mixed cultivations still survive in the Italian agricultural landscape in spite of industrialization of the sector, is not an indicator of backwardness or an inability to innovate. Rather, it bears witness to the set of values expressed by these practices, which have held up against attempts to replace them with technical and productive models that have not stood the test of history as well.

Mixed cultivation is known long before the Roman period. Due to their long persistence, these systems play a fundamental role in landscape conservation and valorisation projects. At the beginning of the twentieth century, Italy had 13.7 million hectares of arable land. Of these, a good 45 % were treed. Unfortunately,

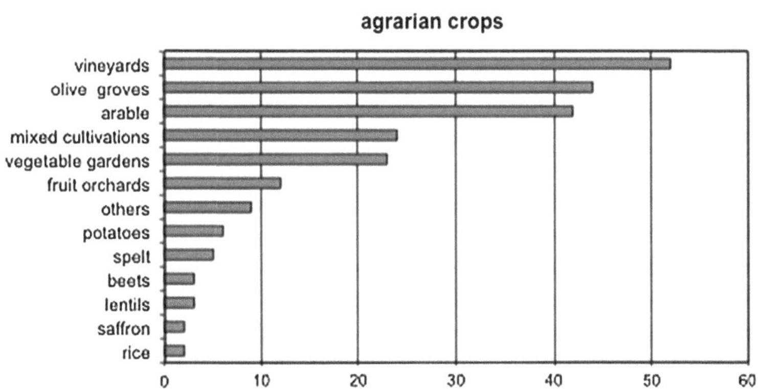

Fig. 5.5 Crops found in the areas surveyed

available statistics do not allow us to follow their evolution down to our day. This is because from 1980 onward, ISTAT, the Italian government statistics institute, integrated them together with specialized woody crops (Toderi et al. 2002). This is an eloquent example of the scarce attention devoted to elements that do not have a significant function from a technical and productive point of view. What we know is that there was a decline from 13,144,000 ha in 1940 to 2,208,000 in 1980. This is a reduction of more than 80 %, which speaks volumes about the importance of this transformation of the landscape (Sansavini 2002). Especially striking is the decline of mixed wine growing from 2,963,000 ha to a mere 455,000 ha. Mixed olive orchards dropped from 1,360,000 ha to 1,088,000, followed by apple, pear, peach, and so on, down to almond, walnut and hazelnut. As to distribution, Sereni's book indicates a higher concentration in the north and centre than in the south. This is explained by differences in socio-economic structures, which favoured prevalence of sharecropping holdings in the north and centre, and of extensively cultivated *latifundia* in the south.

The best-known mixed cultivation systems are the *piantata padana* in the north (Gambi 1995; Cazzola 2008) and the *alberata* in Toscana, Umbria and Marche in the centre, found in association with land-shaping systems such as the *porche* in Tuscany or the *baulatura*—a water drainage system—in Emilia, with many variants. Nevertheless, we should not overlook landscapes such as the terraced almond orchards of Gargano or the carob orchards of the province of Ragusa, which have miraculously survived the transformation of the sector and are documented in our catalogue. It is also important to consider the environmental significance of these cultivation systems as an expression of traditional knowledge intimately connected to the local features of the areas where they occur. One only needs to consider that the more abundant presence of humid areas in the north led farmers, ever since antiquity, to train grapevines on trees to preserve them from the humidity of the soil and improve their exposure to sunlight. The higher fertility and higher percentage of irrigated areas in the north (85 % at the beginning of the twentieth century versus 9 % in the south), allowed much more extensive mixed orchards. In dry lands farmed ever since antiquity in Emilia, Romagna, Polesine, the Ferrarese, the Otrepò Mantovano, and the lower plain of Veneto (Finzi 1998; Landi 1994), the holding system was based on mixed cultivation and, hence, the food self-sufficiency of peasant families. Grape was grown by training it on live supports, usually elms, field maples, willows or poplars arranged in rows at the edges of fields. Tree spacing and field width varied according to local tradition, so that tree densities typically ranged from 90 to 180 per hectare. The arable field with trees and vines is hence the paradigm of a mixed cultivation system achieving optimal energy efficiency.

It is worth remarking that vegetable gardens were an important part of the agricultural landscape, often associated with urban areas. Sereni ascribes a great importance to them as early as the Middle Ages. They show a remarkable continuity expressing their still strong rooting in popular culture. Apart from the areas included in the catalogue, they are widespread in many peri-urban areas, also on publicly owned land. Their small-scale production is used for self-consumption.

There exists a broad range of traditional systems to lay them out, which give rise to surfaces that are very heterogeneous, but of great social importance, especially in some vast peripheral areas where urban expansion is encroaching on agriculture.

5.5 Terraced Landscapes

The high number of terraced areas in Italy, and their importance, also deserve separate treatment. Terracing is possibly the most important system of landscape organization in the Mediterranean area. A terrace is never an isolated landscape element, but rather part of a system of land-shaping works that are the fruit of traditional knowledge of construction and farming methods and a perfect understanding of hydrogeological and climatic factors, allowing local populations to make the most of their environmental resources. These are thus self-regulating systems characterized by high aesthetic quality and capable of modelling the landscape by integrating with its natural characteristics (Laureano 2004). The earliest evidence of terracing in Italy dates back to Neolithic times (Gorfer 1988). Among the terraced landscapes documented in our catalogue, in this case too, there is a prevalence of those whose origin can be traced to the Middle Ages. It is not surprising, however, that in general statistics most terraced landscapes date from the nineteenth century, when demographic pressure led to a remarkable increase of agriculture on mountains and high hill slopes. Terracing, in the form of grass-covered contour terraces and terraces supported by dry-stone walls, is one of the main forms of adaptation to difficult environmental conditions developed by Italian agriculture. This is especially important since mountains and hills cover about 76 % of the Italian national territory. Sixteen percent of the terraced areas selected for our catalogue are in the mountains and 50 % are on the hills, but with some also found in the plain areas. These evidence the fact that this form of land shaping was found even in areas that were not especially hilly, as a widespread and suitable technique to deal with the geomorphological conditions of our country. Terraces principally house vineyards and olive orchards, but also fields, pastures and terraced woods. Chestnut groves, in particular, are often grown on terraces of various shapes, from structures held by dry-stone walls to "lunettes"—that is, half-round walls surrounding each individual tree. Terracing was also widely employed in the reforestation of the Italian mountains by the State, as attested by an area at the mouth of the Sele River in Campania, included in our catalogue.

Many terraced areas host mixed cultivations, such as grape with olive, and very often fruit orchards, especially of hazelnut and citrus. Terraces offer a number of advantages, especially as regards land stability, by reducing erosive phenomena and preserving the valuable soil. In many areas in Italy, building terraces required cutting into the rocky matrix with sledgehammers and pickaxes, and carrying earth up from the valley on people's shoulders or mules' backs to fill the terraces. These are sometimes Cyclopean works of monumental character, both for their size and in consideration of the large work forces required to build them. They have given

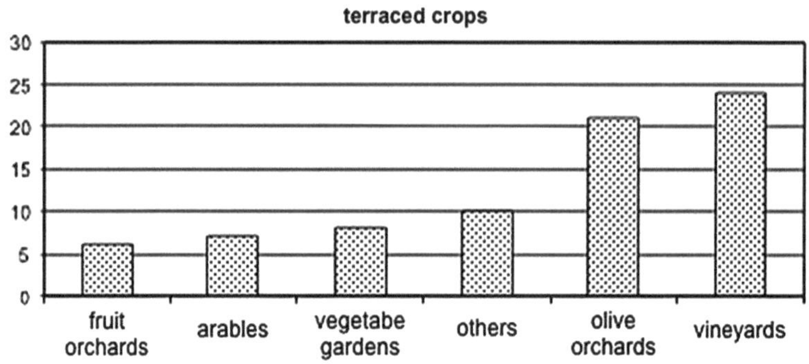

Fig. 5.6 Main terraced crops surveyed

rise to "stone landscapes" found in many areas in the catalogue. Their value as a historical testimony of the work of generations of farmers alone would make them well worth preserving, but in addition to this, they are an element of exceptional scenic value, both on mountains and in coastal areas. Since farming machines are now sufficiently powerful to work on steep slopes, many terraces, especially in wine-growing hill areas, have been replaced over the years with *rittochino* arrangements. The rows of plants follow the maximum acclivity lines of the hills. The employment of mechanization has also called for larger plots obtained by totally or partially removing earlier works, without replacing them with equally effective ones, suitable to house the size of the new fields.

These techniques, besides having low scenic value, cause strong erosion and a reduction of soil fertility that needs to be compensated by chemical means. The results of a research conducted in an area in the Chianti region in Tuscany are emblematic: In the period 1954–1976, erosion increased by 900 %. In vineyards now arranged on the slopes, annual erosion was especially high, about ca. 230 t/ha, versus the 2–12 t/ha regarded as acceptable (Zanchi B., Zanchi C. 2009). Today, to ensure sustainability and landscape protection, we need to combat erosive phenomena and hydrogeological deterioration. This can be both by employing conservative agronomic techniques adapted to each situation, and by restoring and maintaining traditional structures with high landscape value such as terraces. To all this it should be added that dry-stone walls bring substantial benefits to crops by absorbing heat during the hot hours of the day and releasing it during the cooler night-time. Thus, although their maintenance costs are higher, terraces guarantee the conservation of the most important resource for agriculture, that is, the soil. A vast range of terracing types are employed in our country. Unfortunately, we are still far from having reached a comprehensive overview detailing the building characteristics and methods to restore and preserve them, although numerous studies and restoration projects are currently underway. Terraces, both for agricultural and forest maintenance purposes, are not merely valuable testimonies of the human cultural heritage, but also instruments of sustainable development worth preserving as an element of traditional knowledge. This applies in both the agricultural and the forestry sectors (Fig. 5.6).

5.6 Integrity

Assessing integrity is essential if we wish to understand the state of our landscape heritage and its potential as a resource. It is a necessary step in the drawing up of conservation and valorisation policies to stimulate local development and promote the "landscape dimension", which is the frontier of modern rural development approaches. However, in our investigation, evaluating integrity turned out to be the most difficult task we had to face. The conditions to be met for a heritage site to be regarded as endowed with authenticity and integrity are detailed in an article in the UNESCO World Heritage Convention (2008). Many different parameters could apply, and these may differ from culture to culture, and even within the same culture. That is why Article 81 of the Convention prescribes that the value of cultural heritage should be mainly assessed within its cultural context. The conditions for authenticity are regarded as met if the cultural values of the site are expressed in a range of features such as shape, design, material, use and function, tradition, techniques, management, location, and environment. The same features can also be found in a historical farming landscape with its crops. Article 88 defines integrity as the measure of how "intact" a cultural asset is. In practice, it is a matter of determining in what measure it:

- includes all the elements required for its value to be expressed;
- has an extension adequate to fully represent the characteristics and processes expressing its significance;
- is affected negatively by development and/or abandonment.

For this first investigation of the state of Italy's rural landscape heritage, we focused on crop management, without going into any detailed examination of settlement. Hence, we mainly gauged integrity in terms of the extent of the areas under scrutiny and the state of cultivations. As regards the former aspect, our original instructions to our investigators were to single out areas with extent between 500 and 2,000 hectares. The choice of this scale was determined by the application of several considerations. In the first place, it allowed the inclusion of several management units—holdings, farms, large estates, etc.—in a single area. This shed light on changes in farm management, possible connections between the landscape and the types of rural contracts such as sharecropping or those applied on *latifundia*, and on relationships with the environment. This choice of area or size also facilitated studies of relationships between various patches making up landscape mosaics from a landscape ecology perspective. Furthermore, it allowed monitoring by remote sensing and ground checks to assess on-going dynamics and the results of valorisation actions. These objectives might call for larger areas, but experimental data from areas in Tuscany with extent of ca. 1,000 ha had proved this average size to be quite adequate. Another element that later turned out to be decisive, and suggested a certain degree of flexibility in minimum surface requirements, was the awareness that, in all probability, setting inflexible criteria

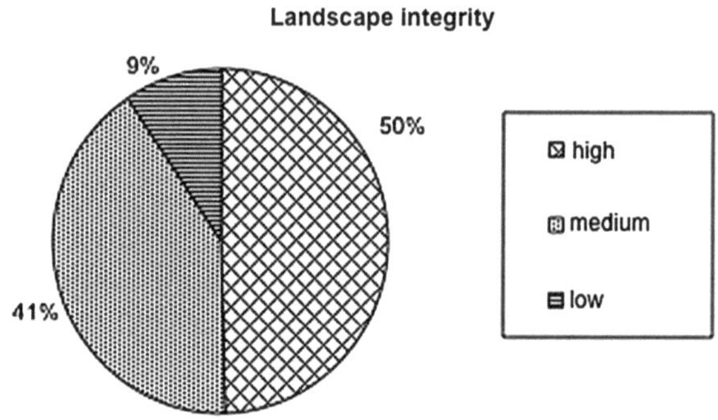

Fig. 5.7 Levels of integrity of the catalogued landscapes

for the singling out of perfectly preserved historical cultivations would bring us up against a scarcity of extensive areas with such characteristics (Fig. 5.7)

The collected data confirmed the problem we had anticipated. We were confronted with a myriad of areas that had preserved their historical characteristics, but over small and fragmented surfaces. Furthermore, historical cultivations were rarely contiguous within a given area. This forced us to modify our initial parameters, extending them to include areas with smaller surfaces—e.g. 100 hectares—or with historical cultivations scattered over broader surfaces. Another important consideration, especially as regards the statistics provided in the present chapter, is the subjectivity of the data gathered by individual researchers on the actual state of preservation of the historical features of cultivations. In general, our researchers estimated integrity based on the survival of cultivations described in historical literature about the area. For example, they regarded the preservation of a terraced vineyard as a sufficient condition for a positive assessment of integrity, even though grape-growing techniques might have changed over time. On the contrary, poor preservation of dry-stone walls or the colonization of terraces by shrub and arboreal vegetation had a negative impact on the assessment of integrity. Likewise, the survival of historical pinewoods was regarded as a sufficient condition for inclusion in the catalogue, even though its colonization by a high number of other tree species due to the lack of maintenance actions aimed at preserving its mono-specificity had had a negative impact on its integrity. (Fig. 5.8).

We interpreted the gathered data based on the contents of the area descriptions themselves, not on qualitative or quantitative statistical data. This can be seen in the first graph showing the degree of integrity of the recorded areas. The second graph highlights the different degree of quality of the landscapes in the main Italian geographical regions. Central Italy appears to show a higher degree of integrity, followed by the south and the islands, and then by the north. Aside from the subjectivity of descriptions, there is undoubtedly a remarkable difference between the condition of landscapes in the Po River Plain and in the rest of Italy. In the Po River Plain, the selected areas often owe their significance to the survival of traditional constructions, distinctive field layouts, or localized cultivations. However, while these elements which are the main focus of the area descriptions, are still intact, historical integrity is low in the rest of the area's landscape.

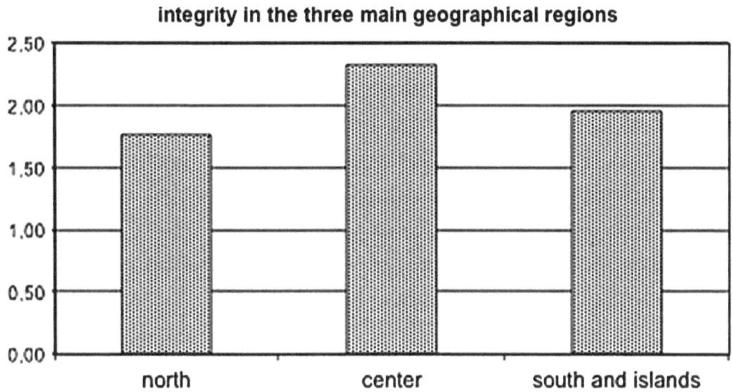

Fig. 5.8 Landscape integrity in Italy's three main geographical regions

The research confirms that the remarkable expansion of agriculture in this part of Italy has deeply undermined the historical character of the landscape. This is as many authors have stressed with regard to phenomena such as the disappearance of the *piantata padana*, the local mixed cultivation system (Gambi 1995). There are, however, important exceptions to this geographical trend. In Trentino Alto Adige and Friuli, there are still some landscapes with high integrity, even compared to their analogues in the central and southern regions. This datum, however, was certainly influenced not only by the subjectivity of the investigators, but also by the lower number of recorded areas: just four in Trentino Aldo Adige, but rather intact, versus eight in Piemonte, but with a lower degree of integrity. In central and southern Italy, historical landscapes are usually in better condition. Especially in the south, this confirms the potential of a resource, which, if integrated into the typical product market and the growingly popular rural tourism, could certainly offer interesting opportunities for economic development. This is so long as the characteristics that make up these landscapes' integrity and uniqueness are given adequate recognition. Unique and intact landscapes of great interest are actually more frequent in the south and the islands than in regions with well-established and widely recognized landscape values, such as Tuscany and Umbria. Examples include the carob groves in the province of Ragusa, or the terraced almond orchards of the Gargano in Puglia.

Another aspect of the gathered data is the distribution of integrity by altitude. If we compare areas in mountains, hills and plains, there is a clear-cut prevalence of hill areas, which also have the landscapes that are the most intact. This depends on two opposite processes discussed more in depth in terms of vulnerability. In the plains, there was a higher intensification of productive processes. Here mechanization and modern industrial agriculture had destructive effects on the historical landscape. On the contrary, in the mountains the abandonment of agriculture and animal husbandry, followed by the exodus of the population and increasing economic marginalization, led to a gradual deterioration of the integrity of the rural landscape, which today is largely being undermined by processes of re-naturalization. In hilly areas, instead, due to the difficulty of establishing extensive, highly mechanized, industrial monocultures, most of the land remained in the hands of small landowners. Besides, and fortunately from a conservation perspective, hills account for most of the Italian land surface. For both reasons, most historical landscapes in good condition are in the hills.

Another element that our area descriptions often present as being connected with integrity, are local landscape restrictions and norms. As we will explain more in detail under vulnerability, the conservation of the historical rural landscape seems to depend much more on the continuity of traditional practices than on landscape restrictions or protected areas. These actually appear to be in the process of deteriorating. Examples such as the historical pinewoods of Ravenna, which shows low integrity in spite of being under several kinds of legislative protection, indicate a certain inadequacy of present legislative instruments. That is why we did not regard the simple fact that one of the selected areas lay within a zone placed under landscape, naturalistic or hydrogeological restrictions as a sufficient condition for assuming the integrity of its historical landscape.

5.7 Vulnerability

As described in the section on integrity, the need to assess vulnerability is already affirmed in the UNESCO World Heritage Convention. Vulnerability studies are commonly carried out as a part of nature conservation actions and environmental impact analyses (Romani 1994). As specifically regards our catalogue, it was necessary to identify threats to the conservation of the selected landscape resources to allow the establishing of guidelines for action and a hierarchy of urgency among sites. Each landscape has an "intrinsic" vulnerability depending on the various types of processes that can affect it. For example, a dry-stone terrace is extremely vulnerable to abandonment, as the absence of maintenance within very few years will lead to the collapse of the terrace and erosive phenomena, including major ones such as landslides. On the contrary, an adult beech wood is relatively much less vulnerable, even if no longer managed. In this perspective, a classification of transformations in land and, hence, landscape uses as related to hydrogeological risk would be extremely useful as a means to monitor and prevent events such as landslides and floods, which in Italy mainly depend on erosion. The presence of erosive phenomena among the causes of vulnerability in the recorded areas bears witness to the destructive effects of processes depending largely on the abandonment of land-shaping and hydraulic works. (Fig. 5.9)

Our data clearly show that the main threat to the Italian rural landscape is abandonment. This is hardly surprising, considering what was stated above regarding transformation and reduction of agricultural lands. Abandonment is directly connected with the third main vulnerability factor, that is, the colonization of pastures and abandoned plots by shrub and tree vegetation. This phenomenon also affects historical woods in terms of modifications of structure. For example, the introduction of other tree species following the abandonment of management

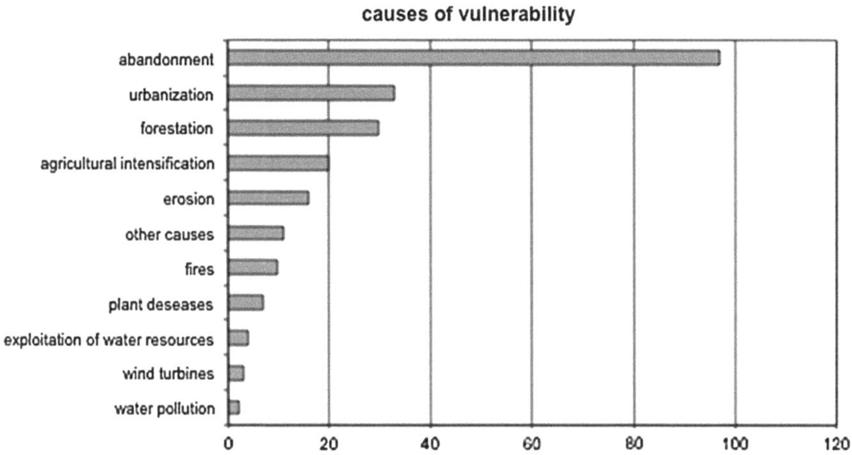

Fig. 5.9 Frequency of vulnerability phenomena in the examined areas

will turn a chestnut grove or domestic pinewoods into mixed woods, changing the specific composition of woods originally designed to be mono-specific. Other major phenomena influencing vulnerability are human pressure and the intensification of agriculture. The former is often a consequence of urban expansion, especially in metropolitan areas, a typical case being the urbanization of the countryside around Naples and Milan. In this case, the conservation of the agricultural landscape, whether historical or not, and independent of its quality, has the important function of limiting building expansion. The catalogue areas in the countryside of Naples are in a situation of great vulnerability that is undermining their integrity and historical character. Intensification, instead, is a result of the industrialization of agriculture following a development model invariably based on mechanization and crop reorganization to increase yields and cut labour costs. This approach, dictated by a vision of rural development where modernization of productive infrastructure is the principal objective, relegates landscape quality and resource diversification to the background. However, the validity of this model is challenged by market globalization and the loss of the economic relevance of industrial agriculture. This is giving way to farming economies based on new relationships between typical products, rural tourism, landscape, and other functions performed by the countryside for society as a whole. Data on vulnerability caused by wind turbines deserve special comment, since this subject is currently very much in the public eye. Present advocates of the development of renewable energy sources in Italy often seem to forget that the contribution of our national landscape heritage to the progress of humanity, the economy, and people's quality of life is much more significant than any contribution wind turbines may give to the solution of the energy problem and the mitigation of global warming. Therefore, although we need to find solutions to our energy problems, the ambitious wind-turbine projects implemented by regions with vast landscape assets, should not limit their costs-benefit analyses to estimates. For example,—invariably

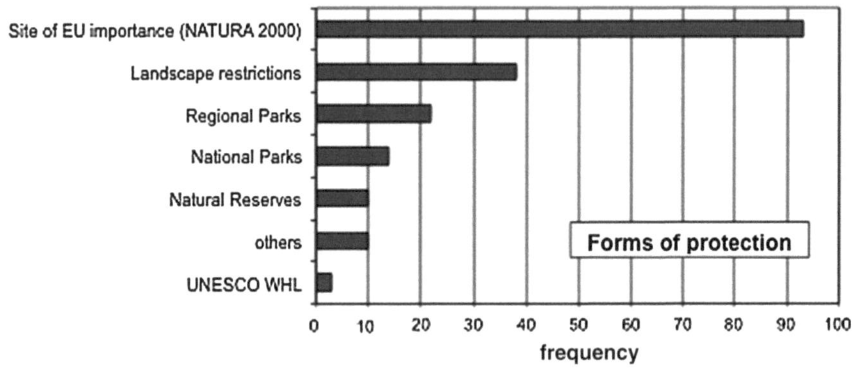

Fig. 5.10 More than 60 % of the catalogue areas lie within protected zones of some kind, and 34 % are under landscape restrictions. However, these forms of protection, insofar as they were conceived for other purposes, have proved ineffective to counter deterioration consequent on abandonment

presented as "facts"—the tonnes of CO_2 saved from being released into the atmosphere, or the energy production of the turbines—nominal, but also presented as a "fact"—, are often quoted. These analyses should also take into account the impact of wind turbines on local rural landscapes. Most of the studies promoting wind as well as photovoltaic power plants overlook the issue of their landscape impact (Martin et al. 2002), just as the rural landscape played a very limited role in the planning of the Italian Environmental Impact Assessment. Even in regions showing a higher regard than others do for landscape values, today there is an extreme reluctance to implement norms taking adequate account of the visual impact of wind turbines, whose height makes them conspicuous over vast land tracts (Fig. 5.10).

Data on the impact of abandonment and increased vegetation cover in abandoned areas gain special significance when crossed with those on legal protection. Among the areas in our catalogue, a high percentage lies in Parks or Sites of Community Importance (60 %). Of these, 51 % lie within SCIs of the *Natura* 2000 network and 37 % in parks or natural reserves. Besides, 64 % are under landscape restrictions under Acts 1497 of 1939 and 431 of 1985 (Fig. 5.11).

If we look at vulnerability statistics in areas under landscape restrictions, we realize that the existing ordinary legislation cannot guarantee the conservation of historical rural landscapes. In fact, in protected areas the trend is often, explicitly or implicitly, to favour re-naturalization. Protected areas should ensure landscape conservation. However, the statistics presented here are perfectly in accordance with present policy orientations in nature conservation at the European and national level. According to these orientations, the habitats especially deserving of protection are natural ones, and animal and plant species not connected to agricultural activities. In fact, natural habitat protection theory regards their fragmentation as something negative, which conservation policies strive to prevent. This obviously clashes with the manmade origin of the Italian landscape and its

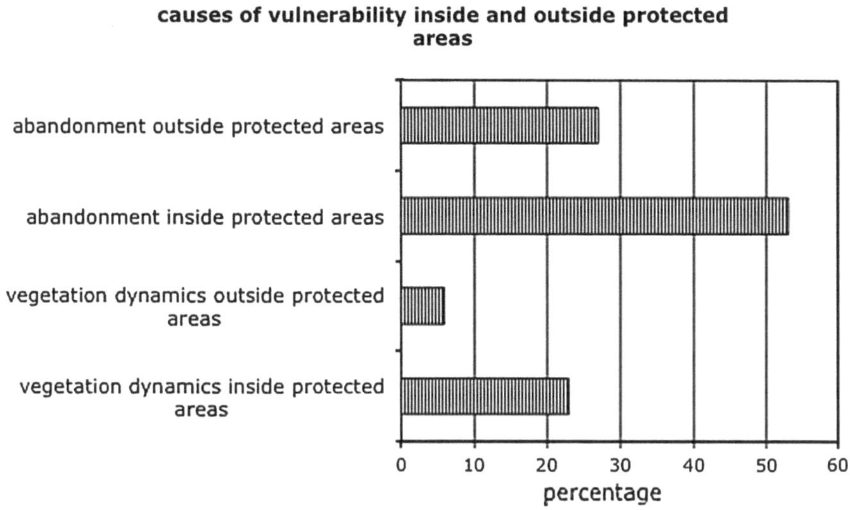

Fig. 5.11 Vulnerability due to abandonment and re-naturalization in the catalogue areas lying within and outside protected areas

very fragmented landscape mosaics characterized by a multiplicity of land uses. In substance, the most commonly accepted interpretation of biodiversity favours the spread of 'natural' tree and shrub vegetation over previously cultivated areas. As regards historical woods—areas, that is, that were already wooded to begin with—management plans tend for example, to encourage the transformation of chestnut groves or pinewoods into mixed woods. This is because the latter are closer to the natural state. As Diego Moreno and Roberta Cevasco explained above, there is a biodiversity associated with historical dynamics that is rarely taken into account. More in general, the scientific theory of biodiversity, at the species and ecosystem level, can be perfectly integrated into the landscape dimension (Baudry and Baudry-Burel 1982; Naveh 1998), which is a fundamental interpretive key for the Mediterranean area. The fact is, however, that norms and applicative directives concerning sustainable management, especially in the forestry sector, are clearly against fragmentation (MCPFE 2003). Nor does the European Habitat Directive of 1992 leave doubt as to which habitats should be protected and which risks avoided. There is thus a problem in the interpretation of the concept of the protection of nature in the Mediterranean area and its relationship with the landscape, a problem that calls for reflection.

An important observation regarding the frequency of occurrence of historical landscapes in protected areas is that, considering the remarkable increase of the latter over the last few years, it was easy to predict that they would be established in economically more marginal zones. These, however, are also those where the most interesting historical landscapes, even from a merely aesthetic standpoint, are often found. It is thus evident that there is a problem regarding the relationship between the preservation of the rural landscape and protected area management, which somehow will have to be addressed.

Different considerations apply to landscape restrictions. These should be theoretically more adequate for the preservation of the rural landscape; however, they have so far failed to achieve this end. In fact, the statistics for our catalogued areas indicates that in those under landscape restrictions the threat of abandonment is even higher (81 %) than in protected areas, and so is the threat from encroaching vegetation (83 %).

This depends in part on the nature of the restrictions imposed by the 1939 Act, which aimed at preserving assets such as parks, gardens, work of art, landscape views regarded as "natural pictures", panoramic viewing spots or belvederes, but not the agricultural landscape. The Galasso Decree of 1985, introduced the category of "environmental assets" among those eligible for protection, including features such as mountain peaks, coasts, rivers, lakes, wooded areas, etc. However, it did not take the rural landscape as such into consideration (Santoloci 2000). Thus, in spite of the important work conducted by several generations of researchers on the agricultural landscape, while natural features were included long ago among categories eligible for protection, agricultural features, which still make up most of the Italian landscape, are still excluded. Actually, restrictions on protected areas and landscape laws appear more effective against phenomena such as the expansion of urban areas, or in limiting the transformation of agricultural into urban land, but are inadequate as a means to prevent the loss of rural formations of special value.

5.8 Conclusions

The dynamic nature of the agricultural and forest landscapes inevitably causes them to change, independently of human action. Thus, in the absence of specific legislation, not only recognizing the origin and significance of cultivations, but also exerting an active influence on their transformation processes and, hence, on farmers, little will be achieved. As we have seen, over recent years there was a further merging of the two approaches, the traditional aesthetic-cultural one drawing on the same inspiration as the Act of 1939, and nature conservation. In a way, landscape restriction legislation tends to adopt "naturalistic" values wholesale, without analysing the real object of conservation and its evolutionary dynamics. Thus, the restoration of a historical agricultural landscape by removing the woodland presently occupying it could appear as a violation both of landscape restrictions and of forest laws, in the absence of specific authorization. The abandonment of a centuries-old pasture or terrace and its occupation by spontaneous vegetation, instead, is not regarded as such. Significantly, our catalogue includes several examples, such as the above-mentioned historical pinewoods of Ravenna, an area under many environmental and landscape restrictions where extensive vegetation dynamics are deeply compromising local historical characteristics with no action taken to prevent this. Among the Italian habitats under priority protection under the European HABITAT Directive of 1992, we find "mixed chestnut-dominated woods", which are nothing but the result of the abandonment of chestnut orchards. Technically speaking, declaring such a habitat a Site of Community Importance and placing it under landscape restrictions would make action to restore the original chestnut orchards—the true habitat at risk of extinction—a violation of EC norms. Thus, if inaction until a few decades ago might be interpreted as a different approach to conservation, today it would be a violation of the law. Another interesting aspect is the fact that the 1939 Act protected panoramic viewpoints as essential elements for the development of landscape tourism, whereas today many hill and mountain roads with great scenic potential run inside vegetation corridors that actually preclude the view of the surrounding landscape. Although this is a widespread phenomenon, rarely actions are taken to remove the vegetation obstructing the view of the landscape to create truly panoramic routes, partly because this would involve violating other laws. It is clear that the issues I have been discussing here concern both existing legislation on natural and landscape conservation, and the subjects involved. At this point, it is important to understand whether the new Italian "*Code for Cultural and Landscape Assets*" will be able to improve the situation. Article 142, comma 3, rules that the State, Regions and Autonomous Provinces can single out historical rural landscape systems to be placed under specific protection for their exceptional integration of the rural landscape and traditional agronomic practices. The purpose of this protection should be to promote their conservation and sustainable development and prevent the deterioration of their land. This article thus offers the Italian regional governments the opportunity to indicate landscapes to be safeguarded, especially through their landscape plans. It still remains to be determined what criteria will be adopted to identify such landscapes, considering that in regions where landscape plans are already under way one finds interpretations that give rise to perplexities and contradictions, such as whether a correct distinction was formulated between "historical" and "natural" landscapes. Besides, the Code provides no indication as to how landscape plans, or at least conservation activities, should integrate with rural development programmes, which may promote activities that are at odds with landscape protection. As regards all these issues, the catalogue intended to provide not only concrete examples of landscapes for which adequate forms of protection need to be developed, but also a reflection on the concept and object of conservation. This was in the hope that a true cooperation among the different subjects involved may arise. In addition, for these reasons, a new law enacted in 2012 assigned to the Ministry of Agriculture, Food and Forestry, the responsibilities for the policies on the rural landscape, so far in the hands of the Ministry of Culture. This too may have implications for future directions and policies.

References

Agnoletti M (ed) (2006) The conservation of cultural landscapes. CAB International, Wallingford and New York

Agnoletti M (2010a) Paesaggio rurale. Strumenti per la pianificazione. Edagricole-Gruppo 24 Ore, Milano

Agnoletti M (ed) (2010b) Historical rural landscapes. For a national register, Laterza, Bari

Agnoletti M (ed) (2012) The Italian Historical Rural Landscape. Cultural values for the environment and rural development. Springer, Dordrecht

Antorp M (1997) The concept of traditional landscapes as a base for landscape evaluation and planning. The example of Flanders Region. In. Landscape and Urban Planning 38:105–117

Baudry J, Baudry-Burel F (1982) La mesure de la diversité spatiale. Relation avec la diversité spécifique. Utilisation dans les évaluations d'Impact. In: Acta Ecologica. Oecol. Applic. 3:177–190

Braudel F (1986) Civiltà e imperi del Mediterraneo nell'età di Filippo II. Einaudi editore, Torino

Cazzola F (2008) Foreste di pianura: espansione e declino della piantata padana (sec XV-XX). In: Marinai V (ed) Paesaggio e sostenibilità. Studi e progetti. ETS, Pisa, pp 19–27

Council of Europe (2009) Landscape and rural heritage. proceedings of the sixth meeting of the council of Europe for the implementation of the European landscape convention Sibiu 20–21 September 2007. Council of Europe Publishing, Strasbourg

Desplanques H (2006) Campagne Umbre: Contributo allo studio dei paesaggi rurali dell'Italia centrale. Quattroemme, Perugina

Farina A (1993) Bird fauna in the changing agricultural landscape In: Bunce RGH, Ryszkowski L, Paoletti MG (eds) Landscape ecology and agroecosystem. Lewis Publishers, Boca Raton, FL, USA, pp. 159–167

Finzi R (1998) Civiltà mezzadrile. La piccola coltura in Emilia-Romagna, Laterza, Bari

Fowler PJ (2003) World Heritage Cultural Landscapes 1992–2002. UNESCO, Paris

Gambi L (1995) Declino o evoluzione della tradizionale piantata in coltura promiscua? Qualche considerazione ricavata dal caso emiliano-romagnolo. In: Ceschi R, Vigo G (eds) Tra Lombardia e Ticino. Studi in memoria di Bruno Caizzi, Edizioni Casagrande SA, Bellinzona

Gambino R (1994) Ambiguità e fecondità del paesaggio. In: Quaini M (ed) Il Paesaggio fra attualità e finzione. Cacucci, Bari

Gorfer A (1988) L'uomo e la foresta, per una storia dei paesaggi forestali-agrari della regione tridentina. Manfrini Editori, Trento

IUFRO (2007) Guidelines for the implementation of social and cultural values in sustainable forest management, a scientific contribution to the implementation of MCPFE – Vienna Resolution 3. IUFRO Occasional Paper No.19

Landi F (1994) Le basi economiche: un sistema ad alta integrazione e bassa produttività. In: Gambi L (ed) Storia di Ravenna. Dalla dominazione veneziana alla conquista francese, Marsilio editori, Venezia, pp 517–582

Laureano P(2004) Il sistema dei terrazzamenti nel paesaggio euro mediterraneo. In: La cultura dei terrazzamenti per la salvaguardia del paesaggio. Atti del convegno internazionale Vietri sul Mare. Maggio, Menabò Edizioni, Salerno

Martin J, Pasqualetti PG, Righter RW (2002) Wind Power in View: Energy Landscapes in a Crowded World. Academic Press, Waltham, Massachusetts, USA

McNeill JR (2002) Qualcosa di nuovo sotto il sole. Einaudi, Torino

MCPFE (2003) Improved Pan European Indicators for Sustainable Forest Management. MCPFE Liaison Unit, Vienna

Moreno D (1988) Il paesaggio rurale fra storia e attualità. In: Monti e Boschi 1, pp 3–4

Moreno D (1990) Dal documento al terreno. Il Mulino, Bologna

Naveh Z (1998) Culture and landscape conservation: a landscape-ecological perspective. In: Gopal B (ed) Ecology today: an anthology of contemporary ecological research. International Scientific Publications, New Dehli, pp 19–48

Piussi, P. (1996) Continuità e trasformazione del paesaggio forestale: problemi e metodi della storia ecologica dei boschi. Istituto Internazionale di Storia Economica "F.Datini". Atti della XXVII Settimana di Studi: L'uomo e la Foresta secc. XIII–XVIII, Prato 8–13 Maggio 1995. In Cavaciocchi S (ed) Collana Atti delle settimane di studi ed altri convegni n 27, Firenze, pp 121–154

Van Der Ploeg JD (2006) Oltre la modernizzazione. Processi di sviluppo rurale in Europa, Rubettino, Cosenza

Rackham O (1986) The history of the countryside. J.M. Dent & Sons Ltd., London

Romani V (1994) Il Paesaggio. Teoria e pianificazione, Franco Angeli, Milano

Sansavini S (2002) Un secolo ed oltre di frutticoltura. L'agricoltura verso il terzo millennio. Accademia Nazionale di Agricoltura, Bologna

Santoloci M (2000) (2000) Diritto dell'Ambiente. Edizioni Ambiente, Milano

Sereni E (1961) Storia del Paesaggio agrario italiano. Laterza, Bari

Toderi G, Baldoni G, Nastri A (2002) Il sistema delle colture erbacee nel XX secolo: aspetti agronomici dell'evoluzione e prospettive. L'agricoltura verso il terzo millennio. Accademia Nazionale di Agricoltura, Bologna

UNESCO WHC (2008) Operational Guidelines for the Implementation of the World Heritage Convention. WHC 08/01, January

Zanchi B, Zanchi C (2009) Le sistemazioni idraulico agrarie collinari quale fondamento della sostenibilità produttiva e della tutela paesaggistica ed ambientale. In: Marinai V (ed) Paesaggio e sostenibilità. Studi e Progetti, ETS, Pisa, pp 195–212

Chapter 6
Severance of a Traditional Grazing Landscape in the Himalayas: Commons and Ecosystems in Crisis?

Minoti Chakravarty-Kaul

Abbreviations

Gaz	Gazeteer
PAR	Punjab Administration Report
PBEI	Punjab Board of Economic Inquiry
Progs	Proceedings
SR	Settlement Report

> *It will be sufficient to speak of original property in land, for among pastoral peoples… In general, property in land includes property in its organic products.* (Karl Marx 1857–1858).

6.1 Summary and Introduction

Nestled in the north-west Himalayas has survived a cultural landscape. Though severely fractured, this is an ancient tradition of pasturing sheep on the grazing *dhars*, not unlike the Swiss alps (Lyall 1865) and *soanas* or grazing runs for cattle, which are in the nature of commons. Contrary to the Hardinian perception of shepherds over-grazing a hillside which is a common property resource, the protagonists of this story—both shepherds and herders—have averted a tragedy of the commons. This has been by adopting "physical mobility to change location" (Scott 2009) not because they are fugitives of a stateless society as in nearby

M. Chakravarty-Kaul (✉)
Formerly Department of Economics, Lady Shri Ram College, University of Delhi, C59 5I.P.EXTN, Patparganj, Delhi 110092, India
e-mail: minoti.chakravartykaul@googlemail.com

I. D. Rotherham (ed.), *Cultural Severance and the Environment*, Environmental History 2, DOI: 10.1007/978-94-007-6159-9_6,

Zomia (which Scott describes), but have rather made it possible for sedentary cultivators to co-survive in extremely isolated locations. Thus have pastoral transhumants reciprocated the needs and services of hill cultivators of "billiard table" (Spate 1963)-sized terraces on steep slopes of river valleys. In the process there evolved ecosystems which required a strategy of rules in use to access and police natural resources which were shared and open to erosion in a fragile and dynamic countryside. Here the 'hill folk' could survive only if they could jointly minimise uncertainty by sharing risk as I describe in the first part of the chapter.

Such a cultural mosaic of commons in the forest and above the treeline of remote mountains was transformed in the nineteenth century. This was initiated by a process of colonial codification which was not like pastoral people in the periphery being shifted "from stateless zones to areas of state control", as James Scott describes in Zomia (Scott 2009) by a process of colonising. It was what happened when the Wasteland Minutes written by Canning in 1861, indirectly opened the issue of settling mobile people where possible, "by replacing open common property land tenure with closed common property" (ibid) and providing boundaries in the waste of cultivating hamlets both in the plains of north India and in the Himalayan uplands. Such a process was furthered with codified rights in land and natural resources legalised by the Revenue Settlements Act of 1871 and a Forest Law in 1878 which ring-fenced settlements and forests in north India. Such codification was extended and deepened by railways and canals which converted the vast open grass lands of the south-west periphery of northern India and the Himalayan countryside into "a fiscally fertile zone" (ibid) eliminating thereby "non-state spaces" (Ibid). It is the ramifications of this process we will describe in the second part.

6.2 Historical Roots of a Culture: Chamba and its People

The cultural landscape of the grazing runs in the western Himalayas from the alpine meadows to the tropical forests of the foothills of the Siwaliks owes much of its diversity to the time when the Indian tectonic plate crashed into Central Asia which resulted in "a diversity of ecosystems from alluvial grasslands and sub-tropical broadleaf forests along the Siwaliks, to temperate broadleaf forest in the middle hills, which then changes to inter-mixed conifer forests and pure conifer forests in the higher hills and alpine meadows" (Mittermeier 2004).

Given this biological diversity albeit risky environment, G.W. Traill, the first Commissioner of Kumaon declared that "the Central and Lower Himalayas were not howling wildernesses, but have been for ages occupied by an industrious agricultural population".[1] Chamba was one such unique Himalayan Kingdom in the

[1] Batten (1856–1864).

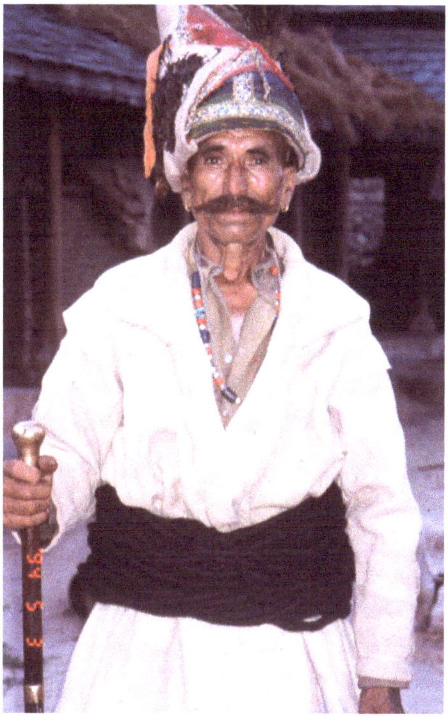

Fig. 6.1 A Gaddi Shepherd. Photograph Minoti Chakravarty-Kaul

north-west which was at once remote and yet open to external influences because its subjects were in the main transhumant shepherds who changed location between mountains and plains not unlike the system in the Swiss Alps. This culture connected two unique land-use patterns of rotating fallows of high mountainsides to those of the village communities in the foothills and plains of the Punjab who rotated village fallows, not dissimilar to the *utmark* of German Teutonic communities. Such a link exposed the traditions of Chamba to the changes on the plains of Punjab when in 1849 the British annexed the Sikh Kingdom.

In Chamba, the British could not fail to recognise the antiquity of both the rulers or the ruled. While its *antiquities* were established by Jean Phillipe Vogel (Vogel 1911) so also did the Gaddis not fit the perception of a hill people who "are best understood as runaway, fugitive, maroon communities… fleeing oppression of states" as was the case elsewhere in the neighbouring hills of Southeast Asia, what James Scott calls Zomia (Scott 2009). Rather, the British administrators recognised the long surviving tradition of transhumant shepherding of the Gaddis (see Fig. 6.1). These were subjects of the Kings of Chamba where the same family ruled for 1,200 years since 700 AD—a record unequalled by any ruler of Europe. The ancient capital of Bharmor was only a village although the kingdom extended to the Padar and Bhadrawah (Map 6.1, Vogel 1911). From this remote area, the

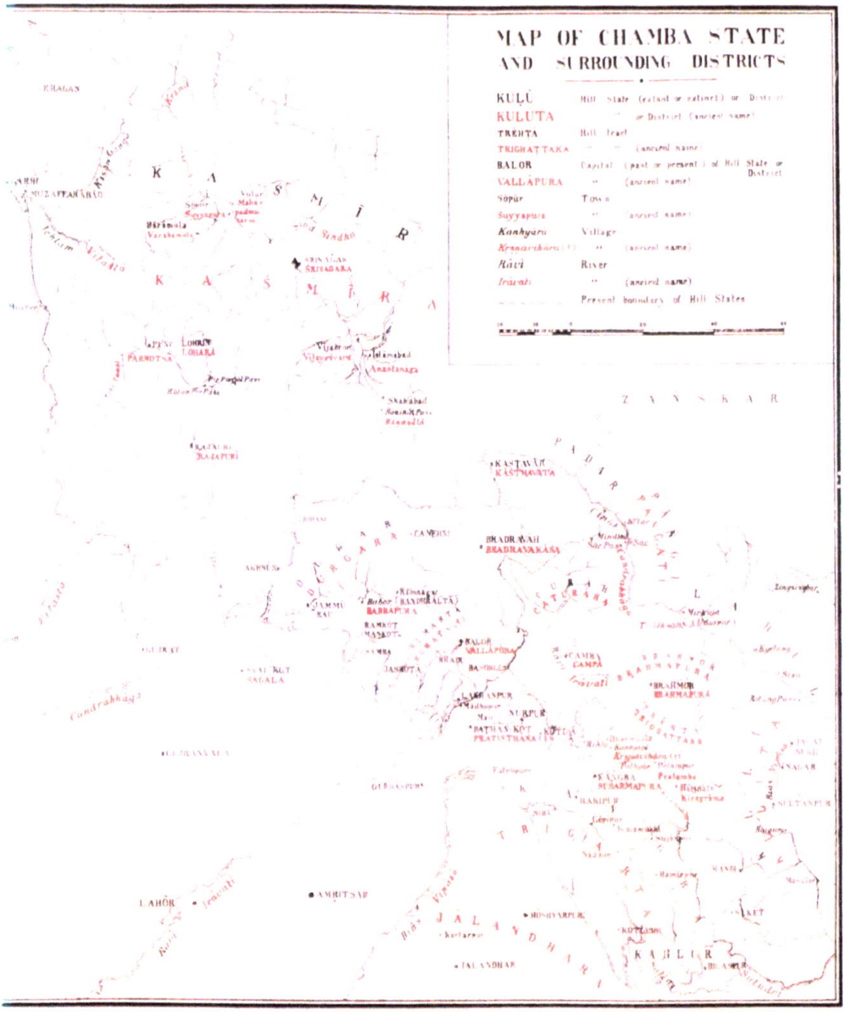

Map 6.1 The Chamba State (Vogel, 1911)

Gaddi[2] shepherds then escaped the harsh winters with their herds of sheep and traversed through the forests on ridges and in the valleys within the mountain system to the foothills and a return journey in the summer (Vogel 1911). Therefore, their homeland Chamba was integrated as part of the whole mountain range and the valleys of Kulu, Lahul and Spiti and Kangra, where lived other people both sedentary and mobile.

[2] Hutchison and Vogel (1933, p. 274) in fn by Vogel. The Gaddis have castes such as Brahman, Rajput and Khatris and they returned themselves in the census under these caste names and not as gaddis. They are principally in the Brahmour Wazarat which is called Gadaran.

Fig. 6.2 *Dhars* or grazing Alps of Pangi. Photograph by Minoti Chakravarty-Kaul

6.3 Part I. A Cultural Landscape

Transhumance has been a critical institution for a pastoral culture, which depends on a tradition of rotating common pastures over space and seasons over the mountain ranges and with those in the plains. In effect, mobile pastoralism has been a support for sedentary cultivators of mountain terraces and the dry tracts of the northern plains of India. Thus did the North West Himalayas become a surrogate commons, which complemented the lowlands of northern India, as we shall see in the next two sections. In Europe, where such vertical migration goes back to very early times in such pairs of regions each of which provides for the climatic defects of the other, and the two climates are regarded as complementary, arising from mere differences in altitude (Carrier 1932).

This story of the pastoral people resonates for a whole region. Transhumance could minimise uncertainty in desert-like locales in the south–west plains of Northern India and icy mountainsides in the north-west Himalayas by horizontal and vertical mobility over space and seasons; while their counterparts—the cultivators could do the same by rotating fallows between cultivation and pasture at the same location. Somewhere along the line, however, both parts of such a culture cannot do it alone without having to share resources which are needed in large scale—namely pasture. Common lands are an integral part of such a cultural landscape and as such could easily be overgrazed and so become victims of the so-called tragedy of the commons. Cheating and shirking of responsibility towards sustaining a resource used in common can always be a real danger. However, as we will see how the compulsions of uncertainty and risk on the mountane, sub-montane and the plains induced institutions of reciprocity to match the Upland-Lowland variations in altitude, rain, temperature and vegetation of grazing tracts.

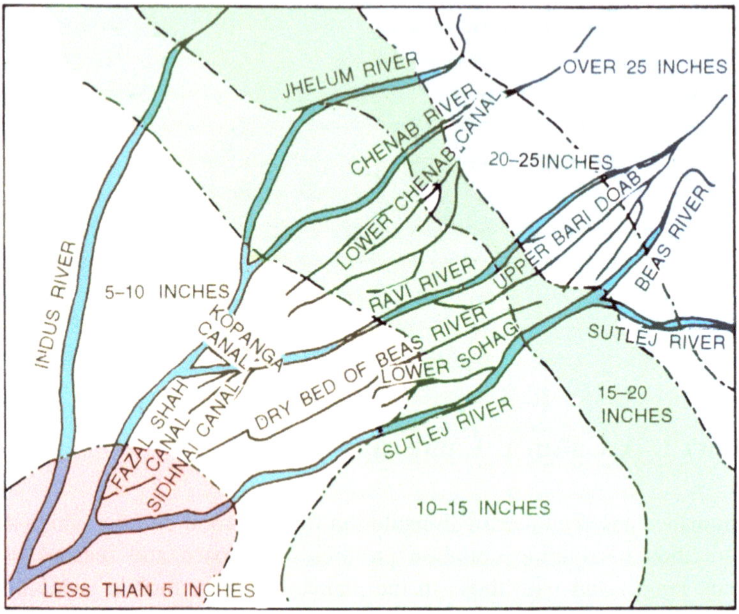

Map 6.2 Complementarity of Periphery and Core in Greater Punjab. (Copyright Chakravarty-Kaul 1996)

6.4 Traditions of Upland-Lowland Commons

6.4.1 The Upland Periphery

The upland periphery of a Himalayan countryside would be impossible for communities like the Gaddi shepherds (see Fig. 6.1) and the Gujar cattle herders. These people cannot afford to stock their herds with the kind of heating necessary in winter, nor can they stock fodder that would be needed during winter, nor grow or stock food for themselves during the extreme weather. Besides this, the scale of economies that are critical for pasturing cattle and sheep would be beyond the means of an average shepherd or herder, particularly working on the mountainside. Transhumance made it cost-effective for livestock herders in Chamba (see Map 6.2) because they could access grazing resources on a scale that is usually referred to as the "open waste" whether this was in the forests of mountainsides, or unenclosed grazing pastures in the alpine meadows, *dhars* (see Fig. 6.2) and *soanas*. The latter are grazing runs synonymous to the Swiss Alps, on steep mountainsides and ridges, used for sheep and cattle respectively. Similarly, grazing was available in the common stubbles of post-harvest cropped terraces in the hills, in the long fallows of the villages in the lowland valleys of river, and in communally held forests in the plains. Such transhumant migration depended on a culture of reciprocal exchange of services between graziers and cultivators on the mountainside and in the sub-montane zone.

6.4.2 The Lowlands

The sub-montane and lowland areas complement the Himalayan landscape in several ways. Political turmoil apart, the whole region is dynamic and so supported two different cultural systems. In the core, there was a strong agricultural tradition of village communities who rotated the short and long fallows between cultivation and grazing. The long fallows were kept as the village commons (Chakravarty-Kaul 1996). On either side of the core in the periphery there were two broad variants of mobile pastoralism, one horizontal from the south-west periphery and the other from the Himalayan periphery through the sub-montane, but neither of them needed to converge on the lowland core at the same time. Here we will concentrate on the second periphery but for details (see Chakravarty-Kaul 1996). In winter, the cattle came down from the forests of the upper reaches of the Himalayas which began after the monsoons and came down to the winter pastures in the sub-montane and sometimes also into the riverain (See Map 6.3).

Thus, the riverine wetlands and the sub-montane provided seasonal pastures for the cattle of all the regions.[3] When the spring or *rabi* crop in the riverine belt was cut, the stubbles of cultivated fields and grass left in the long fallow waste, was available as in common grazing. In winter, this was repeated after the *kharif* crop was cut. In the summer, the cattle would go to the riverain zone, onto the hills, the duns or the valleys (Ibid).

6.5 Mapping Rights and Use in the Grazing Landscape

Traditional grazing trails (see Map 6.3) followed not only grass and water but food for graziers too. In this section, we map out the pattern of access, use and monitoring of the seasonal and unseasonal calendar of transhumance. Pastoral institutions were influenced first by the natural ecology of the sub-montane and the montane tracts; second, by altitude of the pasture; third, by the seasonal/unseasonal variations in risk; and finally by the need to conserve grazing resources in the landscape.

---

[3] J.M. Douie, District Collector, Karnal to Commissioner and Superintendent.

Complementarity and Regional Patterns

Map 6.3 Upland-Lowland Grazing Patterns. Copyright: (Chakravarty-Kaul 1996)

6.5.1 Natural Ecology and Grazing Ecosystems

The grazing pastures of the sub-montane zones held in communally held forests and village common lands fused together two distinct patterns of grazing and forged a tradition of ecosystems. Common property resources as long fallows were key to (i) seasonal nomadic grazing needs of two tracts from the uplands in winter

Table 6.1 The Sub Montane Grazing Fallows

Districts	Grazing wastes	Season of grazing
Hoshiarpur	1. Common lands [Grass fallows] in villages	Monsoon and post harvest
Gurdaspur	2. *Banjar*: or long fallow plots in Andhar[a] circle (Pathankot) and Bharari circle	Winter
	3. Grazing *chhambs* or wetland of Gurdaspur and Jullundur: Khanuwan Chhamb, Magar Mudian Chamb (wetlands)	Winter
	4. Riverain tracts: Belas[b] (islands) in the Ravi, Beas, Sutlej, Kurari and Nahar ki Bir in Pathankot	Winter and summer
	5. *Shamilat* Forest Fallows:	Winter
	Gurdaspur: 16 village *shamilat* (common) forest and Hoshiarpur: 17 village *shamilat* forest[c]	Monsoon
	6. Forest fallows:	Summer
	(a) Lower Siwalik, Hoshiarpur, Jaswan Dun, Sola Singhi Range[d], Mangowal Range, Panjal and Lohara;	Summer
	(b) Gurdaspur: Shahpur Kandi	

Source
[a] Gurdaspur Gazeteer (1914, p. 102)
[b] Siba Jagir Settlement Report (1882, p. 23)
[c] Una Tehsil, Hoshiarpur Settlement Report (1914, p. 27)
[d] P.S. Melville, Comm. and Supdt. Trans Sutlej to D.C. Hoshiarpur, Rev.& Agri. Forests, Progs. 3–5B, Oct. 1887

and from the lowland plains of the Punjab in the summer; and (ii) a year-round grazing of village herds in the sub-montane area itself. It is here that shepherds interacted with a unique combination of two distinct aspects of commons in the long fallows. One was the *banjar kadim* (long fallow) which were *shamilat-deh* or common lands of individual villages in the tract, and the other was in the long fallows and grasslands in the forests (see Table 6.1).

The sub-montane zone could support this double-dependence because of its two unique characteristics. First, the *shamilat ban*[4] or forest held as commons were cultivated in patches and so the grazing on the crop stubbles in these, and the grass fallows within the forests, were shared with the Gujars and Gaddis.[5] These customary obligations were sometimes recorded. In return, the nomads provided manure on post-harvested fields. We can give three examples. (i) The Panjal Tappa of the Hoshiarpur district had eleven villages[6] which had *shamilat* (common) grazing area in the forest and the Gaddis were using the forest along with the villagers; (ii) the Lohara forests of Hoshiarpur[7] had scattered village habitation and the uncultivated forest areas were shared with migrant cattle from as far off as

[4] *Shamilat* meant common land belonging to the entire village proprietary body of the individual villages. This was distinct from the open access forests.

[5] Gujars and Gaddis were pastoral tribes in the main but they were not always nomadic.

[6] Una, *Hoshiarpur Settlement Report* 1876. para 103.

[7] Ibid, para 108.

Chamba, Lahul and the Dhaula Dhar in winter. The tract also received distress movements from as far afield as Hissar and Karnal, the main cattle breeding tracts of the plains.[8] In periods of famine, (Famine Report 1880) cattle were sent up from the breeding tracts of Karnal and Sirsa, and also during the dry season when the water holes and the grass dried up; (iii) finally the Shahpur Kandi tract of Gurdaspur virtually belonged to the zamindars[9] by long usage of grazing and fallow cultivation. Even when the forest area was extensive in the proximity of the village there were patches of cultivation scattered throughout the forest. There were sixteen villages, which held forests in *shamilat*[10] (see Table 6.3). Second, the sub-montane zone also had unique customs of alluvion and diluvion to manage the riverine fallows or *belas* for grazing. These consisted of land thrown up by deposition of alluvium on the banks of the rivers Sutlej and the Beas in the Hoshiarpur district. These became common fallow for grazing. This was a customary way of sharing risk from river action (see Map 6.3).

The grazing runs in the montane zones, called *dhars* and *soanas*, illustrate the use of the principle of complementarity in sharing mountain slopes at different heights of montane and sub-montane forests (for details see Chakravarty-Kaul 2000). These customary arrangements took advantage of the varied vegetation types for providing specific preference for succulents within the forest range itself and with those in the forests outside them. Such a pattern of forest ranges provided pastures in rotation just as fallows do for cultivators of terraces and valleys. However, rights to them differed between the forested tracts of the montane and those in the sub-montane forest districts of Gurdaspur and Hoshiarpur or those of the plains. Illustrations of these are best obtained from British settlement records in 1849–1852.

G.C. Barnes settling the montane district of Kangra observed grazing rights of Gaddi shepherds were claims they laid upon certain beats of forests, which "they regard as their *warisee*, subject to the payment of pasturage tolls. The forests of the lower Hills are apportioned out among the Gudees or shepherds of the Snowy range, who, in the winter season bring down their flocks to graze. In the same manner, the Goojurs with their buffaloes will take up divisions on the hillside and carefully respect their mutual boundaries. Often, as buffaloes relish different shrubs and grasses from those that sheep and goats take, a Gudee and a Goojur will possess a concurrent claim upon a certain tract of forest. Either would instantly resent the intrusion of another of the same tribe, bringing in the same class of animals to graze, but as their respective herds delight in different esculent matter, the rights of the two are perfectly compatible" (Barnes 1849–1852).

Herders like Gujars, who had buffaloes used varied tactics of leasing grazing in closed forests for three months since they could not rely solely on the grass

---

8 Lowis' reply to questions asked by Rai Bahadur Shyam Sunder Lal, *Famine Inquiry Report* 1901, p. 306.

9 *Shahpur Kandi Tract, Gurdaspur Settlement Report* 1877 p. 15.

10 *Gurdaspur Settlement Report* (1912, p. 100).

Table 6.2 The Complementary Montane Grazing Fallows

District	Grazing wastes/fallows	Season
Lahul	1. *Gahar* and *Thatch*[a]: land near the highest cultivated land;	June
	2. *Nigahars*[b]: alpine meadows in sources of rivers Parbati, Sainj, Tirthan;	July
Kangra	3. *Bahan*[c]: fallow fields in the terraced hills and valleys;	Post-harvest
	4. *Kharetars*[d]: grass preserves in enclosed field on hillside;	Post hay-cutting season
	5. Intermediate waste between the hamlets;	Summer
	6. *Soanas*: exclusive pastures used by Gujars in forests;	Spring and summer
	7. *Dhars*[e]: grazing runs on hillside for sheep;	Spring and summer
	8. Forest *bartan*[f]: hillside valley forest rights	Spring and summer

Source [a] Diack (1898, p. 36). [b] Ibid. [c] Lyall (1865–1872. p. 43). [d] Ibid, p. 45
[e] Ibid, p. 19. [f] Ibid

growing on fallow fields. The *soanas* were thus exclusive runs for three months and then when the *thak* or closed grazing was removed all the cattle of the village grazed over the whole forest indiscriminately. The Gujars right was much like the right of the individual to his hay field or *kharetar* which is exclusive for a season. These *soanas* or grazing runs were like *warisi* or heritable rights to the forest commons (Lyall 1865–1872) (see Table 6.2).

6.5.2 Altitude and Grazing Trails

Pastoral acumen in creating rights in the seasonal grazing commons in alpine meadows is perhaps a good example of using altitude to achieve (i) scale efficiency and access to (ii) quality of pasture. Nevertheless, here food could become problematic.

The examples of Kangra in the lower hills and Lahul in the upper ranges illustrate. Here some of the shepherds planned duplicate villages one in the lowlands and the other in the uplands not unlike the Minoan herdsmen in Crete (Carrier 1932). For food, the Gaddi shepherds "cultivate the winter crop of wheat in Kangra in the lower ranges and returning with their flocks to grow the summer crop or a rain crop in Bharmour as the province on the other side of the snow is designated" (Barnes 1849–1852).

Altitude also tested traditional rights, which respected *complementarity of scales* (see Table 6.3). Flocks moving up to these alpine meadows were numerous,

Table 6.3 Scale of Operations (Diack 1898)

	The number of sheep grazing in Lahul in 1890
Kulu flocks	51,665 sheep and goats
Chamba flocks	58,048
Kangra Gaddi flocks	63,205

but they respected each other's customary grazing runs. For example, the Gaddis came down from Bharmor in Chamba into the warmer valleys of Kangra while people from Lahul similarly moved into Kulu with their families for the cold weather (Punjab Census 1881).

Altitude and Access to high quality of grasses were likewise a tribute to shepherdic institutions. The Gaddis got *pattahs* (or deeds) from the Raja of Chamba to access the rich grasslands with species like 'niru', which is dull and bluish in colour and the favourite food of the sheep. Others were '*mat*' and '*morar*' (Lyall 1865–1872), and they occurred in high altitude alpine meadows known as (*gahar* or *thach)* which they entered on the same level as the highest lying cultivation for a month, and then in July moved on to the *nigahars* of Chamba (see Table 6.2). Such a move protected them on the leeward side of the mountains, which was very healthy for sheep and helped to avoid 'foot-rot' that attacked them in the summer. Rights of access to these meadows were seasonal, as some of them were really fallows; but taking advantage of such resources involved transaction costs (see below), by way of food and pasture required for a complex transhumant calendar. The Gaddis did this by keeping their summer and rainy season *dhar* or sheep runs (see Fig. 6.2) in the Chamba territory and "later went in autumn and winter and spring to Kangra Proper" (Lyall 1865–1872). Such a cycle of mobility began in March when the flocks from the lower ranges of Native States and the foothills of the Siwaliks were driven home to spend the spring lambing season near the village of the proprietors and remained there till June (Diack 1898).

6.5.3 Seasonal Variations and Risk Avoidance in Pastoral Customs

Minimising risk through customs of (i) rotation of montane grazing fallows and (ii) in reciprocal rights to pasture in the sub-montane areas, reflect, the means to reduce loss through counter measures to tackle sudden risks in the Himalayan countryside. Two examples illustrate this.

Lahulis are a good example. To escape the freezing winters the Lahulis keep their sheep in the lower hills of Kangra. Similarly, in winter the Kulu sheep and goats of the higher *kothis* or settlements were taken to the pastures of the lower *kothis* of Mandi and Suket (Diack 1898). At the same time, "the Chamba mountaineers descend, with their flocks upon the valleys of Kangra, the people

Table 6.4 Summer Grazing 1891 (Diack 1898, p. 36)

Sheep and goats belonging to	But grazing in native states	Kulu	Total
Kulu Proper	18,948	76,617	95,565
Rupi	21,697	22,750	44,647
Saraj	5,598	76,337	81,925
Total	46,433	222,137	268,570

contest with each other, who shall house the shepherd and his flock and a cultivator will give two or three rupees a night, for the advantage of having sheep folded upon his land. Night after night the shepherd changed his ground and before the harvest is sown, reaps a little fortune without the smallest exertion or cost" (Barnes 1849–1852) (Table 6.3).

Sharing of risk through customs like common of shack, as it was known in England, involved folding and penning of sheep of nomadic herders in winter on post-harvest cultivated stubbles of villages in the foothills of the Siwaliks during the winter months. In the process, organic manuring of the land (Calvert 1933) and fodder were exchanged minimising transaction costs (Table 6.2) of individual shepherds and cultivators while countering seasonal risk (Table 6.4).

6.5.4 *Pastoral Traditions of Conservation*

Finally, customary traditions fended off what is known today as 'tragedy of the commons' since there always existed the biggest risk of all—free-riding—on the mountain pastures which functionally served like rotating common fallows. Shepherding mobility was kept visible by a custom of *lango karu* or what shepherds had to pay for the crossings over the seasonal runs. Each *dhar* or grazing *alp* paid one or two goats and the fleece of a sheep. A village official or the drirkar, who was always a Gaddi and was entitled to take certain perquisites from the shepherds, collected this fee. However, after the settlement of the British all the taxes on the *dhars*, the *lango karu*, were abolished. There was a variation in the titles to these alpine pastures and so also to the claims of those to the *dhars* who were not shepherds. "All the flocks when they descend into the valley in the autumn spend some time in sitting on the fields, but, except in certain cases the shepherd is free to agree to sit on any man's land he pleases; whether he is also free to leave the village at once without sitting on any land is a moot point; the general feeling is that he ought to halt a certain time for the good of the village and with rare exceptions he always does so. In going up in the spring the *dhars* are all free even in Nirwaneh and Kamiara there was always this distinction between spring and autumn" (Lyall 1865–1872)

6.6 Part II. Severance of the Cultural Landscape

The cultural landscape of the Himalayas began to be disturbed, oddly enough, when political conditions in the Punjab plains settled down in 1849 with British annexation. A new source of uncertainty appeared when a modern Government codified institutions of rights to land and natural resources of village communities, which then altered the customary arrangements, which we have described above. The process as we shall see below was initiated by enactments like the Punjab

Land Revenue Act of 1871, which gave boundaries to villages in the plains and the Punjab Laws Act of 1872 that recognised customs of villages in the common lands. This process as we shall see below meant that the Punjab Government virtually 'captured the waste' outside the villages. In these wastelands, the Government also began construction of railways and canals, which went right across the pastoral trails. In the mountains too, something similar happened as we shall see below, even though caution was exercised in the first settlements by G.C. Barnes. He was one of the most able revenue officials who was aware that the nomadic pastoralists had their own special runs in the forests.

Settlement in the Mountains 1849–1852 by G.C. Barnes 'created' communities out of scattered hamlets being clubbed together as co-parcenary bodies. These then had to pay the revenue jointly, a responsibility to which they were strangers, and to balance this Barnes gave the new communities the right to collect certain miscellaneous items, the produce of the waste or their 'created' common lands. The question of the forests in general, was left vague. Unwittingly Barnes created a 'revolution' as his successor Lyall (1865–1872) averred. However, he went a step further and demarcated the boundaries of the hamlets within the *mauzas* or revenue estates. In their eagerness to get solid property in the waste, the people wanted to sub-divide and this allayed their fears that the "Government was about to take away the land"[11] (see Chakravarty-Kaul 1996). This had consequences.

Therefore, *the historic 'capture of the waste'* followed. The village communities, the Government and the foreign tea-planters all became contenders for the waste. It confirmed settlement lore of that time that "limited" land within demarcated villages provided valuable common property in the waste within villages. At the same time, the State acquired rights over the waste at large. In 1852, it was re-iterated that outside the village boundaries was to be "the property of the Government".[12] It was a greater 'revolution' than was officially acknowledged, because it opened vistas of opportunities for the Government to acquire the vast unreserved and unoccupied forests and scrub lands of the Punjab. The Wasteland Minute of Canning recorded in 1861 confirmed this. Given this fiscal security in Kangra[13] the Government could then embark on enclosing the forest commons as we shall see below.

[11] The iqrarnamas (statement of rights) frequently differ greatly on important points. For example "... the unmeasured waste in one is not declared to be shamilat or joint property of the landholders, and a literal rendering of the text would limit their proprietary right to the measured rights only; in another no mention is made of the right of Government to the timber of trees on waste lands etc." Kangra Settlement Report 1865–1872, p. 29.

[12] Board of Administration's circular No. 15 of 1852, prescribing the steps to be taken to encourage plantations and conservancy, Circular Order issued by the Board of Administration in the Revenue Department during the years 1849–1853, Manual for Arboriculture, 1905, para 3.

[13] Edmonstone in his letter No. 2554 of Nov 9, 1853, para 12, "*It may be encouraging to the local officers to know that in the Simla Hills this revolution has within the last year or two actually effected... and joint responsibility fully recognised and established.*" Jhang Settlement Report, 1860, para 12.

Institutional uncertainty in access to common pastures followed Government fiscal settlements and weakened customary reciprocity. First, shepherds suddenly found their customary grazing runs overlaid in some sections by the 'created' common lands for village estates, which had been cut out of the intermediate hill wastes lying between hamlets. Second, friction was caused when Gaddis also had to re-negotiate grazing rights. They objected to paying dues to the landholders when their grazing runs were not co-incident with the village commons. Further, reciprocity on the stubbles of the cultivated fields were not likely to be honoured by cultivators who had newly acquired joint property in the waste. Third, these grazing runs were hampered by forest rules, known as Bayley's rules of 1855. And so these restrictions led to greater pressure on the 'created village commons' in the hills on which increasingly village cattle and those of the transhumant pastoralists had to rely.[14] Fourth, decisions regarding cultivated fallows in denuded forests were restricted by permission from the Deputy Commissioner who intended "to prevent the clearance of the forest". (Ibid) Such a policy strained the institutional patterns as we see below.

Environmental risk was heightened in the mountains from mounting pressure at the end of the nineteenth century. Clearly, rules to protect forests had contrary and opposite impact on the grazing commons of landholders and created friction between landholders and shepherds. First, famines at the end of the nineteenth century, brought herds from the plains so far away as Karnal and Hissar.[15] Second, there was unplanned concentration of cattle, both of the villages and of the nomads, in the *shamilat-deh* or common lands of the hill villages, as was observed in village inquiries of the Punjab later (PBEI 1933). Ironically, it was not this pressure of cattle on grazing commons but the decline of cattle numbers that drew attention to the forest policy!

Enclosing forests for conservation however drew attention to shrinkage of grazing which was suspected to have caused a decline in the number of sheep in Kangra. A new conservation policy emerged. Lyall (1865–1872) reserved trees for the Government, which grew on village common lands and second inserted binding conservancy rules in the village administration papers. Finally, the Government took over the right to collect grazing dues from the Gaddis who objected to payment of grazing fees to the landholders whose estates were not co-incident with sheep runs. All of this increasingly gave the Government a greater hold over users of grazing resources in the uplands.

Reciprocity was thus weakened in a sequence of events, which altered land-use patterns in the lowlands and sub-montane zones with serious consequences for the mountain commons. First, grazing fallows were shortened in the lowlands as

---

[14] P.S. Melvill, Commissioner and Superintendent, Trans-Sutlej States, to District Collector Hoshiarpur, 1/3/1860, Revenue and Agriculture (forests), Proceedings 3-5 B, October 1887.

[15] Revenue and Agriculture (Famines) Proceedings 3-4 A, September 1885, p. 324.

canals encouraged multi-crop cultivation. This increased pressure on village common lands and induced partitions, furthering shortages. Mobile pastoralists started pressing on the montane forest fallows. Second, Government responded to such pressure by formalised enclosure of forests by the 1878 Forest Act. Conservation was strict in the reserved category of forests; but this meant increasing pressure on the protected category of forests, which admitted rights of pasture, fuel and timber. In years to come, it also took measures under the Punjab Laws Act 1872, to achieve conservancy of forests in communal lands in the hill districts of Punjab.[16] Finally, the principles of non-interference in common lands adopted in the settlements were given up in 1875 after the Forest Conference in Simla. Government was to administer common property for the good of all.[17] As we see below.

Upland-lowland complementarity was most specifically modified by what happened to the sub-montane grazing in the three communally held forests which were buffers between the plains and mountains (see Map 6.3). For brevity's sake, we illustrate this from one example. In the Shahpur Kandi Forest, (see Map 6.3), the Government demarcated 2,339 acres as State Forest and the rest undemarcated portion being 5,099 acres was "all the common lands of 17 villages in the tract" (Gurdaspur 1912). Villages from which no land had been taken for the State Forests, "could not exclude from their lands the Gaddis who had hitherto enjoyed the right of grazing there".[18] Consequently, Gaddis could graze here by right as they did earlier. It was however "optional for the others to open their lands or not" (Ibid). The Gaddis paid the Forest Officer their dues and at the end of the season, the money was collected and divided "amongst the various villages", (Ibid), but evidently it only distanced the customary relationships.

Free riding emerged since reservation of forests entailed severe policing beyond a point. The availability of forest per capita was ridiculously low being only a quarter of an acre in the hills and lesser still in the plains. Comparison to European standards made it appear even more so.[19] It consequently resulted in forest offences. Evidence of this were the "protests" by burning forests and other forest offences which were on the increase. In 1909–1910, the Conservator of Forests remarked that "The increase in the number and value of cattle which combined with curtailment of wastelands make it more worth the while of graziers to run the risk of detection, when grazing their animals in the prohibited areas" (Report 1909–1910). There was also an increase in corruption among the officials who were supposedly the "guardians" of the forests. Whilst this can never be proved unofficial discussions suggest that it was indeed so.

---

[16] Punjab Gazette, 1907, Notification No. 179 for the Rawalpindi Forests.

[17] C.F. Amery, at the Advance Conference on forestry in Simla, 1875.

[18] C.A. Roe, Settlement Officer Hoshiarpur to Commissioner and Superintendent, Jalandhar Division, 13/11/1872. Agriculture and Revenue (Forests), Proceedings 3–5 B, Oct. 1887, para 18.

[19] Report of the Royal Commission on Agriculture in India (1927, p. 683).

6.6.1 Tell-Tale Signs of Erosion in the Landscape

The end of a tradition seemed imminent but it threatened also to end in tragedy almost like what Garrett Hardin had suggested. Late in the nineteenth century, soil erosion, or *Chos*, as the phenomenon of hill torrents is known, developed in the submontane areas. It certainly highlighted the problem of reconciling conflicting interests of the increasing demand of the graziers with Government anxiety to arrest the deterioration of the forests in the Punjab. A vicious circle emerged. Just as shortening of fallows induced the pressure on the long fallow commons, so did reservation of forests in parts to conserve only helped to sharpen over-use in others. It was an issue of insufficient comprehension of the importance of complementarity of private–public governance of resources which are diverse, can be shared and have alternate use. Individual and common property rights complement but the colonial government did not appreciate this, at least not at the time.

Extensive de-forestation attended settlements in the hills of north-west Himalayas through the nineteenth century and in the first decade of the twentieth century (PAR 1919–1920). Consequently, forest policy increasingly turned towards reservation of forests and grazing as measures for conservation, which only added pressure on forested and grass covered soil on unreserved portions of the forests. These were not unrelated issues since conservation had externalities too. Yet the Punjab Government took several years to work out the complicated rights involved in the case of the lands in the sub-montane district of Hoshiarpur affected by the hill torrents. This was before legislation in 1901 could enable conservation of the eroded land (Hoshiarpur Settlement Report 1885). Curiously, the Conservator of Forests was to conclude that this was due "to the well-meant but mistaken motives of officers who object to placing any restraint on the user of forest defeat themselves".[20]

The tradition innovated in the early decades of the twentieth century to counter pressure in the uplands as in the Haripur and Mangarh taluqas in Kangra where landowners increasingly reserved hayfields or *kharetars*, which by 1931 represented 14.2 % of the total area in 1931. This was in addition to the *banjar kadim* or long fallow and common forests (PBEI 1933). However, the Government almost gave up. W. Mayes, Chief Conservator of Forests, Punjab, wrote in his Memorandum to the Royal Commission on Agriculture in 1927 that forest conservation in the hills was no longer sufficient with an increase in human and cattle population. He recommended restriction in pastoral pursuit until such time as "people have learnt to breed better stock, to grow fodder crops and cut hay instead of grass and to stall feed where pasture is insufficient".

In the lowlands, the period post-1947 brought the curtain down on the drama, again by institutional change brought about by the Land Reforms Act of 1947, which created open access to the village common lands. Nevertheless, the battle cry on the commons did not come from the whole region of north India but from

[20] Conservator of Forests, Punjab to Officiating Secretary Government Punjab, 4/12/1876, Revenue, Agriculture and Commerce, Government of India, Proceedings 1–3 A, March 1878.

the villagers of Kanjhawala, who trace their lineage back to the twelfth century. The whole cluster of twenty villages with a strong pastoral heritage protested the distribution of 123 acres from their village common grazing land to fulfil the target of a poverty elimination programme of the Government. It indicates the alienation of communities like Kanjhawala from the traditions of a commons heritage (See Chakravarty-Kaul 1996). The Kanjahwala tragedy of the commons is in effect a shared experience of a cultural severance, which was set in motion in the nineteenth century. A few illustrations are given down below.

In the uplands, post-1947 period, the pressure on the grazing landscape in the Himalayas has been exacerbated by the reduction of common lands and erosion of grazing resources in the entire region, which then reduces the opportunity of hill graziers to use their winter pastures in the plains. First, land reforms of 1947 created open access to the village commons in the plains by transferring them to the entire body of village residents that enabled their partition (Ibid). Second, by conversion of grazing fallows to alternate use dictated by market incentives, such as extension of the village residential areas and setting up non-agri-pastoral activities (see Chakravarty-Kaul 1999). Third, this was by the reduction of large areas of grazing fallows, which mountain shepherds used, in the foothills of the Himalayas, which are now reservoirs of several dams (see Map 6.1). Fourth, the effects included enclosure of the open waste in the plains for special economic zones. Fifth, the forest commons were discounted by a policy, which in 1988, very briefly made a move to hand back forests, particularly in the sub-montane zone, to the community. This was by a Joint Forest Management Policy with the Forest Department as a partner. The initiative only succeeded in elbowing out the transhumant graziers from the forest commons landscape.

Once again, at the beginning of the twenty-first century we see the conflict between conservation and institutions of human ecology. We need to ask whether a tradition of commons has necessarily to give way to the setting up of National Parks. Why is it necessary to oust the indigenes from their grazing commons to conserve threatened wild species, like the *bharal* or blue sheep in the Great Himalayan National Park in North West Himalayas and the Marco Polo sheep in the Nature Reserve of the Hindu Kush mountain region? Why is it necessary to forget history, and that both the Nature Reserves above were carved out of the grazing commons designed by shepherding tradition?

Thus, *the loveliest haunt of ancient peace* (William Morris) stands shattered.

References

Barnes GC (1849–1852) Kangra settlement report

Batten JH (1856–1864) Garhwal District settlement report. p 49

Calvert H (ed) (1933) Punjab Board of Economic Inquiry, An Economic Survey of Haripur and Mangarh Taluqas

Carrier EH (1932) Water and Grass. Christophers, London

Chakravarty-Kaul M (1996) Common lands and Customary Law. Oxford University Press, New Delhi, Institutional Change in Northern India in the past Two Centuries
Chakravarty-Kaul M (1999) Market success or community failure? common property resources in colonial north india and a case illustration from a cluster. Indian Econ Soc Hist Rev 36(3):356–387
Chakravarty-Kaul M (2000) Risk uncertainty and collective action on the mountain forest. Lecture at University of Trento
Diack AH (1898) Final report on the settlement of the Kulu Sub-Division of the Kangra District
Famine Inquiry Report 1901
Famine report (1880)
Gurdaspur Gazeteer (1914)
Gurdaspur settlement report (1912)
Hoshiarpur settlement report (1885)
Hutchison J, Vogel JP (1933) History of the Panjab Hill States. Government Printing, Lahore. I
Lyall JB (1865–1872) Report of the Land Revenue Settlement of the Kangra District, Panjab
Marx K (1857–1858) Pre-Capitalist Economic Formations
Mittermeier RA, Gil PR, Hoffman M, Pilgrim J, Brooks T, Mittermeier CG, Lamourex J, Da Fonseca GAB (2004) Hotspots Revisited. Cemex, Mexico
Punjab Administration Report (1919–1920)
Punjab Board of Economic Inquiry (1933) An Economic Survey of the taluqas of Haripur and Mangarh of the Kangra District
Punjab Census (1881)
Report on the Administration of the Punjab (1909–1910)
Report of the Royal Commission on Agriculture in India (1927). VIII p. 683
Revenue and Agriculture Forest Proceedings (1887)
Scott J (2009) The art of not being Governed. Yale University Press, New Haven and London, History of Upland South East Asia
Spate OHK (1963) India and Pakistan. A general and regional Geography. Methuen and Company Ltd., London
Siba Jagir Settlement Report (1881–1882)
Una Tehsil Settlement Report (1914)
Vogel JP (1911) Antiquities of Chamba State, part I: inscriptions of the pre-muhammadan period. Superintendent Government Printing, Calcutta

Chapter 7
Early Wood Commons and Beyond

Della Hooke

7.1 Early Medieval Wood-Pasture Commons

Early medieval documents provide the first written evidence of the folk territories recognised at the beginning of the period. Some are noted in the tenth-century Tribal Hidage, a document which comes from the days of the Mercian supremacy, some are recorded in pre-Conquest charters, while others can be detected through place-name evidence. Charter, Domesday Book and later ecclesiastical sources also reveal estate linkages within these territories (Hooke 1985). It becomes clear that in many of them a focus of activity can be recognised, often in a riverine heartland, with links to distant, more marginal regions, whether the latter were woodland, upland or wetland. Such marginal zones served, at some stage, as distant seasonal pastures for domestic stock—a form of resource management which was also implemental in defining regional administrative divisions. The pattern is particularly clear in south-eastern England where some of the earliest charters surviving show the links between coastal and vale estates with the wood-pastures of the Weald (Hooke 2011a), but a similar pattern has also been studied by the present author within parts of Greater Mercia.

Figure 7.1 illustrates the estate links that can be reconstructed within the kingdom of the Hwicce from charter and Domesday evidence. Ford (1976) has argued that the linkages and the territories they suggest may be extrapolated back into the late Iron Age. Late Iron Age hillforts are found along parts of the Hwiccan kingdom boundary and on the spine of upland that separated the two folk groups of south Staffordshire, the Pencersæte and the Tomsæte, but many hillforts could also have served as collecting points for territorial produce. Interestingly, it is links to the area of the Birmingham Plateau that seem to suggest a degree of

D. Hooke (✉)
University of Birmingham, 91 Oakfield Road, Selly Park, Birmingham B29 7HL, UK
e-mail: Della.Hooke@blueyonder.co.uk; D.Hooke@blueyonder.co.uk

I. D. Rotherham (ed.), *Cultural Severance and the Environment*,
Environmental History 2, DOI: 10.1007/978-94-007-6159-9_7,

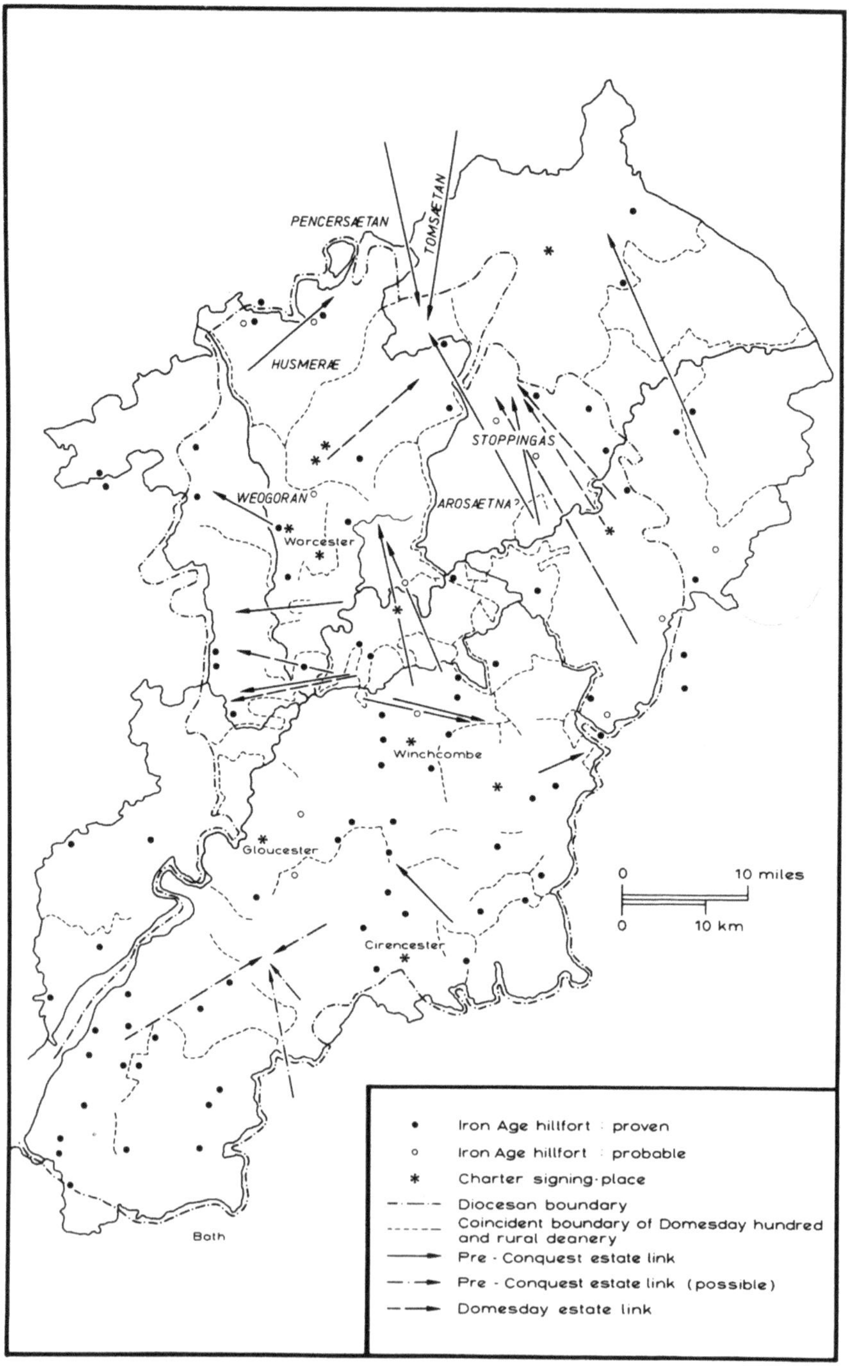

Fig. 7.1 Territorial links within the Anglo-Saxon Hwiccan kingdom (from Hooke 1985)

intercommoning in this wood-pasture region in the early Anglo-Saxon period—here the boundary of the Hwicce seems to have been pushed northwards at some stage, perhaps to compensate for the loss of land on its southern borders. In this northern frontier region, land had earlier been claimed, apparently, by the two south Staffordshire folk groups, on the upland east of the Lickey Hills (part of the main central watershed of England) meeting the territory of a Worcestershire folk group. The Church of Worcester's estate at Shottery near Stratford-upon-Avon to the South-East was also granted woodland at *Hellerelege* in the same upland region (Sawyer 1968, S64; Hooke 1985, pp. 16–17, 80–85). However, detached areas of woodland might reflect early medieval adjustments, as several Kentish charters show, and the evidence confirming intercommoning is not unequivocal.

While all kinds of domestic stock might be pastured in such marginal zones—and the Kentish charters mention cattle, sheep, goats and plough oxen—it was pigs that were particularly important (Hooke 2010, 2011a). Herds of pigs would be taken by swineherds into the woods, and especially in early autumn would forage on available acorns or beech-mast. The term *wudulǣs* 'wood pasture' was used in Worcestershire but other terms for swine pasture were *denbǣra* or *wealdbǣra*, 'den' being the name given to the seasonal pastures of the Kentish Weald. This resource was so important that in Domesday Book the value of woodland in some circuits was reckoned according to the number of pigs it could support.

Such pastures were the origin of many of our later commons. The word 'common' implies an area over which a number of people have legal rights and at this early stage it seems that it was the folk of these territories that enjoyed the pasture rights, although there may have some intercommoning along territorial frontiers before boundaries became precisely demarcated, as noted above. Only at a later stage were the areas of seasonal grazing appropriated by individual manors. The Kentish charters cast more light upon this transitional phase. In a charter of 724 (Sawyer 1968, p. S1180; Kelly 1995, pp. 162–163), Æthelberht, son of King Wihtred, granted land by the River Limen to Mildred, abbess of Minster-in-Thanet, together with additional land which included additional swine pastures at *Brentingesleag on Limenwearawalde* and *Brunesbeam on Weowerawealde* (although these were probably added later to the original grant). Reference here to both *Brentingesleag on Limenwearawalde* 'in the wood of the men of the *Limen* region/district' and *Brunesbeam on Weowerawealde* 'in the wood of the men of Wye' seems to indicate areas of wood-pasture that had previously belonged to the folk of these particular districts (*regiones*) in south-eastern Kent.

As the early medieval period progressed, changes in land tenure altered the administrative pattern of England. As individual estates within the larger folk territories were granted out to lesser lords, the concept of usage by 'folk groups' thus declined, until, by medieval times, rights such as common of pasture were usually held only within an individual manor. Boundaries of estates, too, became precisely drawn and although those in marginal zones appear to have been the latest to be delineated and 'fixed', virtually every inch of land was so allotted by the time of the Domesday survey. The Wealden commons had been appropriated by individual manors within their respective territories well before this.

Frontier regions were often sparsely populated zones of seasonal pasture. Evidence of economic organisation and resource management may be provided not only by estate links but by the routeway patterns within such territories. Often a series of parallel routeways can be reconstructed which seem to have originated as droveways linking the more intensively cultivated areas with the seasonal pastures. Such routes remained in use in the Weald long enough to influence later road patterns and can still be reconstructed from the modern pattern of roads and bridleways within Warwickshire (Hooke 1998, p. 161, Fig. 55) (Fig. 7.2). Linear patterns linking riverine areas with watershed zones used mainly for grazing have also been identified in East Anglia by Oosthuizen (2003, p. 48, Fig. 2).

Roberts aptly summarises the process of appropriation of common rights: 'The history of common rights in England may be viewed as that of an increasing limitation of rights to a more sharply defined class of user' (Roberts 1965, p. 157). Even in medieval times, neighbouring townships might cooperate in using and managing a common that lay upon their borders. Conflict could however arise

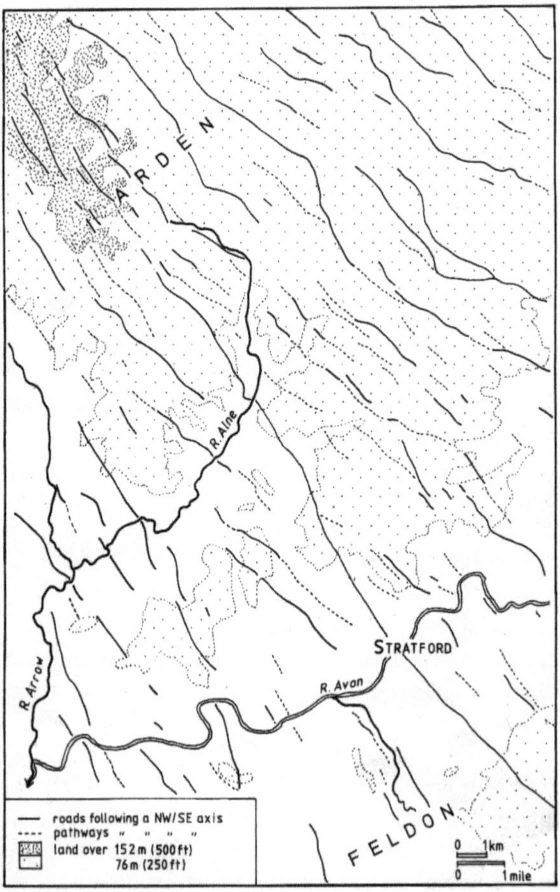

Fig. 7.2 Warwickshire, possible droveways between Arden and Feldon

between a manorial lord and his tenants over the use of the common waste (see below), while on occasions a lord might allow communities to expand when and how they wished (encroachment intakes are a common feature around most commons, having increased in number in post-medieval times). Historical details vary from region to region and will not be further examined here.

7.2 Wood-Pasture Woodland

The character of the wood-pasture zone is now rather better understood following the studies by Vera and others. Generally, the vegetation cover would have been characterised by a mixture of open woodland and heath—rather similar to that of the New Forest today. Vera (2000) has shown how grazing by herbivores helped maintain open woodland, with denser stands confined to more inaccessible places such as steep slopes or within enclosed woodland, thus producing a mosaic of open land, shrub and woodland, with all phases of the cycle present in different areas (Kirby 2003). Some species of trees also resisted grazing pressure better than others—especially the oak, ensuring that these remained characteristic of wood-pasture regions (Rackham 2003, p. 293; Hooke 2010, pp. 193–200). The Old English term *lēah*, especially, seems to have been applied originally to regions characterised by wood-pasture and it retained this meaning when later associated with smaller woods used in this way (Hooke 2008). Many such areas were placed under forest law by Norman kings after the Norman Conquest and for some few hundred years this helped to preserve them from development, restricting both settlement and enclosure (Hooke 2011b). With the growing desire of medieval lords to enclose their own private deer parks, characteristic features of wood-pasture/forest, including veteran trees preserved within grassland, were often preserved in such parklands, especially as such trees, if pollarded, ensured a ready supply of timber grown out of the reach of browsing animals (Fig. 7.3).

The *lēah* type woodlands of early medieval England had also been cropped for timber and this remained an important resource of wood-pasture commons throughout the medieval period. While the commoners of a manor might generally be limited to taking away anything more substantial than underwood and loppings for fuel, fencing etc. (for taking material for making hedges, the right was known as heybote), higher quality timber was not readily available and as more and more woods passed into private ownership, so such timber was reserved for their owners' usage. Wood-banks carefully marked the bounds of such woodland and fines were imposed for trespass.

Fig. 7.3 Pollard oaks in Moccas Park, Herefordshire (photo: D. Hooke)

7.3 The Loss of the Commons

Pressure for development inevitably led to the loss of much of the wood-pasture landscape. As the designated areas of some royal forests were pushed back, so they became of diminished interest to Norman kings interested in hunting and the proceeds to be gained from the taking of venison; money could be made instead through fines imposed for assarting (forest enclosure and clearance). Cannock Forest in Staffordshire, for instance, its core area already granted to the bishops of Lichfield by the twelfth century, became restricted to a number of small separate hays. Nevertheless, the infertile nature of the Cannock Hills—with soils derived from Triassic Bunter Pebble Beds over much of the area—and the continued maintenance of the hays, particularly rich sources of oak timber, helped to ensure that many of the wood-pasture commons of this region survived into the final stages of parliamentary enclosure (Fig. 7.4).

Other areas were less fortunate: many medieval parks, enclosures which often preserved wood-pasture and veteran trees, were enclosed for stock pasture during the late Tudor period when stock rearing became so lucrative and repaid the investment of larger landowners. Wetlands were being drained and enclosed as early as the medieval period but large-scale drainage increased as the sixteenth and seventeenth century progressed. This process had led to the gradual loss of carr or mossland in Cumbria and Lancashire and the enclosure of more limited wetland areas elsewhere, such as the Weald Moors in Shropshire, where 1,100 ha had been enclosed and drained by 1650. There was a surge of enclosure in East Anglia as Dutch engineers were brought into drain the Fenlands. In 1629, the 'Great Level' of the southern peat fens of Cambridgeshire, under the direction of Cornelius Vermuyden, was sanctioned (Williamson 2006, pp. 202–206). Vast intakes were also made along lower upland slopes and the ring-fences along the margins of the hillside commons, belonging to the farms established on the valley sides, can clearly be identified in, for instance, the valleys of the Black Mountains in south-

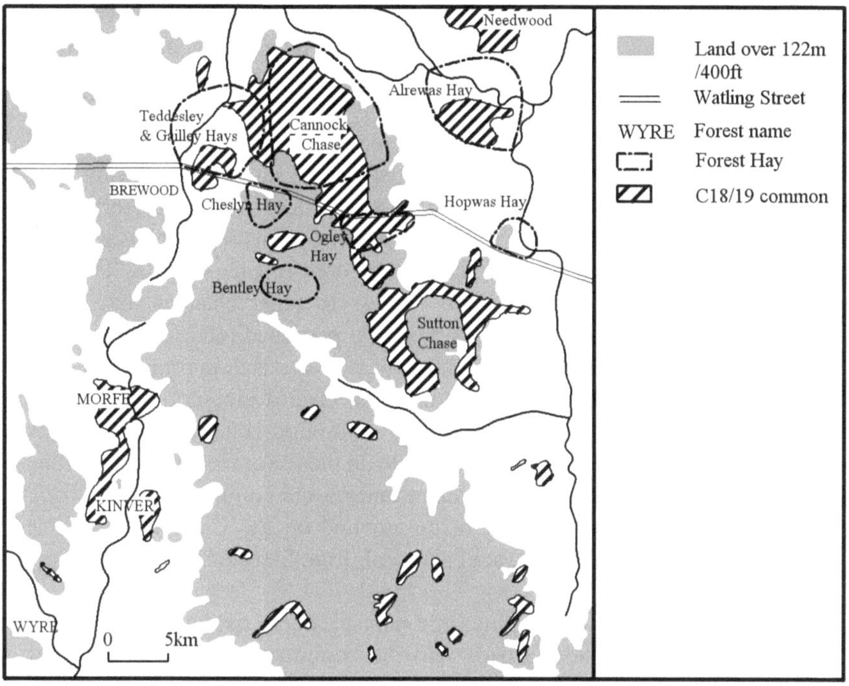

Fig. 7.4 The hays of the former Cannock Forest and the commons of Cannock Chase in south Staffordshire (commons after William Yates' map of the County of Stafford 1775)

west Herefordshire (Hooke 2006, pp. 94–95, Fig. 5.7). Upland commons had long been essential for the traditional rural economy and a form of transhumance had developed in north Wales whereby stock was moved to higher pastures each summer. Initially, long-distance movement may have been involved: the medieval *hafodydd* of the royal *maerdref* of Ystumgwern in coastal Ardudwy, on the west coast of Wales, lay 7 km away in the valley of Nantcol. In 1326, the inhabitants of one coastal parish were still moving in early May with their animals to the mountains (Jones 1972, p. 298) but by the sixteenth century, landownership patterns were changing. Huge areas of open *ffridd* (pastures) were enclosed by individual farms, pushing upwards the margins of the hill commons, the same process seen along the hill ranges of western Herefordshire. The amount of woodland, too, was also to diminish (Hooke 1997). In Cumbria, the unenclosed common 'waste' was also a vital resource and Winchester also argues that here a process of 'manorialisation' took place in medieval times—here the waste appears before that to have been technically the possession of the overlord's demesne hunting preserves. As this became treated as manorial waste, partition of the commons proceeded and each township eventually usually embraced both farmland and moorland with funnels or 'outgangs' leading into the moor, although manorial boundaries do not seem to have been rigidly respected (Winchester 1987, pp. 83–88, 2006, pp. 76–77). It was, however, the increase in the numbers of sheep kept on the hills that was to change the character of much of the wood-pasture, with trees and undergrowth giving way to closely cropped grassland over much of the upland area.

The hills of south Shropshire also offered summer grazing for surrounding communities and driftways or straker routes can be identified, used by animals being driven onto the open commons; these are particularly well preserved on the Long Mynd and Brown Clee. On the latter, grazing rights were enjoyed by all the surrounding townships that had lain within the Forest of Clee (Rowley 1972, p. 52, Fig. 3; p. 242). The Clee Hills were witness to other changes that were not uncommon in regions where rock type not only produced relatively infertile soils characterised by areas of moor and heath, but was also rich in minerals. Coal was already being mined around the Clees in 1235, taken from simple bell-pits scattered around Titterstone Clee. This method of mining continued into the seventeenth century and spoil heaps are still visible on the lower slopes near Doddington where the bellpits spreads onto the area of unenclosed common (Hooke 2006, pp. 106–107, Fig. 5.20). Iron ore was being smelted on the Clees in the sixteenth century (at first using a wind furnace) and coal, iron, limestone and basalt were all being extracted.

It was the influence of developing rural industry in the sixteenth century that led to the some of the greatest inroads into the common pastures. At Broseley in Shropshire, close to the developments that were taking place in the Ironbridge ironworking region, the commons had all but disappeared by about 1600 as immigrant coal miners moved in and built primitive cottages in Broseley Wood. As on the Clees and at Broseley, whole communities erected squatter settlements around the margins of the commons in many developing industrial areas of the west midlands. In southern Herefordshire, overlooking the valley of the River Wye and on the margins of the iron-working area of the Forest of Dean, such settlements around the commons in Walford, Whitchurch, Great and Little Doward were typical of such settlement spread and, like many of the labourers elsewhere, many of the occupants of these encroachment settlements combined mining, quarrying and nail-making with small-scale crofting (*ibid.*, pp. 106–107). Such development was typical across England and Wales, from Cornwall to the north. In Cornwall, tin miners settled in a sprawl of hamlets over the moors around St Agnes and St Austell, for instance, often enclosing tiny fields around.

Some former forests, like the forest or chase of Needwood in Staffordshire and Wychwood Forest in Oxfordshire, even remained unenclosed as late as the nineteenth century. Within Needwood, the detached portions of parishes recorded on the first edition of the six-inch Ordnance Survey map probably represent the pastures claimed by the surrounding parishes within the forest (Hooke 1983, p. 35, Fig. 11b). However, here the pressure to 'improve' argued for by the eighteenth-century agriculturalists resulted in the almost total destruction of the wood-pasture landscape, giving way to a countryside dominated by enclosure fields; only a few parklands retained anything like the earlier forest character. While animals would be excluded in the early years of a coppice cycle within managed woodland, another way in which wood-pasture was lost was by the enclosure of private woodlands as sporting reserves, especially as the nineteenth century progressed, although attempts to regulate wood-pasture within woods increased earlier in the sixteenth century in parts of Eastern Anglia (Rackham 2003, p. 147). With stock excluded, undergrowth and sapling trees rapidly filled in the open space—a feature of many estate woodlands today. No longer used as fodder, holly has choked some former coppice woods.

It was in the later stages of Parliamentary Enclosure that the greatest loss of common land occurred. In upland areas, thousands of hectares of former common waste passed into individual ownership. Encouraged by agricultural 'improvers', enclosure on a large scale had been recommended and implemented since at least the eighteenth century, allowing landowners to control and improve their herds and flocks, just as enclosure of the common fields had enabled them to introduce new crops, fertilisers and more flexible cultivation techniques, greatly raising their profits. Across the uplands, straight drystone walls carved up the open pastures, although land use was not generally greatly affected. Even crop growing on former moorlands was, however, encouraged by higher grain prices during the Napoleonic Wars of the late eighteenth/early nineteenth century, leading to a frenzy of reclamation in the Lake District (Winchester 2006, pp. 106–109). In the west midlands, the revolution in landownership was similar: woodlands, heaths and upland sheepwalks were enclosed across the region: on the South Shropshire Hills, the Long Mynd lost some 2,734 hectares of upland common between 1788 and 1822, but in the old Clun Forest area 5,513 hectares were enclosed by private act before 1845; in the final 'tidying up' process in the Parliamentary Acts of 1845 and later, a further 2,702 hectares (over 20,000 acres in total) were enclosed (Hooke 2006, pp. 100–102). Across Herefordshire, many commons survived enclosure but as tiny patches compared with their medieval counterparts. In Warwickshire, the surviving patches were more limited. These commons had provided free pasture for labourers and the poorer members of the community—a few cattle, sheep or geese feeding on the free pasture available, allowing such families to survive. With the loss of these in the nineteenth century some such communities merely disappeared—often drawn into the slums of the growing industrial towns. At Morton Bagot, on the southern fringes of Arden, for instance, enclosure in 1807 mean that at least ten cottages were abandoned, together with other roadside cottages that had been erected upon the roadside waste (Hooke 1987, pp. 60–71).

In the Warwickshire Arden, many of the commons were historically wooded and bore 'wood' names; although the word 'common' today often conjures up an image of an area of open grassland it is clear that some of the Arden commons, at least, were still dotted with trees in the sixteenth century, when the commoners of Shrawley parish, for instance, pastured their pigs upon mast available from the oak-trees growing on the commons. On others, however, woodland was giving way to heath in many places by medieval times. Felling may have been a contributory factor but grazing on the woodland commons increased and was probably a greater cause. Probably reflecting this process, 'heath' names had become frequent across the region by the time of later mass enclosure, irrespective of the underlying geology and soil type. Despite intakes into the commons themselves, large tracts are still recognisable on eighteenth- and nineteenth-century maps, many lying along parish boundaries. Although many were enclosed in the later stages of Parliamentary Enclosure, surviving commons are still characteristic of the region today (Hooke 1993).

7.4 Landscapes of Conflict

It was the loss of pasture that aroused the greatest conflict with the poorer members of the community. On Cannock Chase in Staffordshire, riots broke out in the sixteenth century when a local landowner, Thomas Paget, attempted to abolish commoners' rights in order to enclose coppices to supply his ironworks (Harrison 1999). Restricted access, even for a number of years, deprived local inhabitants of free pasture and tree loppings for their few sheep in winter. A series of well-planned riots were orchestrated in 1581 on the 20,000 acres of Cannock Chase: hedges were burnt and fights broke out between foresters and protestors. This remained an area where the peasantry continued to feel oppressed. Already hated because of the ferocity of their game laws, the Paget family aroused new anger when they made rabbit warrens on the waste in the late seventeenth century and early eighteenth century, a practice copied by other local lords. Burrows were destroyed by attacks on the Cannock Chase warrens in the winter of 1753–1754 but the Chief Justice upheld the rights of the lord against the complaints of the people: 'the estates of the aristocracy were paramount, and the rights of commoners were beginning to decline to extinction' (Hay 1977, pp. 234–235). Enclosure and the consequent loss of pasture rights was anathema to the people and there were general bouts of unrest or riots at this time in many places, with new fences and hedges pulled out and burnt at Welland and Upper Howsell on Malvern Chase in Worcestershire in 1776. The rioters blackened their faces to escape recognition and ventured forth armed with guns and other offensive weapons (Smith 1978, p. 167). Indeed, riots were not infrequent following the large-scale enclosure movement of the late eighteenth and nineteenth century. Again, there was further unrest at Ogley Hay on the southern fringe of Cannock Chase, especially as the access roads used by those formerly enjoying pasture rights had been stopped up at the time of enclosure. Less dangerously, lawsuits following enclosure often lingered on for many years.

Perhaps the greatest conflict was aroused by the enclosure of the Peak District and Pennine moorlands, long the 'lungs' of the growing industrial towns of the region. Huge areas became private grouse moors and although sheep continued to be pastured, as their grazing stimulated the growth of young shoots eaten by the birds (a process known as swaling), after enclosure the moors were in private ownership and access was seriously limited. Historically, there was no such thing as what is known today as 'public access'. Common rights were usually restricted to those living within the manor. As the available common pasture diminished, the practice of stinting was extended, whereby manorial tenants (copyholders) might only pasture a defined number of animals on the common. Nearly all commons were stinted by the end of the sixteenth century and stints were being reduced by the end of the century due to pressure of population. The concept of *public* access gained ground with the protests by ramblers and other groups about the loss of the commons, especially in the Peak District where mass trespass was carried out in 1932 in order to demonstrate 'for the rights of ordinary people to walk on land stolen from them in earlier times' (Rothman 1982). Only recently has the 'right to roam' over undeveloped land, however, become established by law: under the Countryside and Rights of Way Act 2000 access was granted over mountain, moor, down land, heath land, and registered common land.

7.5 Conclusions: Conservation of the Commons

The Commons Preservation Society, founded in 1865, was struggling to preserve public rights around major cities, and the National Trust, founded in 1895, sought to preserve some of the country's most beautiful landscapes. This was a concept, which by that date included 'wild' areas of countryside. Some individual areas of common were fortunate—the Malvern Hills on the Worcestershire/Herefordshire border had become part of Malvern Forest in medieval times, its wood-pastures used for swine since early medieval times (above), eventually becoming a private chase. Here, too, the commoners might gather windfall wood for the repair of their houses, ploughs etc., although it was the right to pasture cattle and swine that was longest and most effectively deployed. By Elizabethan times, a man holding 30 acres of ploughland might pasture 30 sheep, a freeholder 10 and a cottager 6, but there was no limit to the number that could be kept on the hills. Disafforestation occurred in about 1630 when Charles I decided to sell the chase (completed only in 1660). As elsewhere, this led to local riots but two-thirds of the waste land was set aside for the commoners, left unfenced, and one-third allotted to the Crown over which only the owners would have rights. Encroachment certainly took place around the flanking lowland commons, although large areas of common survive, but the nature of the upland prevented much settlement upon the hills themselves. Continuing encroachment on the commons throughout the nineteenth century led to the Malvern Hills Act of 1884 and the creation of the Malvern Hills Conservators, preventing further encroachment or destruction of the trees. The powers of this body were extended by further Acts in the twentieth century, and in 1959 the Hills were designated an Area of Outstanding Natural Beauty (Bowden 2005, pp. 47–51; Smith 1978, pp. 39–40, 274–278). Large areas of common still survive below the hills, such as Castlemorton Common, noted for its many fine black poplars (Fig. 7.5). These trees, often growing close to cottages ringing the common, were maintained in the past by pollarding to produce young branches for fencing, tools and basketry, etc. This is just one more instance of the way that so many commons in historical times were dotted with trees useful to the community (Barnes and Williamson 2011, p. 114).

Wood-pasture is a precious resource that has undoubtedly been diminishing over the last few hundred years. Today it is realised that commons not only help to preserve some of our most traditional of landscapes but also some of our most valuable traditional habitats for both native fauna and flora—little gems of 'ancient landscape'. Wood-pastures are especially valuable, whether upon common or publicly owned land or within private parkland. A varying structure of tree, scrub, heath and grassland provides habitats for birds and insects. Wood can also be taken for all manner of purposes. Wood-pasture can also absorb a great deal of public use without loss of ecological value, as in the New Forest, Epping and Burnham Beeches (Rackham 2003, p. 201). Centuries of grazing by commoners' ponies and pigs has helped to preserve the landscape of the New Forest in Hampshire as a traditional wood-pasture habitat, a mosaic of woodland and open lawns. Veteran trees, particularly valuable features in their own right, abound. In the Lake District, it is some of fells managed by the National Trust, such as those at the head of Langdale or around Patterdale, where some of our best maintained wood-pasture survives. In the midlands, Sutton Park, the remnant of an ancient hunting chase, has been recognised as a National Nature Reserve although most of its ancient trees have been lost. Several thousand such trees were taken in the mid-eighteenth century when timber was required to erect a new workhouse, but it still conveys an impression of what an ancient wood-pasture may have looked like (Fig. 7.6). Its hollies had been preserved because they were recognised as valuable sources of fodder for stock, although today they are no longer used and consequently form overgrown thickets.

Techniques for managing wood-pasture have often been neglected within modern-day farming landscapes. In wood-pastures, old pollards need to be maintained and grazing continued or reintroduced. Without grazing, gorse and other shrubs can overwhelm open spaces and heathland; woods can become

Fig. 7.5 Castlemorton Common below the the Malvern Hills (photograph: D. Hooke)

Fig. 7.6 Sutton Park near Birmingham—an ancient wood-pasture landscape (photograph D. Hooke)

choked with brambles and bracken. To maintain these as traditional landscapes, conservation bodies now introduce cattle, sheep and ponies. Controlled grazing by sheep and cattle helps to preserve the Malvern commons; Longhorn cattle will be encountered in Epping Forest and Windsor Forests; Exmoor ponies and various breeds of cattle are found in Sutton Park, and ponies on the cliffs owned by the National Trust in Devon and Cornwall. Of course, large numbers of ponies on the moorlands of the South–West have often given rise to particular hardy breeds that have evolved over the centuries. Providing stocking numbers are controlled, and winter poaching of soils is avoided, such continued grazing maintains habitat diversity and ensures that local flora and fauna survive. This fauna and flora has adapted over many hundreds of years to such an environment, and the grazing maintains a more dynamic environment than artificially cleared or mown land.

Patches of wood-pasture landscapes are also preserved in many parks, as noted above, especially those first established in medieval times with continued lives as ornamental parks. Moccas, in Herefordshire, is one such example and the ancient pollards it contains are another valuable feature of many early wood-pasture regions (Fig. 7.3)—for optimum nature conservation, a wood-pasture can have a tree cover of 30–90 %. Others such as Dinefwr, Calke and Hatfield also have outstandingly good habitats preserved within them (Cox and Sanderson 2001, p. 40, tale 6). As Rackham points out, 'The historical character and biological importance of wood-pastures reside in grassland, heath, and scrub, and in old and dead trees' (Rackham 2003, p. 202).

Wood-pastures are fragile landscapes requiring specialised management, vulnerable because they are now rarely part of the general farming economy. They offer glimpses of an ancient past but have a strategic part to play in maintaining the health and diversity of today's rural environment.

References

Barnes G, Williamson T (2011) Ancient trees in the landscape: Norfolk's Arboreal Heritage, Windgather, Oxford

Bowden M (2005) The Malvern Hills. An ancient landscape. English Heritage, London

Cox J, Sanderson N (2001) Livestock grazing in national trust parklands—its impact on tree health and habitat. Jonathon Cox Associates, Lymington

Ford WJ (1976) Some settlement patterns in the central region of the Warwickshire Avon. In: Sawyer P (ed) Medieval settlement. Continuity and change. Edward Arnold, London, pp 274–294

Harrison C (1999) Fire on the chase: rural riots in sixteenth-century Staffordshire. In: Morgan P, Phillips ADM (eds) Staffordshire histories. Essays in Honour of Michael Greenslade. Staffordshire Record Society & Centre for Local History. University of Keele, Keele, pp 97–126

Hay D (1977) Poaching and the game laws on Cannock chase. In: Hay D, Linebaugh P, Rule JG, Thompson EP, Winslow C (eds) Albion's fatal tree. Crime and society in eighteenth-century England. Penguin Books, Harmondsworth, pp 189–253

Hooke D (1983) The landscape of Anglo-Saxon Staffordshire: the charter evidence. Department of Adult Education, University of Keele, Keele

Hooke D (1985) The Anglo-Saxon landscape. The kingdom of the Hwicce. Manchester University Press, Manchester

Hooke D (1987) The Arrow Valley Project 1. Morton Bagot, a parish survey. Department of Geography, University of Birmingham, Birmingham

Hooke D (1993) Warwickshire's historical landscape—the Arden. School of Geography. University of Birmingham, Birmingham

Hooke D (1997) Place-names and vegetation history as a key to understanding settlement in the Conwy valley. In: Edwards N (ed) Landscape and settlement in medieval Wales. Oxbow Monograph 81, Oxbow, Oxford, pp 79–95

Hooke D (1998) The landscape of Anglo-Saxon England. Leicester University Press, London

Hooke D (2006) England's landscape. The West Midlands. HarperCollins/English Heritage, London

Hooke D (2008) Early medieval woodland and the place-name term *lēah*. In: Padel OJ, Parsons DN (eds) A commodity of good names. Essays in honour of Margaret Gelling. Shaun Tyas, Donington, pp 365–376

Hooke D (2010) Trees in Anglo-Saxon England. Literature, lore and landscape. Boydell Press, Woodbridge

Hooke D (2011a) The woodland landscape of early medieval England. In: Higham NJ, Ryan MJ (eds) Place-names. Language and the Anglo-Saxon landscape. Boydell Press, Woodbridge, pp 143–174

Hooke D (2011b) Royal forests—hunting and other forest use in medieval England. In: Ritter E, Dauksta D (eds) New perspectives on peoples and forests. Springer, Dordrecht

Jones GRJ (1972) Post-roman Wales. In: Finberg HPR (ed) The agrarian history of England and Wales, I.ii, A.D. 43–1042. Cambridge University Press, Cambridge, pp 283–382

Kelly SE (1995) Charters of St Augustine's Abbey Canterbury and Minster-in-Thanet. Anglo-Saxon charters IV. Oxford University Press, Oxford

Kirby K (2003) What might a British forest landscape driven by large herbivores look like? English Nature Research Report No. 530, English Nature, Peterborough

Oosthuizen S (2003) The roots of the common fields: linking prehistoric and medieval field systems in west Cambridgeshire. Landscapes 4(1):40–64

Rackham O (2003) Ancient Woodland, its history, vegetation and uses in England. Castlepoint Press, Dalbeattie, Kirkcudbrightshire

Roberts BK (1965) Settlement, land use and population in the western portion of the forest of Arden, Warwickshire, between 1086 and 1350. A study in historical geography. Unpublished PhD. thesis, University of Birmingham, Birmingham

Rothman B (1982) The 1932 Kinder trespass. Willow Publishing, Altrincham
Rowley T (1972) The Shropshire landscape. Hodder & Stoughton, London
Sawyer PH (1968) Anglo-Saxon charters: An annotated list and bibliography. Royal Historical Society, London
Smith B (1978) A history of Malvern, 2nd edn. Alan Sutton/The Malvern Bookshop, Stroud
Vera FWM (2000) Grazing ecology and forest history. CABI Pub, Wallingford
Williamson T (2006) England's landscape: East Anglia. HarperCollins/English Heritage, London
Winchester A (1987) Landscape and society in medieval Cumbria. John Donald Publishers, Edinburgh
Winchester A (2006) The farming landscape. In: Winchester AJL (ed) England's landscape. The North West. HarperCollins/English Heritage, London

Chapter 8
What, How, and Why? Collecting Traditional Knowledge on Forest Uses in Switzerland

Matthias Bürgi and Martin Stuber

8.1 Introduction

During the nineteenth and twentieth centuries, forest use and management in Switzerland underwent radical changes (Bürgi 1999). Before this period, traditional forest uses, such as woodland pasturing, wood hay and litter collection and even crop production on temporary fields in the forest were a common practice throughout the country. In the late nineteenth and early twentieth century, these non-timber forest uses have been increasingly abandoned and/or banned from the forests. This development has been paralleled by an increasing interest in wood as an industrial good and the introduction of the science of forestry together with the implementation of forest laws. Furthermore, agricultural modernization and a rapidly growing infrastructure after the Second World War facilitated importing resources from abroad and consequently took pressure from the forests to supply resources for the local demand. Lately, uses, such as woodland pasturing and litter collection, have gained attention from various scientific disciplines due to their importance for cultural history, ecosystem development and carbon sequestration in forests (Perruchoud et al. 1999; Gimmi, Bürgi and Stuber 2008).

To assess the characteristics, extent and the intensity of mainly agricultural uses of forests in Switzerland, an extensive literature review has been conducted (Stuber and Bürgi 2001; 2002; Bürgi and Stuber 2003). This literature review illustrates the diversity and variability of human impacts on forest ecosystems and implies a massive change in forest uses especially after the Second World War.

M. Bürgi (✉)
Department of Landscape Dynamics, Landscape Ecology Group, Swiss Federal Research Institute WSL, Zürcherstrasse 111 CH-8903 Birmensdorf, Switzerland
e-mail: matthias.buergi@wsl.ch

M. Stuber
University of Berne, Bern, Switzerland

I. D. Rotherham (ed.), *Cultural Severance and the Environment*, Environmental History 2, DOI: 10.1007/978-94-007-6159-9_8,

However, literature and archival material show specific limitations as main sources for reconstructing forest uses, as not all activities are adequately represented in written over-delivery. One reason for that is the social context of many non-timber forest uses, which often have not been conducted by forestry officials, but by the local population, especially women and children. These parts of society are often not adequately reflected in written sources and so are activities, which do not lead to marketable goods, but serve self-supply. Thus, the knowledge remained just with the people who performed theses uses and has rarely been documented otherwise. Under these circumstances, conducting oral history interviews is most suitable to complement written and other archival sources in order to get a more complete picture of former forest uses and to document the connected traditional knowledge (Roth and Bürgi 2006; Gimmi and Bürgi 2007). As many practices have been abandoned in the last 50–70 years, now is the last chance to collect and document the connected traditional knowledge, which otherwise will be permanently lost and deleted from the collective memory.

To gain a more complete picture of traditional forest uses in Switzerland, a project "Hüeterbueb und Heitisträhl" (i.e. "Shepherd boy and blueberry comb") has recently been finished and its results published together with the release of a documentary movie (Stuber and Bürgi 2011, www.youtube.com/watch?v=rr3WYXJ-pQ8). The results are based on 56 oral history interviews conducted in five study regions across Switzerland (Fig. 8.1). The regions were chosen based

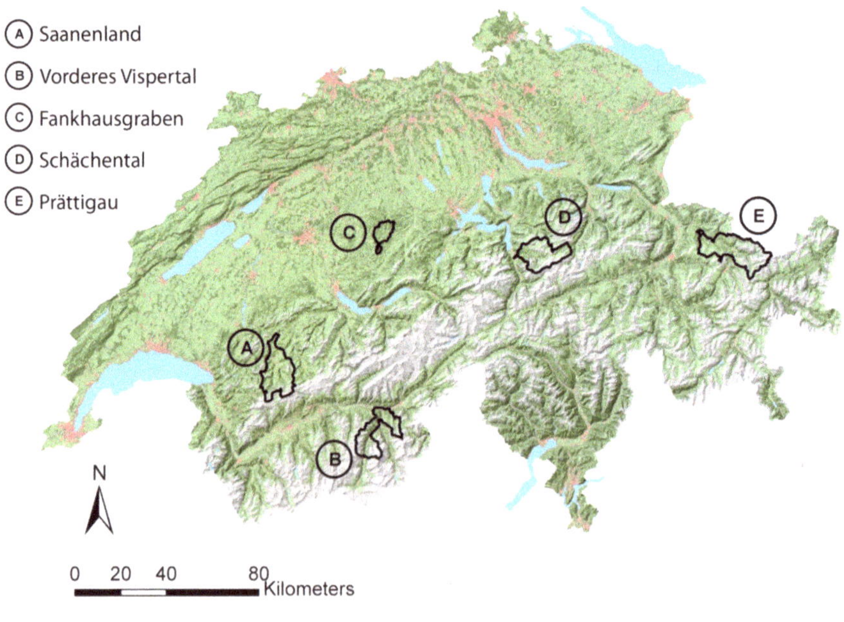

Fig. 8.1 Location of the five study areas of the overall project. This paper reports of the results from the region "C Fankhausgraben" (*Data* Arealstatistik 1992/97, BFS GEOSTAT; swisstopo (DV033492.2); DHM25 © 1994 Bundesamt für Landestopographie)

on the literature review mentioned above, i.e., in regions, where we knew that such practices might have been performed up to the second half of the twentieth century. In this chapter, we report on the results from one of the study regions, i.e. the Fankhausgraben.

By preserving the traditional knowledge on forest uses, this study aims at: (a) preventing the loss of cultural heritage, (b) documenting the human impact on forest ecosystems, (c) assessing the regional variability in forest use, and (d) contributing to the understanding of the human impact on forest ecosystems.

8.2 Study Area

The region Fankhausgraben is located in the Emmental, a hilly area of the Canton Bern in central Switzerland. The Fankhausgraben ranges from 792 to 1250 masl and it is part of the municipality of Trub, which covers an area of 6,201 ha. This municipality today is still dominated by agriculture and includes about 1,450 inhabitants and 140 working farms. The number of inhabitants has declined since the beginning of the twentieth century, when 2,606 people lived in Trub. Worldwide, about 50,000 people trace their roots back to Trub (www.trub.ch), illustrating a long history of emigration due to difficult economic conditions and limited availability of employment. The region belongs to the German speaking part of Switzerland, in which various regional dialects persist.

A core feature of the landscape in Fankhausgraben is the topography, consisting of steep, often wooded ravines cutting into soft rolling hills of which the flatter part is mostly grassland. The farms are surrounded by their farmland and therefore distributed throughout the landscape, leading to a dispersed settlement pattern ("Streusiedlung") (Fig. 8.2).

8.3 Materials and Methods

In the Fankhausgraben, a total of eleven interviews were conducted in January and February 2008. The interviewees include three women and eight men, who were 60–85 years old and long-term inhabitants of the area. The same interviewer (Susan Lock) conducted all interviews. She used an interview guideline, addressing the diversity forest uses, the context in which they were performed, the specific practice, and its temporal and spatial development. Towards the end of the interview, historical photographs displaying different forest uses were shown to the respondents. This semi-structured method assures enough freedom for the interviewees to tell their own stories based on their recollection, but at the same time guarantees that a core set of forest uses are discussed in all interviews. The list of forest uses addressed was compiled based on the literature study mentioned and a pre-test. The interviews lasted 60–120 min and were tape-recorded. For the

Fig. 8.2 Single farmes dispersed throughout the landscape, which is characterized by a pattern of forests along the ravines and on steeper slopes and pasture and meadows on slopes suitable for farming (*Source* Hof Wüthrich, Fankhaus. Reproduced from Uetz, 1948)

analysis they were transcribed into standard German, only special terms or names related to forest uses were put in quotations marks and remained in the regional dialect. The interviewees agreed that a picture, their age and address were published in a monograph (Stuber and Bürgi 2011). For the present paper, all information taken from the oral history interviews are referenced as "(interview XY)", with XY standing for the respective informant.

We aimed at collecting information on forest uses, which the individual interviewees performed themselves. At the same time, we aimed at drawing a complete picture of the forest uses performed generally in an area. Therefore, the questionnaire also included questions regarding different actor groups, i.e. uses which have been performed by other members of the interviewee's family or which the interviewee just had heard of. For the data analysis, we compiled a new, specific list of forty-four different forest uses mentioned in the interviews in Fankhausgraben.

Every single forest use was defined by three terms, which name the relevant activity (e.g. mowing, collecting, hunting), a resulting product (i.e. leaves, berries, fuelwood), and the connected use of this product (e.g. fodder, fertilizer, tools). This approach enables to distinguish between, for example, leaves being collected as bedding for the cattle ("Streuelaub", i.e. raking/leaves/litter) from leaves being collected as stuffing for mattresses ("Bettlaub", i.e. raking/leaves/mattresses).

For all 44 forest use categories, we counted how many times they were mentioned and specified the user group (numbers range from one to eleven, i.e. the total number of interviews conducted). All uses were classified according to prevalence and specialization to develop a region-specific profile of forest uses. For prevalence, uses, which were mentioned by just one to three people, were subsequently called "rare". Uses mentioned four to eight times, where called "known", and uses called nine to eleven times, were called "common". We then calculated a weighted count, in which counts of own performance were multiplied by three, counts for uses performed by the family by two, and counts of uses heard of were not multiplied (numbers range from one to thirty-three). The ratio between the weighted number of counts and the simple number of counts was taken as a measure for degree of specialization. If the ratio for a certain forest use was three, i.e. all people mentioning the use did also perform it themselves, this use fell in the category of "no specialization". If the ratio was between 1.5 and 2.9, an "average specialization" was stated. A ratio below 1.5 was interpreted as "highly specialized or abandoned", as theses uses were nearly always reported as "heard of", i.e. no direct experience was recorded.

Generally, reports on forest uses, which are classified as "rare", respectively as "highly specialized or abandoned", have anecdotal character and might not have been collected and reported with the same consistency as uses listed in the categories "common" and "known" regarding prevalence and "no specialization" and "average specialization". Information on most widespread and known forest uses is much more likely to be representative for the study area.

8.4 Results and Discussion

For Fankhausgraben, collecting firewood (i.e. logging/wood/fuel in Table 8.1, Fig. 8.3), using needles and leaves for cattle bedding in the stables (pollarding/needles/litter and raking/leaves/litter), collecting bark for leather tanning (logging/bark/tanning), and collecting mushrooms for food (collecting/mushrooms/food), were forest uses, which almost all people knew and performed themselves (Table 8.1). Whereas some of these uses are widespread and do not require additional explanations, the specific use of needles for cattle bedding seems to be a local specialty, as it has been reported in no other case study region assessed in Stuber and Bürgi (2011) (data not shown). For this use, the needles were not raked together as reported otherwise, but cut from the branches, i.e. pollarded from trees that had been felled. The branches were then bundled together and used as firewood. As trees standing in open spaces provide more branches with needles, they were especially sought after. This left the open areas open, which an interviewee recalls as being beneficial also for the capercaillie (*Tetra urogallus*) (interview MH). The resulting manure from needle litter was kept on a separate pile and then mostly used to fertilize

Table 8.1 Forest uses in the Fankhausgraben, Switzerland, based on 11 oral history interviews.

	Common	Known	Rare
No specialization	collecting/mushrooms/food logging/bark/tanning logging/wood/fuel pollarding/needles/litter raking/leaves/litter	collecting/moss/div. logging/wood/pulp & construction mowing/fern/litter mowing/sedges/litter	collecting/clematis/smoking collecting/lichen/div. collecting/lycopod/div. collecting/resin/chewing gum collecting/snails/food logging/bark/fuel removing ant hill/needles/fertilizer removing ant hill/resin/fumigation
Average specialization	collecting/berries/food collecting/cones/fuel collecting/herbs/medicine collecting/resin/pig slaughtering logging/wood/tools etc.	mowing/sedges/mattresses	collecting/forest earth/fertilizer collecting/juniper/fumigation collecting/juniper/schnapps collecting/resin/medicine mowing/fern/pillows raking/leaves/fodder
High specialization or abandoned	logging/wood/shingles	clearing/wood & food & pastureland/div. logging/ash/cleaning logging/wood/fencing pasturing/grass & leaves/fodder cattle	hunting/venison/food mowing/sedges/fodder pasturing/grass & leaves/fodder goats pollarding/brushes/litter pollarding/leaves/fodder

The forest uses are defined by three terms, addressing the related activity, resulting products, andtheir use. The criteria for classifying the uses according to specialization and prevalence are explained in the text

potato plots. The manure had to be ploughed into the soil and it was said to have a positive effect on soil structure, as it brought air into the otherwise heavy loamy soils. There even is a saying that manure from the forest makes the soil "proud", i.e. productive ("Mist von Holz macht Boden stolz") (interview LZ).

The interviews do not only provide information about the specific regional practices, but also about their temporal development. For example, the widespread use of pollard needles for bedding was said to be abandoned in the course of increasing availability of straw in the 1950s and 1960s, which was bought and transported to the Fankhausgraben from grain producing areas, i.e. the lowlands. Apart from financial considerations and transport facilities, new hygienic requirements for milk production and the increasing shortage of workforce, were reported to be additional triggers for the abandonment of pollarding needles for litter. Leave litter collection lasted for longer, but ended, when stables became

Fig. 8.3 Not surprisingly, collection of firewood has been reported as an important forest use also for the study area Fankhausgraben. Smaller branches and twigs traditionally were bundled into so-called "Wedelen" (*Source* Clemens Schildknecht. Reproduced from "Schweizer Familien Wochenblatt", Nr. 9, February 26, 1949, page 21)

equipped by automatic dung cleaning machines, as these mechanisms are chocked by the leaves all the time.

Also common, in the sense that almost all interviewees mentioned the use, is the recollection of using wood for the production of shingles (logging/wood/shingles in Table 8.1). However, only very few people actively performed the use, i.e. it has been classified as highly specialized. Shingle making was almost a professional activity in a sense that only a few farmers learned the skills which were often passed along from father to son (Fig. 8.4). Thus, the classification procedure applied seems to capture the traits of this forest related activity correctly. In other parts of Switzerland, charcoal making has been classified similarly

Fig. 8.4 Production of shingles, Trachselwald (Oberemmental) (Source: Ernst Brunner, Schweizerisches Institut für Volkskunde, Basel [Serie LY No 63])

as a highly specialized forest use, but charcoal making has not been reported on in Fankhausgraben.

Comparing the results from Fankhausgraben with the results from the other study regions (overall results published in Stuber and Bürgi (2011)), reveals that the use of pollarded needles for bedding is not the only specialty of Fankhausgraben. The use of bark from spruce for tanning (Fig. 8.5) also has been named in all interviews in this study region, but by none in any of the other areas, despite the use of oak bark for tanning is known from other areas of Switzerland (Bürgi and Stuber 2003). This lack of information from other study regions is because they are all located outside of the oak growing area. Spruce seems to have been a core resource in Fankhausgraben, since using its needles, bark and timber all rank very high among the common uses reported. This example illustrates how information on the ecological characteristics and constraints of a region has to be taken into account in interpreting the findings from local studies on forest use.

The comparison with forest uses reported in other study areas reveals also that some otherwise widespread forest uses seem to have been underrepresented in Fankhausgraben—at least during the period captured with the oral history interviews. For example, woodland pasturing, one of the most widespread and known forest uses in other parts of Switzerland (Stuber and Bürgi 2001), seems to be not very widespread in the Fankhausgraben (Table 8.1). The interviews reveal that woodland pasturing has been abandoned due to the steep terrain (interview FB), but also due to the influence of foresters, which did not want any more goats grazing in the forest (interview MH). These statements are supported by statistical data that reveal a decline in numbers of goats in the late nineteenth and early

Fig. 8.5 Rüsten der Baumrinde, bei Eggiwil (Oberemmental) (*Source* Ernst Brunner, Schweizerisches Institut für Volkskunde, Basel [Serie HZ No 19])

twentieth centuries, from a peak of 919 in 1866 to just 205 in 1939 (Bernhist 1994–2006).

The long list of forest uses classified as rare (Table 8.1), regardless if they are reported as conducted by the interviewee him- or herself (i.e. "no specialization) or just heard of ("high specialization or abandoned"), are harder to interpret. Whereas some of them might actually have been rare, others might just not have received appropriate attention during the interviews. Still, it is unlikely that common uses remained unnoticed to an extent that they finally ended up in this category of rare forest uses. Thus, this list of rare uses has at least anecdotal character and illustrates the wealth of traditional knowledge on forest uses.

Such detailed and systematic studies of forest uses inevitably lead to problems in defining "traditional forest uses", as obviously not all forest uses were performed continuously. For example, the use of pollarded leaves for fodder (pollarding/leaves/fodder) has been reported only for 1947, when extremely dry weather conditions caused a shortage of grass. Where did the knowledge to use this resource come from all of a sudden? Similarly, during times of shortage, also leaves were raked as fodder for goats (pasturing/grass & leaves/fodder for goats), but also this use was classified as rare (interview CW). Of course, such practices might have been more common earlier or always been restricted to times of scarcity. However, it might also be possible that they were adopted from other areas or added to the regional set of forest use as an innovation. Some for example, may have been promoted by the agricultural extension service. This example raises the question, how long and frequent a certain use has to be performed to become a local "traditional" forest use.

8.5 Conclusions

Our study reveals a great diversity of forest uses. Classifying all information on forest uses by activity/product/use worked well. The procedure does not only provide valuable insights in the characteristics of a forest use, but is also a first step to interpret the use in the socio–economic context of a given region. In addition, the classification system developed addressing various degrees of specialization and prevalence (Table 8.1) seems to be a good way to separate common and rare forest uses and thus structuring the information collected from the oral history interviews. Of course, the interviews provide a wide range of additional information on specific uses, which cannot be depicted easily and which in itself refer to again a much greater wealth of experiences and traditional knowledge.

Developing regional forest use profiles as proposed in this chapter illustrate the important role of forests as a core provider of various resources for the local population. The regionally distinct profiles can be interpreted in their regional natural, economic, political and cultural context. This analysis will help generating hypotheses regarding the human-nature interaction in (pre-) alpine environments. Furthermore, if we will be able to determine general spatio-temporal pattern of forest uses, we might attempt to assess the ecological impact of humans on forest ecosystems regionally (Wohlgemuth et al. 2002; Bürgi and Gimmi 2007) but even outside the study regions, i.e. by extrapolating the findings to areas which are similar in relevant context.

Studies as the one presented here help to preserve knowledge and values attached to woodland landscapes and to understand long-term developments in the relationship between societies and forests. As such, they promote the dissemination of valuable knowledge of wooded landscapes and provide the bases for the development of management schemes for sustainable future forests. At the same time, they challenge simplistic views on so-called "traditional" forest uses by revealing a wealth of temporal and spatial dynamics and variability in how the local population used the resources provided by forest ecosystems.

Acknowledgments This research was generously funded by the Bristol-Foundation.

References

Bürgi M (1999) A case study of forest change in the Swiss lowlands. Landscape Ecol 14:567–575

Bürgi M, Stuber M (2003) Agrarische Waldnutzungen in der Schweiz 1800–1950. Waldfeldbau, Waldfrüchte und Harz. Schweizerische Zeitschrift für Forstwesen 154:360–375

Bürgi M, Gimmi U (2007) Three objectives of historical ecology: the case of litter collecting in Central European forests. Landscape Ecol 22(suppl. 1):77–87

Gimmi U, Bürgi M, Stuber M (2008) Reconstructing anthropogenic disturbance regimes in forest ecosystems — a case study from the Swiss Rhone valley. Ecosystems 11:113–124

Gimmi U, Bürgi M (2007) Using oral history and forest management plans to reconstruct traditional non-timber forest uses in the Swiss Rhone Valley (Valais). Environ Hist 13:211–246

Perruchoud D, Kienast F, Kaufmann E, Braeker OU (1999) 20th century carbon budget of forest soils in the Alps. Ecosystems 2:320–337

Roth L, Bürgi M (2006) Bettlaubsammeln als Streunutzung in St. Galler Rheintal. Schweizerische Zeitschrift für Forstwesen 157:348–356

Stuber M, Bürgi M (2001) Agrarische Waldnutzungen in der Schweiz 1800–1950. Waldweide, Waldheu, Nadel- und Laubfutter. Schweizerische Zeitschrift für Forstwesen 152:490–508

Stuber M, Bürgi M (2002) Agrarische Waldnutzungen in der Schweiz 1800–1950. Nadel- und Laubstreue. Schweizerische Zeitschrift für Forstwesen 153:397–410

Stuber M, Bürgi M (2011) Hüeterbueb und heitisträhl–traditionelle formen der waldnutzung in der schweiz 1800-2000. Bristol- Schriftenreihe 30

Wohlgemuth T, Bürgi M, Scheidegger C, Schütz M (2002) Dominance reduction of species through disturbance — a proposed management principle for central European forests. For Ecol Manage 166:1–15

Chapter 9
The History of Utilization and Management of Commons and Consequences of Current Social Change in the Alpine Region of Austria

Elisabeth Johann

9.1 Introduction

Europe's roof garden has a length of 1,200 km long and a width of 300 km, and reaches from the Mediterranean up to an altitude of 4,808 m (Hamberger et al. 1998) (see Fig. 9.1). As defined in the *Alpine Convention*, the region of the Alps is home to some 13 million people (Alpenkonvention 1991). The 190,912 sq.km contained within the Alpine arc comprise a high cultural and linguistic diversity. The region is characterized by eight countries, about 100 regions and about 6,200 communities, four world languages, seven further languages countless dialects and traditions. Human activities can be traced back for perhaps 100,000 years, and permanent settlements for around 7,000 years. The landscape is home to a wealth of eco-systems and cultural diversity. With their unique combination of natural and cultural history, the Alps have become a living space, an economic area and a recreational playground of eminent importance at the heart of the European continent. The experiences of nature-adapted management have by some means been maintained until the present day.

Austria plays an important role among the Alpine states because in relation to the national territory the share of mountain areas is the highest (70 %) within the European community. 67 % of Austria's total territory complies with the EU-definition of 'mountainous area'. Fifty-two percent of the farms are situated in mountain regions, 80 % of the national territory is located in disadvantageous regions. Nearly half of Austria's 2,351 communities live and work in this area. The maintenance of the settlement areas is therefore of great importance to more than three million people.

E. Johann (✉)
Austrian Forest Association, Oberdoerfl 9, A-9173 St Margareten, Austria
e-mail: elisabet.johann@aon.at; elis.johann@utanet.at

I. D. Rotherham (ed.), *Cultural Severance and the Environment*, Environmental History 2, DOI: 10.1007/978-94-007-6159-9_9,

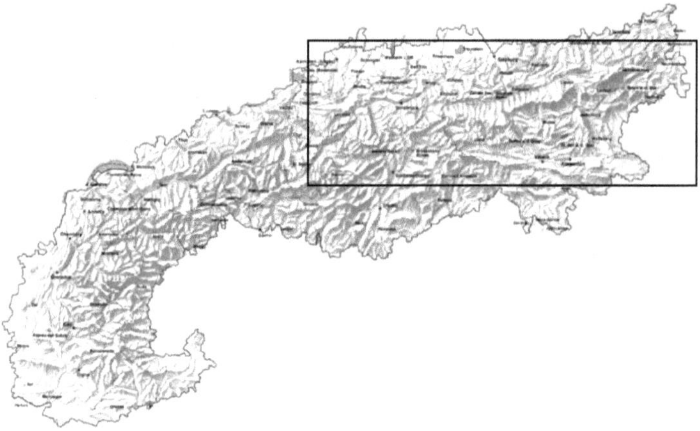

Fig. 9.1 **Area of** ***Alpine Convention*** (N.N. 2010)

Without mountain forests, which stabilize mountain slopes on a lasting basis, a large part of the Austrian territory would become inhabitable and substantial infrastructure jeopardized. 82 % of the forest surface is privately, 18 % publicly owned. Half of the privately-owned forests are small-scale forests; about 10 % have joint ownership structures (Pro: Holz Austria 2001–2012) (see Fig. 9.2).

Rural and craftsman's small-scale enterprises are contributing remarkably to biodiversity and livelihood of the local people. In Austria, the importance of small-scale forestry is proved by the number of about 280,000 human beings (around 5 % of the total population) earning their living in forestry and timber industry.

It has been assumed, that the precondition for the sustainable development and utilization of cultural landscapes is based on the development of the entire society including nature and social agreements. At the heart of this chapter lies the question how the society placed in the mountain area is related to its surrounding.

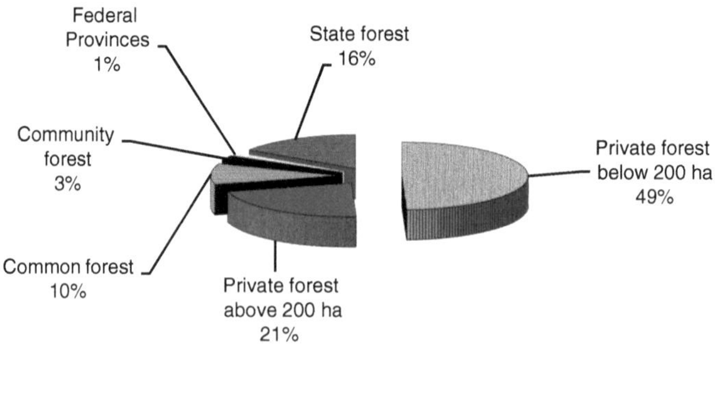

Fig. 9.2 Structures of forest ownership in Austria (2001)

It highlights the measures that were taken by the commons in their entirety to secure the available natural resources on a long-term basis for the benefit of subsequent generations. The study also addresses the contemporary role commons play in the rural society and discusses their future perspectives within the framework of migration and urbanisation.

9.2 Material and Study Area

The study relies on diverse sources. On the one hand, the evaluation refers to unpublished administration-material, stored in the archives of big private and public forest enterprises such as *Lainach* and *OBfAG* and in the public archive of Carinthia (Kärntner Landesarchiv Klagenfurt). These documents give information about the engagement of the local people to manage the land autonomously. On the other hand information with regard to participatory management of common property in particular concerning forests and woodland can be obtained from publications tackling the history of forests and woodland of several Federal provinces *(Bundesländer)* such as Tyrol (Oberrauch 1952), Salzburg and Upper Austria (Koller 1975a, b), Styria (Hafner 1979), and Carinthia (Johann 1968, 2004). Historical conflicts referring to utilisation rights are documented by cases at Court stored for instance in the public archive of Carinthia (Kärntner Landesarchiv Klagenfurt). Present conflicts are documented by contemporary journals and newspapers, TV and radio. Several publications deal with the former utilization rights in State forests (Schindler 1885) and the relief of utilisation rights such as Schiff (1899). The legal framework has recently also been discussed and illustrated by Herbst (2004). Within the last years the history and present status of common property *(Agargemeinschaften)* has also been addressed by several diploma theses carried out at the University of Life Sciences, Vienna (Krondorfer 2005; Buerg 1998). Also the diverse declarations and conventions based on the agreements of the European member countries and the European Community and aiming at a sustainable development of the rural areas have been taken into account (Alpenkonvention 2006; MCPFE 2003; UNESCO 2003).

The scope of investigation covers the alpine region of Austria, located in the Federal provinces Styria, Salzburg, Vorarlberg, Tyrol, Upper Austria and Carinthia. The Austrian landscape is characterized by a high percentage of forests. It amounts to 46.8 % (EU 27.1 %). The forest area *per capita* comes to 0.49 ha (EU 0.26 ha). By nature steeper mountain forests show a higher share of protection forests: their economic importance in terms of commercial timber use is only secondary. Compared with technical control measures protection forests entail relatively low costs and are therefore considered highly important. The economic value of protection forests is in many cases underestimated and only realized in the event of natural disasters. However, the commercial use of mountain forests still represents a major source of income for the population of Alpine regions. Forests still provide a secured source of livelihood to many people, especially in areas distant from easily accessible labour markets. Moreover, mountain forests allow visitors to experience and enjoy nature and to recreate and are a famous destination for tourists from abroad and Austrians alike. Thus, in addition to the commercial use of mountain forests also tourism constitutes an important source of income for the rural population in alpine areas. In Austria's current account balance forestry and timber management ranks second just behind tourism (BMLFUW 2000).

9.3 Results

9.3.1 Living Condition of the Rural Population

Austria's contemporary landscape has been shaped by time with changes through various climatic periods, and by man. In the late Neolithic Age, as various artefacts show, settlers advanced into the Alpine regions. Their existences at numerous different places of settlement are clear indications of an expansion of the community around 4000 B.D. Here we also find the beginnings of Alpine pastures. However, major changes in the landscape did not occur before the exploitation of the Alpine mineral resources set in, and after permanent settlements were established. The enormous quantities of wood needed in the copper, gold, iron and salt mines most likely led to the development of an efficient timber industry as well as the development of the cultural landscape (Pittioni 1974).

On the one hand, the living conditions of the rural society are determined by natural factors such as topography, climate, exposure and altitude, and on the other hand by social factors such as ownership structures, legal framework, and the demographic evolution.

9.4 Natural Factors: Landscape Structure

Although the alpine region offers a high variety of locations suitable for the survival of plants and animals, the locations fitting for human settlements are very sparse. They are limited by a large number of rocks, glaciers, steep slopes and gorges. The upper settlement boundary is influenced by numerous natural factors, such as the exposition to sun and shadow, altitude, gradient and soil type and also by economic activities such as mining and traffic. Farming is determined above all by slope orientation: e.g. on the slopes exposed to the south grain can be cultivated at a very high altitude, sometimes almost reaching the upper timber line. The slopes being exposed to the north are wooded more or less to the bottom of the valleys (Johann 2003).

9.5 Social Factors

Possession of land, cultivation methods and the colonization process are closely interrelated. Austria's commons and forest rights are rooted in the undivided village-owned cultural land of the Middle Ages. It was one of the essential elements of colonization that forests, pastures and alpine pastures were jointly used, whereas fields were distributed among the individual farmers. Depending on the region, the common land was called *Nachbarschaft, Gemein* or *Allmende*. These communities of land-owning farmers often managed large areas of woodland (commonland). Many farmsteads also owned extended rights in forests belonging to private and state-owned estates. They were allocated to the farmsteads in order to ensure a sufficient supply of the farmstead concerning timber, fuel wood, litter and pasture. The extent of the respective utilization rights was determined by the size of the farmstead. Due to the general lack of grazing land (pastures and alpine pastures) these rights were essential. Moreover each farmstead owned a private forest parcel situated in its close proximity the so-called *home forest* for its own disposal.

From the fourteenth century onwards increasing controversies, due to a variety of reasons (shortage of available cultivated land, buying and selling of land, increasing population), caused remarkable changes in the ownership structure. This was the reason why villagers asked for the common land to be distributed among them. This development was due to the fact that their increasing uncertainty about forest ownership and utilization rights had led to forest devastation by forest users. Over the course of centuries, the allocation of former common land to individual farmers was practised in ways that varied with time and place.

The Imperial law from July 1853 gave way to the regulation and suspension of former forest rights. In many cases the transference of the forest area happened with regard to the benefit of commons, communities, villages or the summary of the villagers being entitled to the utilisation of the former common land. In 1871, when land records got rearranged in Austria, it was necessary to formally install these ownership rights and attribute them to specific bodies. The Imperial skeleton law of July 7 1883 enabled common land to be allocated and forest rights to be regulated. Wherever it seemed to be favourable, the commons were divided and distributed to all co-owners, according to their shares of rights; wherever this turned out to be unfavourable for ecological or economic reasons, the commons were regulated and became "*Agrargemeinschaften*". Such regulations specified the rights and duties of all members (shareholders) and the organizational structure of each *Agrargemeinschaft*, and provided a clear distinction of the scope of duties of the local communities (political self-governing bodies) on the one hand, and that of the *Agrargemeinschaften* on the other (Herbst 2004). This development varied from Federal province to Federal province. In Carinthia, for instance, most of the commonland was divided during the following decades, whereas, in some provinces such as Tyrol and Vorarlberg, the government passed public forests into the ownership of communities or commons to guarantee the livelihood of the

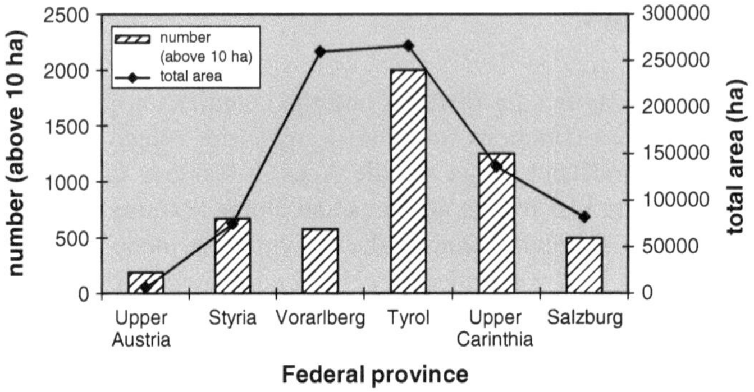

Fig. 9.3 Commons in the Alpine region of Austria (status 2010)

villagers (Johann 1983). Today the size varies between one and thousands of hectares and also the number of commons varies between 100 in Salzburg to 570 in Vorarlberg. Common rights are usually attached to the farmstead except in Vorarlberg where they are attached to the person. Today Vorarlberg has the highest proportion of commons, which cover over 50 % of the land (see Fig. 9.3).

At present, common property is one of the prevalent categories of rural landownership and therefore a major factor in land use planning and policy in Austria. Their legal status is that of corporations under public law (*Agrargemeinschaften*, rural common property). All regulations are found in the already mentioned Austrian Federal skeleton law 1883 and the detailed legislation of the Federal provinces. *Agrargemeinschaften* are self-governing bodies entitled to decide all their internal matters autonomously and self-responsibly. A special supervising authority (*Agrarbehörde*), installed in 1883, safeguards the sustainable management of the commons, which consist mainly of pastures, alpine pastures and forests (Herbst 2004). Due to the historic development and socio-economic structure, the importance of common property and the roles commons play within rural society differ markedly from Federal province to Federal province.

9.6 Benefits and Obligations of Commons : Historical Tasks

The idea of sustainable management of forests and pastures is a many centuries old tradition in the Austrian Alps. However, it is in the commons-related legislation that this idea was first laid down. The members of commons have participated in the administration and management of local resources since the thirteenth century. They accepted responsibility for managing the woodland, the waters and the village infrastructure.

The general aim was to secure the sustainable maintenance of the local resource for future generations. Standard was the annual consumption of the farmstead. Villager's acts tackled fuelwood and timber harvesting, the collection of litter, and forest grazing. The allowable cut and the number of cattle sheep and goats tolerated in the forest depended on the available resources which means the size of the common land and the number of villages entitled to rights.

The protection of cultivated land and settlements against natural hazards, such as avalanches and torrents, was one of the main tasks of the commons. Agreements based on local customary laws give evidence of the cautious handling of the resources and the respect towards nature. The first protective activities aimed at safeguarding cultural land along streams and torrents and date back to the thirteenth century, common laws aiming at the protection against avalanches by careful forest management have been proved from the beginning of the sixteenth century onwards such as in Tyrol and Upper Carinthia (Johann 2004). In the course of centuries, commons developed several techniques to avoid floods, avalanches and other impacts on common property including bans on parts of the forest surface with protective functions. The general aim was the safeguarding of the local resource for future generations.

9.7 Changes in Land Use Management and the Present Role of Commons

Traditional farming remained relatively unchanged until the middle of the twentieth century, when industrialization of agriculture was introduced in alpine valleys. This resulted in a shift from arable land to pastures, stock breeding in the valleys, stable feeding throughout the whole year and the abandonment of alpine pastures. Consequently, unprecedented marked structural changes and modifications in the living conditions of the rural population took place. These also resulted in the abandonment of a considerable number of farmsteads (Johann 2003).

However, *Agrargemeinschaften* are still important economic factors in the rural areas of the Austrian Alps. They generate considerable income that is mostly reinvested in conservation and possible improvement of the common property. The Plenary Assembly may, however, distribute income to the shareholders according to the shares held (Herbst 2004). Although regulations and structures of these *Agrargemeinschaften* in most cases are more than a 100 years old (though regularly adapted), their economic and cultural values are important today. Large areas of land even of low economic value are still managed because they are commonly owned.

At present, the most important kind of common management concerns Alpine pastures. The Federal province Vorarlberg is an example for this development (see Fig. 9.4). The careful cultivation of Alpine pastures contributes markedly to the attractiveness of the landscape, being also the basis for low-impact tourism.

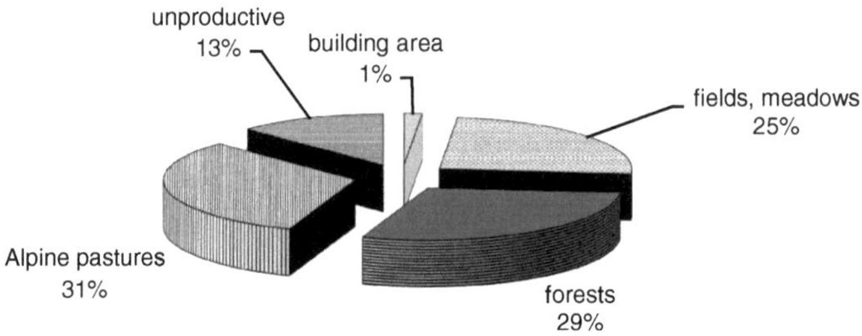

Fig. 9.4 Commons in Vorarlberg: land use categories 2010

Although only 19 % of the area of Austrian Alpine pastures is in the ownership of *Agrargemeinschaften,* commonly owned Alpine pastures provide 48 % of the fodder producing area (see Fig. 9.5). About a half of it is situated in the high mountain region. The percentage of commonly owned and managed Alpine pastures is the highest in Austria's mountainous west (44 % in Vorarlberg, 36 % in Tyrol) .

One of the main contemporary tasks of commons is to take care of the appropriate utilisation of the common-owned land by the entitled shareholders, particularly with regard to timber and fuelwood harvesting and forest grazing. There are two main goals: the sustainable maintenance of the common property and the maintenance of the cultural landscape by adequate management. These also take into consideration public interest. At present, for instance there is a variety of social demands forests have to fulfil, such as recreation and protection, which can only be provided by sustainable and careful forest management without exploitation and over-utilization of the given natural resources.

Agrargemeinschaften are generally aware of this increasing responsibility. There are excellent examples of sustainable forest management as it has been carried out for instance by the *Agrargemeinschaft Schnann*, and recognised by an award from the Austrian government.

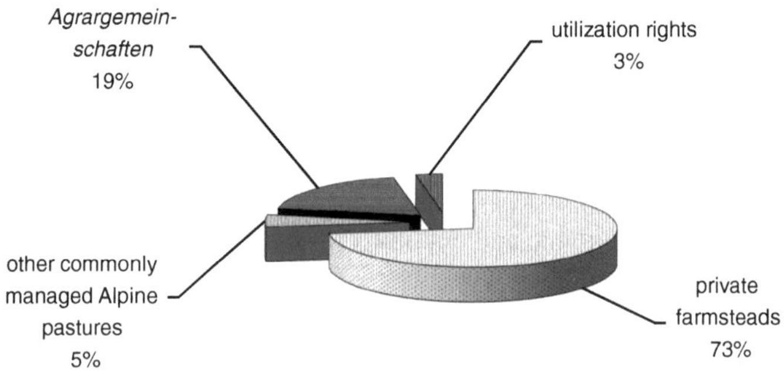

Fig. 9.5 Managed Alpine pastures percentage of ownership 2004 (Parizek 2006, modified)

9.8 Discussion: Future Perspectives in the Context of Migration and Globalization

Studies carried out in the Bavarian and Austrian Alps prove that traditional mountain farming with diverse intensities of land use led to a diverse mosaic of considerable natural or close to nature landscape elements and a high diversity of plants. In the Alpine area, contemporary economic trends cause severe problems in mountain farming. These issues are characterized by the abandonment of former agricultural land situated on difficult sites and increasing intensity of utilization at locations that are easier to work. Until today, agriculture was an important pillar of the cultural identity and rural society and the crucial driving force for the maintenance of the artificial ecological balance of the Alpine cultural landscape. Changes in landscape management therefore also cause severe ecological shifts and consequent loss of biodiversity. Furthermore the affected ecosystems become increasingly unstable.

The cultural and traditional identity of the Alpine population is stamped by projections from outside. That is why the Alps are perceived as an intact, healthy and social undisturbed living space. These myths are also strengthened by the commercialization of tourism, music industry and business. Due to globalization in all fields, the cultural and natural heritage of the alpine region is endangered. This development is mirrored by the alteration of architecture as well as changes of the cultural landscape. The urbanization of the countryside as well as the migration being noticed in remote areas also results from this development. Urban lifestyle is increasingly influencing everyday life even in remote areas.

Natural as well as cultural landscapes are today disappearing rapidly. However, there is an increasing social demand for the safeguarding and renewal of this living space. Therefore the Alpine population in its entirety, bears a shared responsibility for the sustainable maintenance of this unique resource. Socio-cultural as well as economic motivations, are the reasons why utilization of the Alpine area has been re-evaluated. Sustainable development can only be achieved if economic, ecological and socio-cultural aspects are effectively considered. Inclusion of socio-cultural dimensions also means active regional networking. In the remote Alpine areas, the contemporary dominant subject of social discussion tackles the question: "Can I make my living in my close surroundings or do I have to go''? The answer depends on the future perspectives in the field of social, economic, and cultural development. Thus multiple opportunities and functioning social networks are just as important as the achievement of a particular economic level.

Traditional farm forest management was based on the available local resources. The farmer acquired the bare necessaries of life by diligence, skill and traditional knowledge related to fields, woodland, cattle and alpine pasture. Under the village-life settings, consensus and acceptability were especially important, where people participated in decisions that determined the development of the society. Traditional knowledge was preserved from generation to generation, and provided the community with a sense of identity and continuity. The maintenance and safeguarding of knowledge and practices concerning nature and the universe developed and

Fig. 9.6 Forest ownership categories in the context of forest functions (Tyrol)

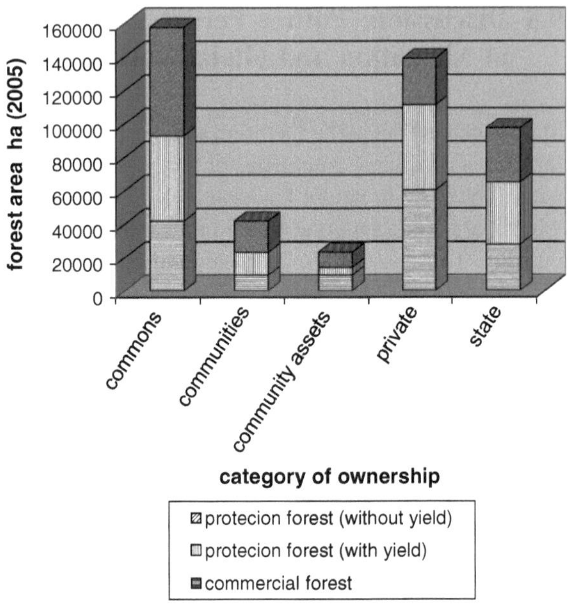

perpetuated by communities in interaction with their natural environment was targeted by the *Convention for the Safeguarding of the Intangible Cultural Heritage*. The 37th Assembly of UNESCO (2003) passed this Convention in 2003. In addition, the *Alpine Convention*, established by eight Alpine states and the EU in 1995, raised awareness of the effects of demographic change on living and working conditions in the Alpine areas. Particularly the Declaration *Population and culture* (Alpenkonvention 2006), attached to the Convention, targeted socio-economic as well as socio-cultural aspects. The ambition is to maintain and further develop the material and intangible cultural heritage, and to secure the livelihood of the local people.

Thus the promotion of regional policy, taking into account the particular demands of mountain regions and the strengthened cooperation of agriculture and forestry, is of essential importance. Through this, the local people are considered the main pillar of the self-determining economic development of the Alpine region. The association with tourism is obvious. In Austria, cultural tourism is one of the prevalent sectors of the tourism industry. Therefore it is of great importance to balance the increasing demands and to safeguard the entire living space in its cultural, natural and historical identity and diversity. There are different initiatives promoting this goal such as *Netzwerk Land*, *Netzwerk Forst und Kultur* or *Allianz in den Alpen*. *Allianz in den Alpen* was established in 1997 and joins communities and regions of seven states of the Alpine arc. Its goal is to develop the Alpine area for the future in a sustainable and participatory way (N.N. 2010).

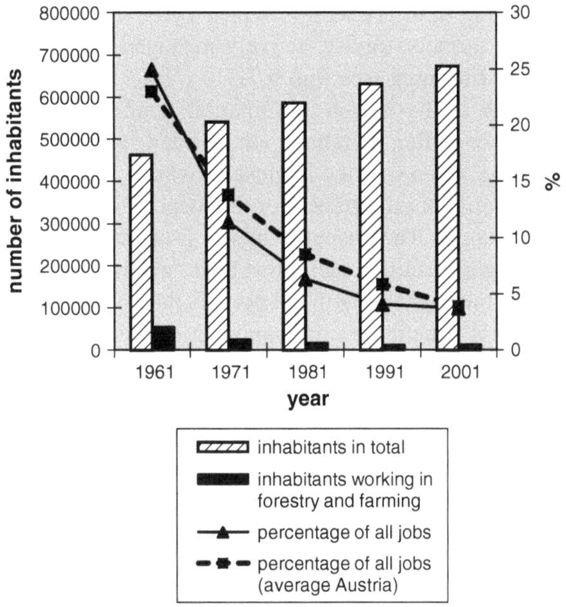

Fig. 9.7 Development of people working in forestry and farming in reation to population development (Tyrol)

Traditional commons are characterized by certain features which can be important for the future. Their structure is still evident and can be illustrated by social equity including cultural, social and spiritual dimensions. There are internal rules to reduce the misuse of power, economic principles which concern the priority of subsistence over profitability, and the priority of local factors over global ones. This model of participatory intercourse with nature makes damaging impacts more difficult, and encourages responsibility for the shaping of the landscape and the sustainable protection of natural resources (Zenkl 2000).

A considerable part of the commonly owned forest area is categorised as 'Protection Forest'. Figure 9.6 gives an example of the Federal province Tyrol. The management of the Protection Forests requires time, energy and high costs, thus diminishing or sometimes even eliminating the financial yield resulting from the eventual selling of wood. Conversely the carefully cultivation of the forests with protective functions, as it is still carried out by the *Agrargemeinschaften,* is of essential importance for the safeguarding of the entire living space in the valleys.

However, because of the changes in social structure and decreasing income, the members tend to switch their professions from farmers to service-rendering employees. In Carinthia for instance, the number of farmsteads decreased by half from 1945 to 2007 (Amt der Kärntner Landesregierung 2009). The decline of active farmers gives way to increasing social conflicts, arising within the commons as well as between commons and communities. These are illustrated by the problems occurring in the Federal province Tyrol. There, about three-quarters or even more of the farmers, having actively managed their land in the 1960 and 1970s have stopped farming (see Fig. 9.7).

This is the reason for internal conflicts between farmers and non-farmers. Commons often own huge areas suited for the local development of industry and housing, often situated in close proximity to the settlements. Therefore they are of high political importance for the local planning and the development of the whole community. The disputes tackle property rights of high economic value concerning agricultural and forest land, but also tourist and other infrastructure as well as hunting rights. The focal point is that the income generated from this ownership does not benefit the community in its entirety but the stakeholders of the *Agrargemeinschaften* (commons). However, communities aspiring to have the common owned land transferred to their own use do not take into account, that the cultural landscape, the result of traditional mountain farming, is the basis of the attractiveness of the living space and offers additional added value. Until the present day, the communities did not address the tasks related to the future safeguarding of traditional values and the material and intangible cultural heritage and the maintenance of cultural landscapes.

9.9 Conclusions

There is an increasing public awareness that recognises that the future of the Alpine living environment must be seen in the context of overall sustainable development. Furthermore, this needs to strive for protection as well as utilization of the cultural landscape, migration and urbanization being important driving factors. Whereas small communities in authentic mountain areas suffer from population decline, there is continuous increase in already densely populated valleys and Alpine cities. With regard to the safeguarding of the cultural landscape and sustainable development of the rural area, commons still have to fulfil important roles. These roles are not restricted to the utilization and the management of pastures, forests and alpine pastures. Traditional commons maintain the material and intangible cultural heritage and the traditional knowledge concerning nature and natural hazards. The safeguarding of indigenous knowledge is of high importance for rural as well as urban society. The feared loss of this knowledge has internationally been tackled within recent decades and was the reason for agreements on a global and European level, such as the *Convention on the Immaterial Cultural Heritage* (UNESCO 2003), the *Alpine Convention* (Alpenkonvention 1991), and the *Ministerial Conference on the Protection of Forests in Europe* (MCPFE 2003). In the Alpine area of Austria, commons and communities have an important role to play in transforming these agreements into working programmes. However, taking into account the increasing social changes within society now living in remote areas the impacts are open to question. Will these efforts be able to stop the current trends, or will this be the end of tradition?

References

Alpenkonvention (1991) Übereinkommen zum Schutz der Alpen. http://www.alpconv.org/pages/default.aspx. Accessed 12 June 2012

Alpenkonvention (2006) Deklaration Bevölkerung und Kultur. AC_IX_11_1_de Verabschiedet von der 9. Alpenkonferenz, Alpbach. 9th November 2006 http://www.cipra.org/de/alpmedia/positionen/87. Accessed 12 June 2012

Amt der Kärntner Landesregierung (2009) Statistisches Handbuch des Landes Kärnten. 54. Jahrgang. Verlag Johannes Heyn, Klagenfurt, pp 102–120

BMLFUW (Federal Ministry of Agriculture, Forestry, Environment and Water Management) (ed) (2000) The sustainable future of mountain forests in Europe. Federal Forest Research Center, Vienna, pp 7–11

Buerg J (1998) Befragung der Anteilsberechtigten von Agrargemeinschaften in Tirol. Universität für Bodenkultur, H44000 Inst.f. Sozioökonomik der Forst- und Holzwirtschaft. HB–DIP: D-8316.en

Hafner F (1979) Steiermarks Wald in Geschichte und Gegenwart. Österr, Agrarverlag, Wien

Hamberger S, Baumeister O, Erlacher R, Zangl W (1998) Schöne Neue Alpen. Raben Verlag, München, pp 1–241

Herbst P (2004): Rural Common Property in Austria. International Association for the Study of Common Property (IASCP) tenth biennial conference. Oaxaca, Mexico http://www.iascp2004.org.mx. Accessed 12 June 2012

Johann E (1968) Geschichte der Waldnutzung in Kärnten unter dem Einfluß der Berg-. Hütten- und Hammerwerke, Verlag des Geschichtsvereins für Kärnten Klagenfurt

Johann E (1983) Die Geschichte des Gemeindewaldes in Österreich biszum Ende des 19. Jahrhunderts. In: Forstliche Bundesversuchsanstalt (ed) Forst- und Jagdgeschichte Mitteleuropas. Mitteilungen der ForstlichenBundesversuchsanstalt Wien, 151, Heft, pp 75–85

Johann E (2003) More about diversity in European Forests: The interrelation between human behaviour forestry and nature conservation at the turn of the 19th century. In: Leos Jelecek et al. (eds) Dealing with Diversity. 2nd International Conference of the European Society for Environmental History Prague 2003. Charles University in Prague, Department of Social Geography and Regional Development, pp 202–205

Johann E (2004) Wald und Mensch. Die Nationalparkregion Hohe Tauern (Kärnten). Verlag des Kärntner Landesarchivs, Klagenfurt

Kärntner Landesarchiv Klagenfurt. Finanzdirektion Klagenfurt, Rechnungsdepartement, Schachtel 29 und 39

Koller E (1975a) Forstgeschichte des Landes Salzburg. Verlag der Salzburger Druckerei, Salzburg

Koller E (1975b) Forstgeschichte Oberösterreichs. Verlag LWK, Linz

Krondorfer A (2005) Soziale und wirtschaftliche Struktur von Agrargemeinschaften in Oberkärnten. Online-Katalog der Universitätsbibliothek, Bodenkultur, AC04716267

MCPFE (Ministerial Conference on the Protection of Forests in Europe) (2003) Wiener Resolution 3: Erhaltung und Verbesserung der sozialen und kulturellen Aspekte nachhaltiger Waldbewirtschaftung in Europa. http://www.foresteurope.org/filestore/foresteurope/Publications/pdf/mcpfe_work_cor3.pdf. Accessed 12 June 2912

Oberrauch H (1952) Tirols Wald- und Waidwerk. Schlernschriften Bd. 88. Universitätsverlag Wagner, Innsbruck

Parizek T (2006) Alm und Almwirtschaft im Überblick. Beitrag zur Alp-Austria Präsentation Großarl 06/2006. Federal Institute for Agricultural Economics, Vienna

Pittioni R (1974) Landschaftsgestaltung in der Urzeit. In: Zitta R (ed) Wildnis, Forst und Ackerland. Verband der wissenschaftlichen Gesellschaften Österreichs, Wien

Pro: Holz Austria – Arbeitsgemeinschaft der österreichischen Holzwirtschaft (2001–2012) Wald und Holz. http://www.proholz.at/wald_holz/wald_zahlen.htm. Accessed 30 Jan 2012

Schiff W (1899) Die Regulierung und Ablösung der Wald- und Weide-Servituten. In: Die Geschichte der Österreichischen Land- und Forstwirtschaft und ihrer Industrien. 4 Bde., Commissionsverlag M. Perles, Wien. Bd. 1
Schindler K (1885) Die Forste der in der Verwaltung des k.k. Ackerbauministeriums stehenden Staats- und Fondsgüter. Staatsdruckerei, Wien
UNESCO (United Nations Educational, Scientific and Cultural Organization) (2003) Convention for the safeguarding of the intangible cultural heritage. http://www.unesco.at/user/texte/docs/conv_saveg_ch.pdf
Zenkl M (2000) Beteiligung der Öffentlichkeit – Probleme und Modelle. In: Mensch und Landschaft. Tagungsband zum Symposium Nachhaltige Entwicklung von Kulturlandschaften im Donau- und Alpenraum, TU, Graz 17-18 February 1994,

Chapter 10
Guided Pollards and the Basque Woodland During the Early Modern Age

Alvaro Aragón Ruano

This research has been carried out within the Inheritance and Cultural Landscapes Research Group (Project IT315-10), funded by the Basque Government, Department of Education, University and Research.

10.1 Introduction

In the early modern age, oak was considered the most useful tree, not only for shipbuilding or house building, but also for charcoal making. This was due to the extraordinary qualities of its timber and the proximity of vast oak forest areas to shipyards and forges. Moreover, this was related to geography, which does not change so rapidly, and the observation that oaks grow in Guipúzcoa up to a height of 600–700 m. As a result, it was the most exploited and re-forested tree in Guipúzcoa during the Early Modern Ages; followed by chestnut, despite the fact that this has a shorter lifetime, and beech, which had not such a good characteristics.

Throughout the Modern Ages, humanity faced the problems in the field of forestry. These were on the one hand, the incompatibility between the rate of forest exploitation and the voracity of production activities, and, on the other hand, the biological rhythms and growth-rate of plant species (Perlin 1989, pp. 16–17). As a result, deforestation and scarcity of raw materials became increasingly the most serious problems. Trying to ensure the sustainability of woodlands, forests and their related activities, from the fifteenth to the eighteenth centuries, the people of Guipúzcoa introduced a number of different forestry techniques (coppice, maiden trees and pollards). This was an attempt to achieve the highest possible level of productivity. These techniques gradually evolved over time, in accordance with the needs and priorities of economic activities and the abundance or scarcity of forest materials and resources.

A. A. Ruano (✉)
Departamento de Historia Medieval, Moderna y de América, Facultad de Letras,
University of the Basque Country, Paseo de la Universidad, 5, 01006Vitoria-Gasteiz, Spain
e-mail: alvaro.aragon@ehu.es

I. D. Rotherham (ed.), *Cultural Severance and the Environment*,
Environmental History 2, DOI: 10.1007/978-94-007-6159-9_10,

10.2 Types of Forestry Techniques

Coppices were the trees cut from the base of the trunk, which needed 5 years to produce hoops suitable for barrel-making, and 12 or 15 years for charcoal. The essential problem with coppices in this landscape was that when the stools were pruned, they could not be protected. This would have required the building of fences or hedges, and because the majority of them were located on common woodland, where everyone had both right of way and the right to graze their animals it was not possible. In his *Ciencia de Montazgos*, written in 1783, José Odriozola affirmed that coppices for firewood could be achieved in many different ways, but that the best way was to sow oak acorns and beeches mixed with chestnuts (Cruz Mundet 1989, p. 149). In the nurseries, it was important to maintain a certain distance between those plants destined for coppices, in order to enable pollarding or the formation of the stools that would later serve as the base, and to permit the branches to spread correctly. According to Villarreal de Bérriz, the best coppices were chestnut one, because they grew more quickly and provided good wood for making charcoal, although for fires the wood was not as good as oak. Following an initial cut half a metre from the ground, when fully grown, it was best to cut them almost right back down to ground level, since in that way they would shoot up anew (Villarreal de Bérriz 1736, p. 152) (Fig. 10.1).

Maiden trees did not have the disadvantage of the coppices. Livestock could be left to graze beneath the trees without fear that they would harm the new growth. Trees were left to grow for between 80 and 100 years, and were used for building houses and smithies, as well as for the shipbuilding industry. However, even maiden trees required a certain amount of care and handling, since as the Marquis of Rocaverde, Superintendent of Ships and Plantations, stated in 1743, the lower branches had to be pruned and the tree should ideally be guided.[1] José Odriozola recommended following the traditional method of planting oaks thickly in these areas, thus

Fig. 10.1 Coppices in Leitza (Navarre)

[1] AMF (Municipal Archive of Fuenterrabía), C, 5, II, 10, 1.

Fig. 10.2 Maiden trees in Urdanibia-Irún (Guipúzcoa)

making the most of the available land and enabling the trees to protect each other from the wind: 2 m for re-populated woodland, and 1.5 m for virgin areas or areas which had been sterile for some time (Cruz Mundet 1989, p. 160) (Fig. 10.2).

Pollarding fulfilled, as the lawyer Arnedo explained in 1662, a large number of different functions, including that of responding to the combined demands of the iron and steel industry, the shipbuilding industry and livestock farming. First, it produced pastureland, but less acorn harvest than coppices. According to bachelor Arnedo in 1662, pollards "*…produce several branches and, even though it takes longer to produce fruit at the beginning, until the tree grows and complete the head, later the cuts come every 8 years or 10 years at the least, and although it does not produce as much fruit as coppice, it comes 4 years before, which compensates each other…*" Therefore, it seems that the pollards produced less acorns than coppices, taking into account that it used to take quite long to made a pollard from a coppice and it could not produce any acorns during the process. Nevertheless, there could be another reason to explain this. Pollards were cut every 8–10 years, whereas coppices were every 12–15 years; therefore, coppices produced more acorns from cut to cut. Moreover, when coppicing not every branches was cut, just some of them, which allowed the tree to produce acorns; on the contrary, most of pollard's branches were removed, which means that the tree would not produce any fruit during the next 5 years. Recent research shows that a hard thinning could damage the acorn production, as it happens with a weak density of trees: the harder the thinning, the poorer the harvest (Gutiérrez Galindo et al 2003, pp. 141–145; Gea-Izquierdo 2006, pp. 346–348). Pollarding also enabled grass to grow so that livestock could graze, while at the same time preventing them from eating the spring growth on the cut trees, because the pruning was carried out at a greater height. Second, it left it two or three main guide branches, which with time could be used for obtaining the curved parts (tuertas or curvatones) so essential for shipbuilding. Likewise, from these main branches a series of smaller ones grew, which could be used for making planks, although they were mainly used for producing charcoal, with the added advantage of increased productivity, since pruning could be carried out 4 years earlier (oak pollards could be cut every 8 or 10 years, and beech pollards every 5–6). Third, the distance between the trees enabled them to expand, thus growing longer branches.[2]

In this sense, Javier Ignacio de Echeverria, author of *Discurso sobre la plantación del roble*, written in 1775 for the *Real Sociedad Bascongada de Amigos del País*, calculated that a guided pollard oak could grow to occupy 60 m^2, while a straight one could only occupy 15 m^2.[3] On this occasion, we are more interested in describing the last technique than the other two, which would be subject of a deeper research in a next opportunity.

10.3 *Ipinabarres* or Guided Pollard Oaks

The *ipinabarros, ipenabarres* or guided or shaped pollard oaks, were created using the *horca y pendón* method, in which one tree was left standing every 4.5 m, and of its main branches, all but two or three were cut off. Those that were left were pruned to a height of 3 m, one at a right angle to the main trunk (the *horca* or fork), and the other at an obtuse angle (the *pendón* or standard) (Fig. 10.3).

10.4 The Beginnings of the Technique

The first references to *Ipenabarres* (a Basque word that in the Vizcayan dialect means *to put or leave a branch*) in the Guipúzcoa region are linked to the Deva and Urola river basins and date from the 1530s.[4] Nevertheless, the Basque Country was not an exception and this forestry technique had been in use in the Castilian region since at least the end of the fifteenth century, mainly in connection with the development of pasturelands. By means of an ordinance dated 28th October, 1496, the Catholic Monarchs ordered that trees should no longer be cut down to the base of the trunk, but rather should be left with two (or sometimes three) main branches (the aforementioned *horca y pendón*) from which new shoots would be able to sprout. The use of this technique could be found in Castilian, Cantabrian and

---

[2] AMH (Municipal Archive of Hernani), C, 5, I, 3, 4.

[3] AMR (Municipal Archive of Rentería), C, 5, V, 1, 5.

[4] AML (Municipal Archive of Legazpia), 167-9.

Fig. 10.3 Guided pollard oak in Artikutza (Navarre)

Spanish American countries. In Castile, for instance, some forest regulations, instead of using that expression it was claimed to leave two *aleros* and one *pendolero*: first was the *horca* and second the *pendón*. This technique has been used to date in the mountains of the Central System (Martín Jiménez 2003, pp. 412–414).

Nevertheless, it seems that this practice did not become common in Guipúzcoa until 1548. In the *Juntas Generales* (or General Assembly) held in Zumaya between 14th and 24th April, 1548, the issues facing timber production were addressed. In light of the increasing deforestation and the general concern over the scarcity of timber in the region for building ships and wood for making charcoal, in addition to ordering the planting of 500 oaks and chestnut trees every year on the commons, the *Juntas* decreed that no tree should be felled. This was with the exception of those trees destined for the shipyards and the construction of buildings. Those trees used for firewood and charcoal should be turned into guided pollards.[5] The concern over forestry resources was not exclusive to the Guipúzcoan authorities. Voices rang out all over the Iberian Peninsula and across Europe also, warning of the problems of deforestation (Aragón Ruano 2001, pp. 62–63; Bechmann 1984, pp. 286–287; Belhoste 1990, pp. 223–224; Rackham 1990, p. 77).

By the middle of the sixteenth century, the scarcity of materials, mainly for the construction and shipbuilding industries, forced the province to issue the aforementioned order in 1548. They also petitioned Carlos I for the confirmation of another order issued by the Elgoíbar *Juntas Generales* on 10 May 1552, regarding

[5] AGG-GAO (General Archives of Guipúzcoa), R2.

the management of coppices. The order recommended that guided pollard oaks be left standing in the coppices at 22-m intervals. Curiously enough, towns like Elgóibar, Hernani, Rentería and Oyarzun, in which coppices were common and where the principal industrial activity was ironworking, rejected the order and voted against it. Consequently, the confirmation was issued but in a modified form, with the principal change being the distance between the pollard trees left standing, established as 33.5 m instead of 22 m.[6]

10.5 The Spread of the Technique Throughout Guipúzcoa

In other areas, outside the aforementioned Deva and Urola river valleys, guided pollards, or *ipinabarros*, were unknown, or at least the technique was not applied until the end of the seventeenth century. Indeed, most of Guipúzcoa's woodland areas were covered with coppices and maiden trees, at least until the seventeenth century. This seems to have been the predominant situation throughout the Late Middle Ages, as indicated also by the municipal bylaws. According to the *Juntas Generales* Díez de Salazar 1991, p. 509), in 1564 coppices occupied at least a third of Guipúzcoa's woodland area[7], and continued to predominate in towns with a strong ironworking tradition, such as Legazpia (1591)[8], Oyarzun (1691)[9]and Hernani (1662)[10]

The survey that Doctor Hernando Suárez de Toledo carried out by royal decree in 1569, aimed to determine the degree of compliance with the 1563 Royal Order commanding the planting of oaks for the Royal Armada. This applied to all areas within two leagues of the sea (an order that Guipúzcoa failed to fulfil). The findings were that two forest models coexisted in the region: coppices, mainly used for charcoal production, and maiden trees, used for shipbuilding and the production of charcoal.

According to the aforementioned survey, the predominance of coppices during that period was most manifest in towns which had strong ties to the iron and steel industry such as Elgóibar, Legazpia, Rentería, Oyarzun, Fuenterrabía and Hernani. In other towns, such as Zarauz, for example, maiden trees were more common, while in Zumaya, pollards existed alongside the coppices. Deva, on the other hand, had a few trees from which bow pointers (known locally as *corbatones*) could be obtained, while in Guetaria, the majority of the woodland areas were dedicated to the production of these elements. We can see, then, that in those towns with a strong iron and steel industry, coppices were most common. Whereas in the coastal

[6] AGG-GAO, JD IM 2/17/4.

[7] AGG-GAO, JD IM 2/17/5.

[8] AML, 170–162, 3.

[9] AMOr (Municipal Archive of Ordizia), leg. 3, 1.

[10] AMH, C, 5, I, 3/4.

villages, which were more concerned with shipbuilding, it was not unusual to find a combination of coppices, maiden trees and pollards used for producing parts for the shipyards. The majority of those surveyed agreed upon the general lack of not straight wood, but rather curved wood. Suárez de Toledo himself affirmed that in Guipúzcoa coppices and *straight* woodland areas (used for plank production) were most common, while in Vizcaya, *twisted* wood (used for shipbuilding) was more abundant. In Vizcaya, it was not customary to cut maiden trees destined for charcoal production from the base of the trunk, but rather from the base of the branches. According to Brad Loewen, based on archaeological sources, during the sixteenth century in the Basque Country, there was a standardized system to obtain pieces for shipbuilding. This means that all pieces used in shipbuilding had the same proportions, though different sizes, which allowed to build the same basic vessel in different sizes. This involves branches that were deliberately shaped using ropes and stakes or making cuts (Loewen 2007, pp. 291–292) (Fig. 10.4).

All the towns along the coast and in the immediate surroundings (Fuenterrabía, San Sebastián, Zarauz, Guetaria, Zumaya, Cestona, Azpeitia, Elgóibar and Motrico) highlighted the lack of trees available for shipbuilding, and the need to import them from Vizcaya, where they were more abundant. Rentería, for example, had been doing this for quite a while. One of the keys to the question was that the aforementioned 1552 ordinance regarding coppices, which specified that a guided pollard be left every 33.5 m, had not been respected along the coast, and was difficult to enforce. The townsfolk complained bitterly that they were forced to go further and further afield for their timber, and pay ever-increasing prices. The supply problems experienced by Guipúzcoa's shipbuilding industry lasted right up until the second decade of the seventeenth century. In 1580, the witnesses asked in different towns of the Guipuzcoan coast reckoned there was from Fuenterrabía to Motrico, enough timber to build 50 vessels of 500–600 metric tonnes. Fuenterrabía and Irun had timber for five vessels, Oyarzun for three vessels, Rentería for 30 vessels, Urnieta for just one, Usúrbil for five, Zumaya for three and Deva and Motrico for two each. These figures were confirmed some years later, when, on

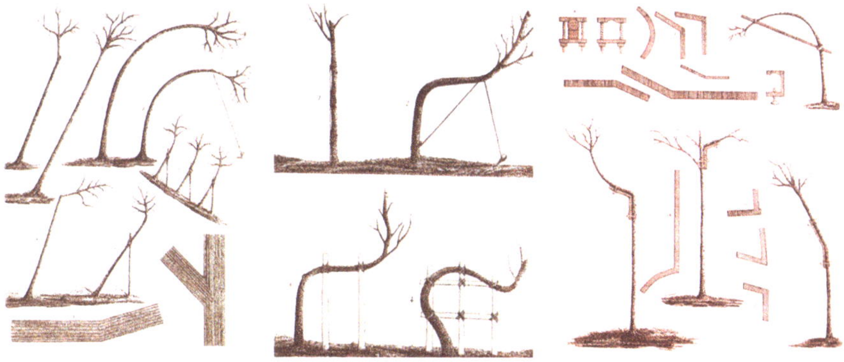

Fig. 10.4 Shaping methods in the eighteenth century

19th of July 1589, the War Council ruled an order to find out how much timber could be used in shipbuilding. According to the witnesses, the village of Zubieta had wood of 10,000 ducats worth and timber suitable to build five–six vessels of 500 tons. In Usúrbil, there was timber suitable to build 14 vessels. In Rentería, traditionally, the main supplier of timber in Guipuzcoa, 30–40 vessels could be built. In San Sebastian, there were some particular forests with enough timber to build four vessels. Deva, as in particular as in common lands, there was timber for four–five vessels and planks for another three. Finally, in Zumaya there was timber for four–five vessels. As a result, in 1589, there was in Guipúzcoa, within the two leagues from the sea timber suitable to build 52–66 ships of an average of 500 tonnes; quite similar to figures in 1580. Nonetheless, proof of the increasing scarcity is the fact that the *Juntas Generales* of Guipúzcoa were obliged to petition Vizcaya for the sale of bow pointers for the manufacture of ships in 1611, 1616 and 1617 (Fig. 10.5).[11]

While by the mid-seventeenth century, towns such as Azpeitia and Azcoitia were obliged to pollard their oaks in the *horca y pendón* style, in the Free Mountains of the Urumea, which belonged to Hernani, San Sebastián and Urnieta, and it was not until 1658–1671, that pollards replaced the hitherto predominant

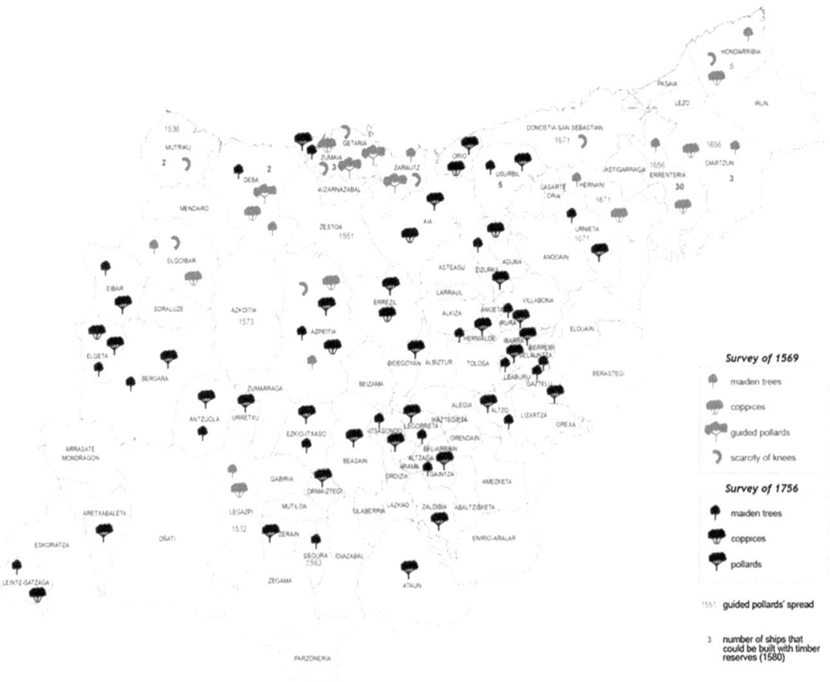

Fig. 10.5 Structure of Guipuzcoan Forest according to surveys of 1569 and 1756

[11] AGG-GAO, JD AJI 2, 4, JD AJI 2, 5 and R 17.

coppices. This change, however, was not without its difficulties. In 1658, a legal dispute ensued regarding the new cutting system, which ended in its generalised enforcement. The practice was definitively established in the aforementioned woodland areas in 1671, after the *Concordia* held in Astigarraga on 21st March that same year between San Sebastián, Hernani and Urnieta decreed that every cut must leave reserves and guide branches. From then on, in the firewood auctions of the Free Mountains of the Urumea, foresters were expressly ordered to respect the rulings of the 1671 *Concordia*, and were prohibited from felling those oaks marked for guided pollarding, known as *ypinabarros*. Meanwhile something very similar was occurring in the nearby towns of Oyarzun and Rentería. Here an order was issued to leave oaks suitable for pollarding at intervals of 6.5 m in the coppices. In Oiartzun, in the forest auctions carried out by the council from 1656 to 1691, cut conditions bound to leave among the coppices oaks suitable to become pollards or *guiones* (guided oaks)—this word was also used in the neighbouring Rentería. At the end of the century pollard oaks were called *suaritzak*, which literally means "oaks for fire". Contemporary Basque language uses the word *zuhaitza* to refer to wood trees, whereas fruit trees are called *arbola*. Probably, and this is our own hypothesis, the word *zuhaitza* comes from the aforementioned *suaritza* (Aragón Ruano 2009, pp. 84–86).

Unlike what had occurred previously, during the eighteenth century, the number of coppices decreased notably, although they continued to be used (mainly) for the basket-weaving and barrel-making industries. The reason for this decrease was that most of the woodland areas and forests in Guipúzcoa were now occupied by pollards and maiden trees. It is more than likely that the needs of the Real Compañía Caracas, from 1728 onwards, and the Royal Armada, following the Royal Order of 1748 (with its specific chapter for Guipúzcoa in 1749), played a key role in this change. Both were major customers of the Guipúzcoan shipyards. In light of all the available data, we confirm that by the middle of the eighteenth century, pollard forests had overtaken coppices, and even maiden woods (Aragón Ruano 2001, pp. 147–172).

10.6 The Results of the Application of the Pollarding System

Despite the efforts put in practice by Royal Army, it could not have at its disposal all the supplies previewed, because the application of the aforementioned techniques had to overcome a series of major difficulties. Firstly, many of the plantations were lost before the wood could be harvested. In places such as Larraul, Ordicia, Legazpia, Hernani, Rentería, Irún and Tolosa, between 1749 and 1808, around one third of all the trees planted were lost. Others such as Asteasu, Segura, Villarreal de Urrechu, Fuenterrabía, Oyarzun and Ataun lost only 10 %. Perhaps the most flagrant case was that of Legazpia, which lost 84 % of its plantations between 1776 and 1805, harvesting only 16 % (Aragón Ruano 2001, pp. 77–79).

Secondly, as Jerónimo Tavern and the majority of the naval officers and forestry experts of the time had warned, due to the ignorance of those responsible for carrying out the pruning operations, there were only a very few pollards whose two guide branches grew in the necessary direction. In the majority of cases, a lot of bad faith underlay these practices, along with the vested interests of the nursery owners, eager to get their hands on the one *cuartillo de real* paid by the Guipúzcoa Provincial Council for every tree planted. The ironworkers and charcoal makers, eager to obtain a quick, safe harvest of firewood and charcoal, and the councils themselves were keen to make a greater profit than that provided by selling the wood to the Royal Navy. All this meant that many trees, despite being marked for shipbuilding purposes, actually became unguided pollards, i.e. no guide branches were left. In light of the Crown's lack of cash, and delays in payment, many towns attempted to make a profile from their forests by felling or pollarding (although without guides) young maiden trees, before they grew large enough to be suitable for building royal ships and were claimed by the Naval Authorities. This practice was reported in 1792 by Bernardino Corvera, the Navy Commissioner for San Sebastián. The practice grew so common that when in 1811, during the Napoleonic occupation, Governor Thouvenot of Vizcaya ordered the Province to conduct a survey to determine the status of the Navy's forests, in the whole of Guipúzcoa only 175 trees were found to be suitable for the purposes of shipbuilding.[12]

Thirdly, given that the forests of Guipúzcoa were responsible, through contractors, for supplying the naval dockyards, such as the one at Ferrol, another factor should not be overlooked. This was the terrible destruction wreaked on them by the needs and requirements of the Royal Navy throughout the whole of the eighteenth century, and particularly between 1749 and 1794. Proof of this is the survey conducted in 1784 by the provincial authorities, in which of the approximately 11 million trees counted in Guipúzcoa, only 1.5 % (156,132) were ripe or suitable for use in the manufacture of Royal ships, either as maidens or as pollards. Of this 1.5 %, the vast majority (97,403) were young trees which would provide the required timber in the future. Only 16,476 were mature maidens and 42,253 mature pollards, capable of providing curved timber (for bow pointers, rib planks, deck beams, etc.) (Otaegui Arizmendi 1999, pp. 481–486). After the disaster at Trafalgar, the pressure exerted by the naval authorities on the forests of Guipúzcoa disappeared, although the needs of the Merchant Navy prompted the Mountain Ordinances drafted by the Province in 1815 to state that at least one third of all woodland areas be dedicated to maiden trees and the remaining two thirds to pollards. The Ordinance also ruled that young woodland areas located in river basins situated within one league of the sea be dedicated to maiden trees, for future use in the shipbuilding industry (Aranda y Antón 1990, p. 83).

[12] AGG-GAO, JD IM 2/13/75 and 81.

Fourthly, another major problem was the shortening of the intervals between cuts, since the recommended 10-year intervals and the 12- and 14-year-long intervals established at the woodland auctions were not respected, and pruning was carried out once every 7 or 8 years. As the eighteenth century advanced, and the demand for firewood for kitchen stoves increased, cutting intervals became shorter, as reported by the Real Sociedad Bascongada de Amigos del País in 1766 and by the Marquis of San Millán in 1788.[13] The harm this practice caused was fourfold. Landowners suffered because many thin, young branches were wasted and much shade was eliminated. Indeed, a forest pruned once every 8 years enjoyed only 5 years of shade, since during the first 3 years the new shoots and branches provided little protection, especially compared to forests pruned once every 13 years, which enjoyed 10 years of shade, which in addition to keeping the soil moist, also provided a greater quantity of dry leaves and fertiliser, causing the trees themselves to grow stronger. The ironsmiths, despite having to pay the same price, got less firewood for their money and needed more labour to remove it, thus increasing the price of each load. The practice also harmed the villagers, who had less firewood for their kitchens and other uses. And finally, the tree itself gradually died as a result of too frequent pruning.

Between the sixteenth and eighteenth centuries, a permanent conflict raged between the naval authorities and the ironsmiths regarding the cutting and pollarding method applied. The naval authorities always demanded that the trees be pollarded, leaving guide branches in order to render them suitable for use in shipbuilding. Despite their insistence, however, the demands of the naval authorities were rarely respected, as shown in the complaints made by the aforementioned Tavern. The charcoal makers and ironsmiths were aware of the demands made by the naval authorities, but were more concerned with their own interests and failed to respect the guidelines established by the Navy Commissioners during their visits. The charcoal makers and woodcutters were not receptive to new ideas which required the shared exploitation of woodlands, and continued to cut wood and firewood in the same way as they had for centuries, i.e. felling from the stump, unguided pollarding and selective felling. The main complaints aired by the naval authorities in their visits to the woodlands of Guipúzcoa focused on the methods used by charcoal makers during the first and second pruning operations. It is true that the methods used by ironsmiths, who cut the branches randomly, in any direction, caused the tree to store water in its trunk, thus causing gradual but unstoppable rot to set in. Cuts needed to be made in such a way as to ensure that the water was allowed to fall to the ground, rather than be retained. During the middle of the eighteenth century, the Commissioner denounced the poor management of the majority of the Guipúzcoan councils, with the exceptions of Tolosa and Ataun. Consequently, following the instructions of the Navy Commissioners, many places appointed an "intelligent person" to make the cuts, so as to avoid leaving the task in the hands of the charcoal makers.

[13] AMT (Municipal Archive of Tolosa), C, 5, II, 1, 3.

Nevertheless, as Commissioner Garmendia stated in 1780, the town councils failed to follow the guidelines which stated that the cuts should be made by those qualified in the building of ships for the Royal Navy.[14](Fig. 10.6).

10.7 Conclusions

Perhaps the model of tree that offered the widest variety of uses and which was able to combine the greatest number of exploitations was the guided or shaped pollard, also known as *ipinabar*, *guión* or guided tree. This might be oak, beech, chestnut or alder. This model of forestry exploitation, therefore, enabled a single tree or forest to respond to the needs of a range of production activities, while at the same time increasing productivity. Nevertheless, despite the fact that, repeatedly and over the course of three centuries, the guipuzcoan authorities and the Navy demanded and ordered, by means of binding legislation, that pruning and pollarding be carried out leaving guide branches in the *horca y pendón* (fork and standard) style, the majority of ironsmiths, woodcutters, foresters and charcoal makers failed to comply. The same, however, was not true of carpenters or shipwrights. This twofold practice resulted in two types of tree, despite the fact that, in theory, a single model was followed, p. pollards and guided pollards, with *horca y pendón* guide branches.

Despite the fact that the technique had also developed in other countries, it had specific characteristics in the Basque Country, mainly due to its ties with shipbuilding. The timber needs of shipyards were the key factor which boosted the generalization of the technique across Guipúzcoa in the seventeenth century. This

Fig. 10.6 Pollard oak in Sara (France)

[14] AMR, B, 6, 1, 1.

was not only along the coast as had happened previously. The practice of pollarding became more widespread and this took place as an attempt to prevent livestock from damaging the coppices. Curiously enough, Guipuzcoan iron men, the most important livestock owners until the sixteenth century, decided consciously to get rid of them and to replace them with flocks, apparently less harmful. This was on account of the damage made by their animals in common coppices, and which slowed down the tree growth; in turn harming the charcoal production. Thanks to their control on the councils, the iron men encouraged people to replace coppices with guided pollards.

Although the practice and use of pollarding without guides has survived almost to the modern day, guided pollarding has been lost to history. The decline of the shipbuilding industry was linked to the Royal Navy following the defeat at Trafalgar. There was also the gradual disappearance of timber-based shipbuilding during the nineteenth century, the import of exotic wood and foreign species, and the voracity of the iron and steel industries. The latter survived right up to the last third of the nineteenth century, and was compounded by the destruction wreaked during wartime (War of the Convention, War of Independence and the Carlist Wars). Confiscation processes and the spread of agriculture and livestock farming, all resulted in the disappearance of this forest husbandry model, the recollection of which now only remains in our collective memory. While everyone knows of the existence of pollard trees, because many beech groves and a few isolated oak groves still survive today with these characteristics, only a very few examples of guided pollards (*ipinabarros*) now survive in the province. Curiously, the majority of them are mistaken for unguided pollards, used basically for the production of charcoal, rather than recognised for what they really are. This view is not only deeply rooted among the general population, but is also prevalent among Basque historians (Uriarte Ayo 1988, p. 92; Carrión Arregui 1991, p. 26).

Currently, the Regional Government of Guipúzcoa is carrying out a programme in which the author of this article is involved. This is part of a European LIFE project, related to the conservation of biodiversity, in order to recover old pollarding system, and involves surveying elderly woodcutters. This action will lead to the publishing of a guide to good practice. The guidance should then be applied by anyone, either public or private, in order to preserve the large areas of pollard beeches that exist today in Guipúzcoa but which are on the verge of destruction. Something similar should also be applied to the few guided pollard oaks that currently survive and are authentic historical and biological monuments.[15]

[15] https://www.lifetrasmochos.net/en. Accessed 23 December 2011.

References

Aragón Ruano Á (2001) El bosque guipuzcoano en la Edad Moderna: aprovechamiento, ordenamiento legal y conflictividad. Sociedad de Ciencia Aranzadi, San Sebastian

Aragón Ruano Á (2009) Una longeva técnica forestal: los trasmochos o desmochos guiados en Guipúzcoa durante la Edad Moderna. Espacio, Tiempo y Forma Serie IV 22:73–105

Aranda y Antón G (1990) Los bosques flotantes: historia de un roble del siglo XVIII. ICONA, Madrid

Bechmann R (1984) Des arbres et des hommes. La forêt au Moyen Age. Flammarion, Paris

Belhoste J-F (1990) *Une silviculture pour les forges, XVI-XIX siècles*. In D.Woronoff (ed), *Forges et forêts : Recherches sur la consommation proto-industrielle de bois*. École de Hautes Études en Sciences Sociales, Paris, 219–261

Carrión Arregui IM (1991) La siderurgia guipuzcoana en el siglo XVIII. University of the Basque Country, Bilbao

Cruz Mundet JR (1989) Un ilustrado renteriano? José Odriozola y la "Ciencia de montazgos". Bilduma 3:141–147

Díez de Salazar LM, Ayerbe Iribar MR (1991) Juntas y Diputaciones de Gipuzkoa III. Regional Government of Guipúzcoa, San Sebastian

Gea-Izquierdo G, Cañellas I, Montero G (2006) Acorn production in Spanish holm oak woodlands. Investigación Agraria: Sistemas y Recursos Forestales 15(3):339–354

Gutiérrez Galindo E, Bravo Oviedo A, Sanchez I, Serrada Hierro R, Serrada Hierro R, Allué-Andrade M (2003) Evaluación del regenerado posterior a diferentes tratamientos de resalveo en un monte bajo de Quercus Faginea Lamk. (Torreadrada, Segovia). Cuadernos de la Sociedad de Ciencias Forestales 15:141–146

Loewen B (2007) Le navire de Red Bay et les structures de la construction navale basque. In: Grenier R, Bernier M-A, and Stevens W (eds) The underwater archaeology of Red Bay, basque shipbuilding and whaling in the 16th century. Parc Canada, Ottawa, 271–322

Martín J, Eva P, Navarro F, Gil Sánchez L (2003) El aprovechamiento tradicional de la dehesa boyal en un área de montaña del centro de España. Puebla de la Sierra (Madrid). Estudios Geográficos, LXIV 252:407–434

Otaegui Arizmendi A (1999) El paisaje forestal de Guipúzcoa en 1784. In: Actas del IX Congreso de Historia Agraria, SEHA, Bilbao, pp 481–486

Perlin J (1989) A forest journey: the role of wood in the development of civilization. Harvard University Press, Cambridge, Massachusetts

Rackham O (1990) Trees and woodland in the British landscape: the complete history of Britain's trees woods and hedgerows. Phoenix Giant, London

Uriarte Ayo R (1988) Estructura, desarrollo y crisis de la siderurgia tradicional vizcaína (1700–1840). University of the Basque Country, Bilbao

Villarreal de Bérriz PB (1736) Máquinas hidraúlicas de molinos, y herrerías, y gobierno de los árboles, y Montes de Vizcaya. Antonio Marín, Madrid

Chapter 11
The Evolution of Forest Landscapes in Spain's Central Mountain Range: Different Forests for Different Traditional Uses

N. López Estébanez, G. Gomez Mediavilla, G. Madrazo García de Lomana, F. Allende Álvarez and E. Sáez Pombo

11.1 Introduction

The main objective of this chapter is to characterize the different phases of forestry uses and their impact of forest dynamics on the northern and southern slopes of the Somosierra and Ayllón mountain ranges (Madrid-Segovia, Spain). Thus, we attempt to define long-term models reflecting forest dynamics. To this end, there is a vital need to break down our large study area into smaller zones, known in Spanish as "montes",[1] and to define in each of these the current features of their vegetation and to identify the historical uses and exploitation systems that have shaped them.

[1] Monte is the most common Spanish term used to define forest areas. There are two problems: one is that it has multiple meanings, as it can also be used to refer to mountain; the other is its lack of accuracy, as it can be used to refer to woodlands and scrublands, and even to pastures, barren lands and dunes. On the other hand, its major advantage is that its flexibility allows it to include all form of forests and woods in the widest sense.

N. López Estébanez (✉) · G. Gomez Mediavilla · F. Allende Álvarez · E. Sáez Pombo
Geography Department, Autónoma University, 28049 Madrid, Spain
e-mail: nieves.lopez@uam.es

G. Gomez Mediavilla
e-mail: gillian.gomez@uam.es

F. Allende Álvarez
e-mail: fernando.allende@uam.es

E. Sáez Pombo
e-mail: ester.saez@uam.es

G. Madrazo García de Lomana
Department of Regional Geographic Analysis and Physical Geography, Complutense University, Madrid, Spain
e-mail: gmadrazo@ghis.ucm.es

I. D. Rotherham (ed.), *Cultural Severance and the Environment*,
Environmental History 2, DOI: 10.1007/978-94-007-6159-9_11,

Our research project integrates different disciplinary and methodological approaches habitually employed in studies on the history of the environment (Rackham 1995; Kirby and Watkins 1998; Agnoletti and Anderson 2000; Métailié et al. 2003; Rotherham 2007), and makes use of converging methods:

- Consultation of different files (local, national, private), and documentary and photographic archives.
- Consultation of available technical plans and interviews with forest managers in order to obtain information on current forest management.
- Field work: inventory and geo-referencing of charcoal furnaces and other heritage-related elements (fences, charcoal burners' trails, canals, etc.), excavation of charcoal furnaces.
- Characterisation of current forest formations and structures by means of stratified floristic inventories and forest characterisation records (100 m^2 plots) (López Estébanez 2003).
- Anthracological analysis: we excavated the charcoal furnaces, following standardised methods (Davasse 1992), in order to identify the forest species that had been charred.

The study area selected (Somosierra-Ayllón forests) is of particular interest from this multidisciplinary perspective (Fig. 11.1). Indeed, we are dealing with a historically very large area of public land presenting very diverse uses, which can clearly be seen in the present rural landscape. It is a Mediterranean mid-mountain environment (from 1,000 to 2,200 m above sea level), lying upon siliceous materials and with a vegetation dominated by formations of Mediterranean oak (*Quercus pyrenaica* Willd.) and pine (*Pinus sylvestris* L.) which have been conserved mainly in the common forests, as well as of scattered birch (*Betula alba* L. and *Betula pendula* Roth.), beech (*Fagus sylvatica* L.) and holly (*Ilex aquifolium* L.). Lastly, the rapid recent transformations (depopulation, increased forest area, abandonment of agricultural and livestock farming uses…) call for studies on the dynamics of vegetation (Sáez Pombo 2000; López Estébanez 2003; Madrazo García de Lomana 2010; Gómez Mendoza et al. 2010).

11.2 Types of Forest Use Over the Last Eight Hundred Years

The ways in which rural societies have used natural resources in the last millennium have shaped the current appearance of forests. Over centuries, there have been different phases in the use and management of these forests, which have affected the latter's appearance and dynamics, and which are diverse according to specific European contexts thus giving rise to time periods that do not necessarily coincide (Agnoletti and Anderson 2002; Clemént 2002; Métailié et al. 2003; Arnould et al. 2003).

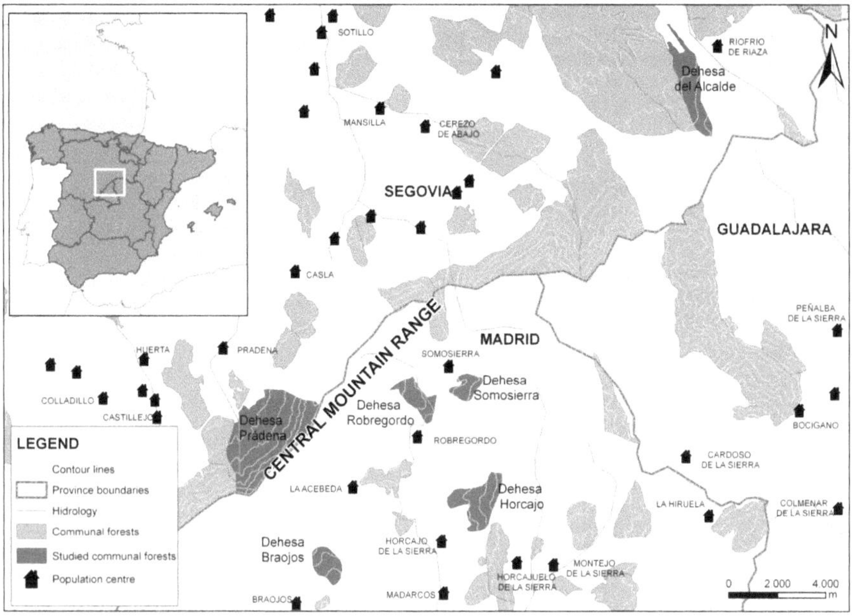

Fig. 11.1 Location of the study area and of the forests studied

Ultimately, to understand current forest dynamics and to appreciate their cultural values, one needs to go back in time. The temporal perspective of this study involves 1,000 years of forest history, which will oblige us to synthesise the phases of the forests, highlighting the points of interest and the specific forces that have played a role at each moment in time in the management and configuration of these landscapes (Fig. 11.2).

From the eleventh century, the Christian Kingdoms reoccupied this part of the Iberian Peninsula (González 1974; Clément 1993), ending two centuries of scant population and little use of resources.[2] Thus, the foundations were laid for a production-based society, which, with certain nuances, remained stable throughout eight centuries. From the eleventh to the twelfth centuries, the municipal governments were established, taking over and managing these territories with all their resources, including forests. The administrative boundaries set up at this time were based upon head of communities, (roughly equivalent to *county towns*), with extensive jurisdiction over all the corresponding villages, lands and forests (Fig. 11.3).

Thus, the jurisdiction of the territory had a notable effect on forestry uses (Sáez Pombo 2000; Madrazo García de Lomana 2010). Basically it would govern

[2] This area was known as *Castile's Extremadura,* which constituted a no-man's-land from the ninth to the eleventh centuries, with little population and which served as a border the Christian Kingdoms to the north of the peninsula and the Muslim ones to the south. But there had been a stable human population in the area (Franco Múgica et al. 1998).

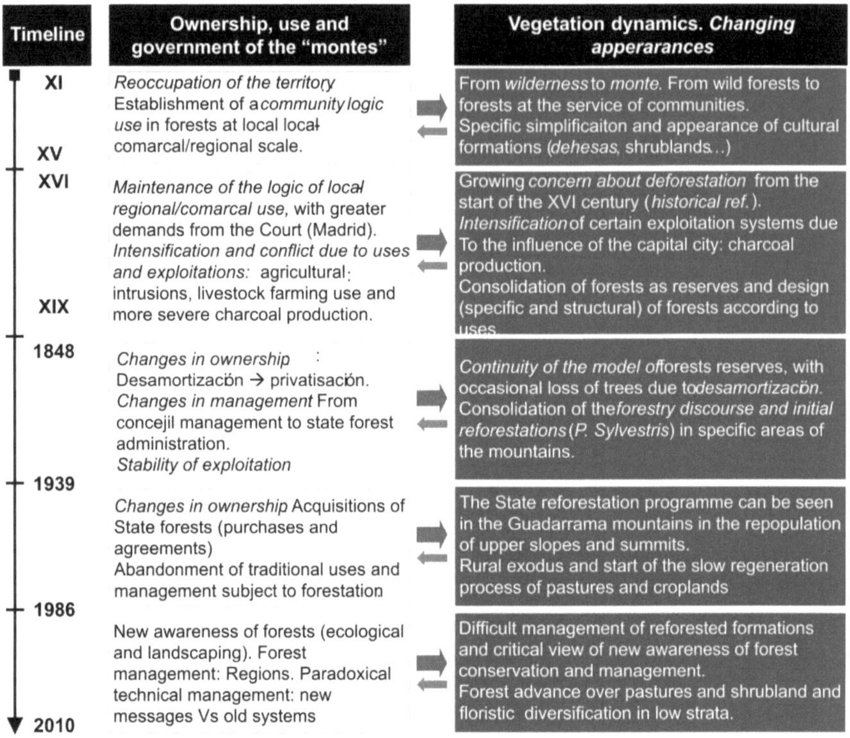

Fig. 11.2 Changes in uses and vegetation in forests of the Guadarrama Mountains (Own sources)

ownership and management of the *montes*, so that management of some forests was transferred to the village, in the form of common lands (*montes concejiles*), whereas others remained under the administration of the communities. The vast majority of the forests therefore remained in public hands, with the exception of small estates and dispersed woods on private fields, and secondly the ancient woodlands were classified into two groups: those controlled by the villages and those controlled by the head of communities.

Common lands (*Concejiles*): With some exception, all villages—regardless of their size—owned an area of forest, ceded by the communities. As owners of this forestland, these villages managed it as they considered fit and defended it from any intrusion of livestock belonging to outsiders and from forbidden uses. This defensive attitude in relation to their forestland heritage reveals two fundamental aspects with regard to characterising the history of these forests:

(1) On one hand, semantic confusion exists, as much of this forestland consists of *dehesas*. This denomination has nothing to do with the kind of forest vegetation, but is rather a simple evolution of the term *deffesa* ("defence" in Latin), referring to the fact that the area is physically demarcated by boundaries.

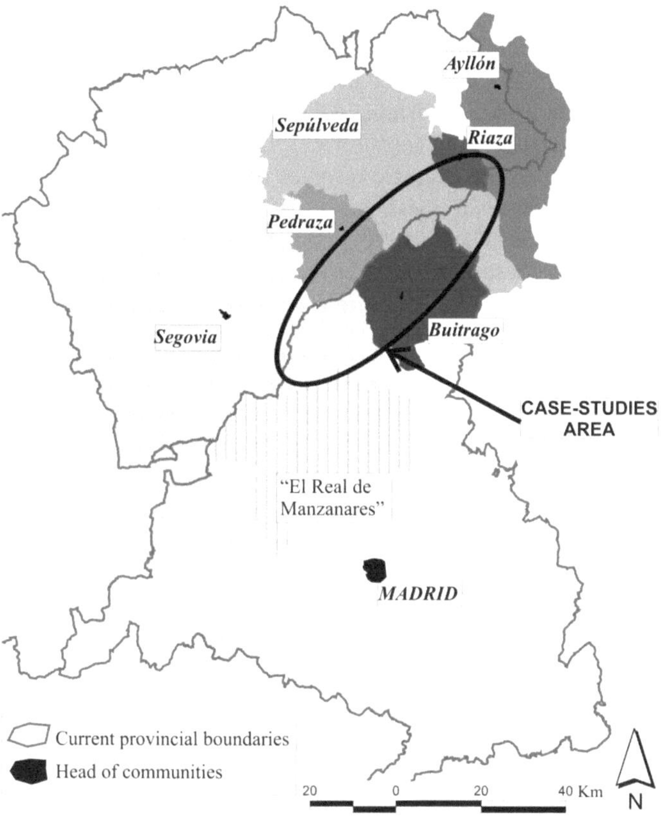

Fig. 11.3 The ancient boundaries of the communities in the Guadarrama Mountains

Thus, these *dehesas* are not always *dehesas* from the point of view of vegetation (open forests), as one is often dealing with shrubland or mixed vegetation (trees and shrubs).

(2) On the other hand, these forests are of great value to local people, due to the fact that they constitute authentic forest reserves, which indicate the importance of protecting and managing them. To this end, some basic rules were established to regulate forestry uses (*charters*), which from the fourteenth to the sixteenth centuries, were extended in Ordinances (*ordenanzas concejiles*), and became more specific and generally stricter (Gómez Mendoza et al. 2009). These common land forests generally constitute a good example of adaptation of specific techniques for management of unique ecosystems (Berkes and Folke 1998).

Community lands: The head of communities continued to own many of the forestland estates in the mountain ranges until well into the nineteenth century, when they were legally revoked. Among the forest lands they owned, we can differentiate two types: pastures, heaths and fields of Spanish broom (*Cytisus)*, which were rapidly deforested, but still called *montes*, and some big forests exploited for timber. The interest of the former lay in the management of these pastures, vital to the livestock farming economy in these mountains (García Sanz 1977), whereas the large pine forests (*P. sylvestris*, L.) in Valsaín, Cabeza de Hierro or Navafría, became strategic reserves of forest products at regional scale. Ancient historic references to these masses of conifers (Manuel Valdés and Rojo Alboreca 1994; Martínez García 1999), along with pollen records (Franco Múgica et al. 1998) identify these masses as *historical pine forests,* differentiating them from other pine forests resulting from reforestation in the twentieth century.

The two models of ownership and administration of forests during the old regime (villages and communities) came to determine the appearance of forests masses. Therefore, considering the forested areas as a whole, one can appreciate a repetition of the distribution patterns (López Estébanez and Sáez Pombo 2002). Thus, the *dehesas concejiles* were usually situated in the vicinity of the villages that owned them, on the lower slopes of the mountains, forming green patches among pastures and croplands. The upper slopes and summits of these mountains belonged to the communities and were mostly deforested, at least as from the fifteenth to the sixteenth centuries (Madrazo García de Lomana 2010). Falling outside this system were the historical dispersed pine forests, totalling approximately 26,000 ha throughout the Mountain Range (Martínez García 1999).

To synthesise, forest landscapes in the Guadarrama Mountains at the end of the Old Regime reflected a livestock farming economy with some agriculture and forest reserves. On one hand, these mountains had long supported a heavy livestock load, although the need for new croplands, arising from population increases, was addressed by occupying common lands. On the other hand, two forces, aimed at protecting forests, attempted to prevent depredation of resources. Firstly, the communities and villages themselves, which, by means of bylaws, promoted rational use of forest resources. Secondly, the Crown, after moving to Madrid (1562), developed an increasingly interventionist and punitive forest policy (Hernando Ortego 2003). The Crown considered it crucial to respect forest boundaries in the surroundings of the Capital, as it relied on forestland for energy and other resources. The inexistence of mineral coal mines in the vicinity of Madrid meant that charcoal became fundamental for energy until the start of the twentieth century, much later than in other European cities (Bernardos et al. 2011).

Halfway through the nineteenth century, the idea began to spread of the need to regenerate forests in Spain, and in the Guadarrama Mountains, this scientific current could be seen in a call for reforestation, founded upon a "forestry vocation" (Gómez Mendoza 1992). These ideas relating to forest management, however, promoted by forestry engineers, would require over one hundred years to take root, with the initiation of an intensive reforestation programme in 1940 (Gómez Mendoza and Mata Olmo 1992).

In the years after 1855, the forests in the Iberian Peninsula suffered a terrible blow with regard to ownership and administration. From 1855 to 1870, most of the land that had belonged to the communities was privatised in a process (*desamortización*) not exempt of opposition by the forestry engineers, who believed that these lands should be public and reforested, and by the communities, who lost ownership of their lands. According to the 1862 Forestry Law, public land covered by pines, oak or beech were not privatised and came to be managed by the public administration through the recently created Forestry Engineering Corps, which meant that the villages ceased to manage them.

But these substantial changes in forest ownership and administration did not immediately affect the physiognomy of forest landscapes. Forests continued to be exploited for one hundred years, and forest area hardly varied, with unchanging vegetation. This lack of change was related to backward agriculture on the Iberian Peninsula: only after 1950 did an exponential increase in inputs (energy, fertilisers, machinery etc.) do away with ancient conditioning factors for agricultural production, thus accelerating the exodus from the country to the city and causing forests to lose their old functionality and intensive use.

In short, one can see that ideas advanced more rapidly than did practice in each territory. This is reflected in the one hundred years that transpired between the spread of the *reforestation* discourse and the definitive transformations in forest cover halfway through the twentieth century.

A reforestation plan was started in the 1940s involving the planting of over 3.3 million hectares from 1940 to 1986 (Gómez Mendoza and Mata Olmo 1992). Prior to the start of this "Forestry crusade", reforestation had already been initiated in high-risk territories (dunes, high-mountain ravines…). Examples exist of this early reforestation in our study area (Lozoya basin), as it contains the headwaters of the rivers that provide Madrid with its water (Sáez Pombo 2000).

But it was between 1940 and 1986, when many of the slopes and summits of the Somosierra and Ayllon ranges were re-populated, after remaining deforested for several centuries. For these reforestations, *Pinus sylvestris* was chosen due to its similarity to the patches of "historical pine forest", although there was occasional testing with other conifers (*Pinus nigra* Arnold, *Pinus uncinata* Mill., *Pinus pinaster* Aiton), some of these exotic (*Pinus ponderosa,* Dougl, *Chamoecyparis lawsoniana* Par, *Cedrus atlántica* Endl.).

Forest planning has changed in the last few decades, due both to the new administrative organisation of the Spanish State and to the acceptation of certain criticism of the reforestation processes under Franco. But in practical terms, discourse and rhetoric apart, one can see that there has hardly been any change in technical interventions in forests (Madrazo García de Lomana 2010). However, present-day silviculture does not fully adjust to the post-industrial discourse (Milbourne et al. 2008) that has emerged, more sensitive to the environmental and cultural values of forests. As we previously stated in relation to the nineteenth century, it seems that discourse precedes reality.

11.3 Characterisation of Forest Vegetation Based Upon Their Exploitation: Charcoal, Livestock Farming and Timber

The different models of forest exploitation in this territory have shaped them in a way that, in many cases, can be seen to the present day. In some of the forests studied, we only found vestiges of the most recent uses, associated with the agro-silvo-pastoral uses from the start of the twentieth century. In others, however, the structure and floristic composition of the canopy enables us to reconstruct the different phases of exploitation.

In order to characterise the different facies that have lasted to the present in these forests, we typified these areas by means of the method of regeneration method which has often occurred over many centuries.

11.4 Exploitation of *Quercus Pyrenaica* Coppices for Charcoal

The most widespread type of forest in this sector of the Central System comprises coppices of *Quercus pyrenaica*, characterised by woodlands with clear signs of degradation caused by long periods of clear cutting in short turns (ten to twenty years). The physiognomy of these coppices and stumps results from intensive exploitation for charcoal (López Estébanez and Sáez Pombo 2002; López Estébanez 2003). This exploitation dates back, in some cases, to 1300 (Fuero de Sepúlveda) (Sáez 1953), when the activity was regulated. After the sixteenth century, the area of these formations is very much associated with demand to supply Madrid, and its royal factories and palaces (Gómez Mendoza et al. 2010).

Based upon the fieldwork conducted in the coppices studied, we drew up detailed maps of the charcoal furnaces that were still recognisable (Fig. 11.4), subsequently analysing some of these. The results obtained from this analysis show that the specific composition of the coppices has varied, and some species found at the base of the furnaces (Braojos 2, Riaza 2) disappeared towards the roofs (*Taxus baccata* L. or *Quercus petraea* subsp. *petraea* Matts. Liebl.), at the same time as the proportional relationship between the dominant species varied (Fig. 11.5). The forests contemporary with the lower level of each charcoal furnace appear to present a multi-stratum canopy, albeit dominated by *Ilex aquifolium* in the case of Braojos 2 and with a noteworthy presence of *Quercus pyrenaica* at Riaza 2. Towards the roofs of these two furnaces, we observed a big change in composition, due to the fact that the possible holly wood in Braojos gave way to masses of *Quercus pyrenaica* with other oak species (*Quercus ilex* Samp. and *Q. petraea* subsp. *petraea*); and the mixed forest of *Quercus pyrenaica*, *Quercus petraea* subsp. *petraea* and *Ilex aquifolium* at Riaza 2, tends to become an oak forest of *Quercus pyrenaica*, accompanied by the remaining species. At present, the

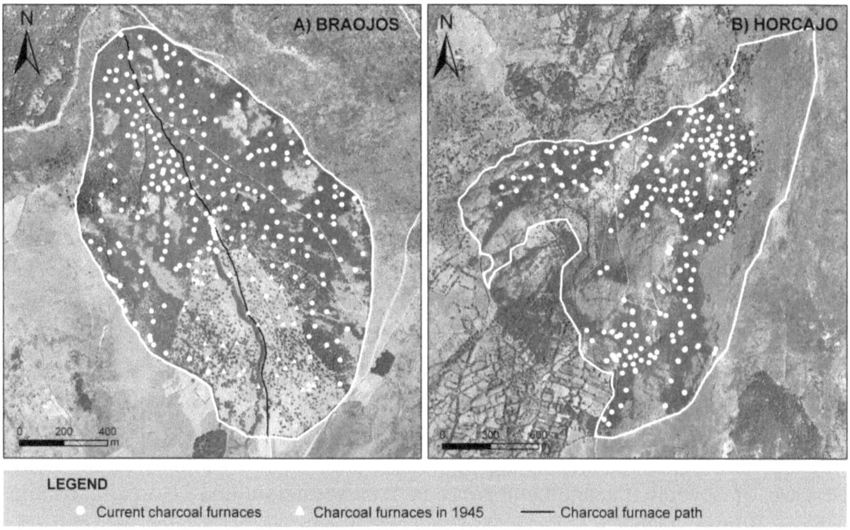

Fig. 11.4 Location of the charcoal furnaces in the Dehesa Boyal de Braojos and Horcajo de la Sierra

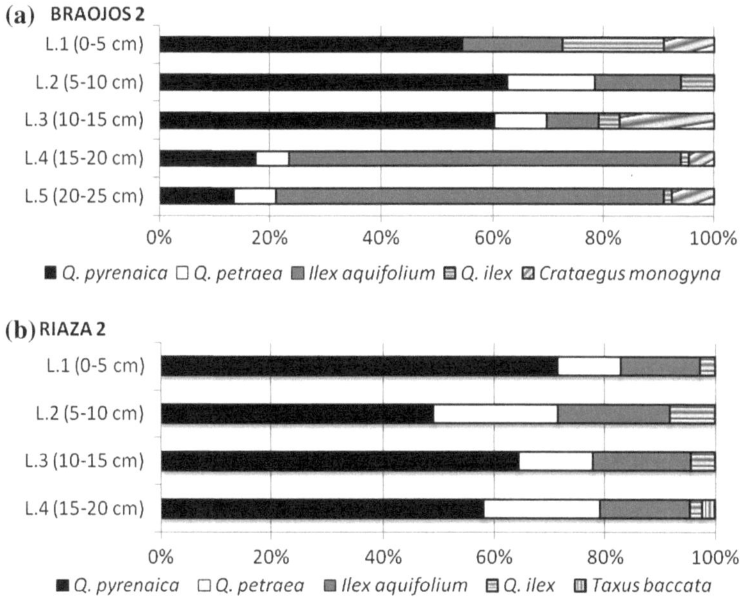

Fig. 11.5 Analysis of the percentage of forest species found in the charcoal furnaces at Braojos 2 and Riaza 2

biogeographic inventories conducted at the two stations sampled (Braojos 2 and Riaza 2) reveal formations of *Quercus pyrenaica*, which is dominant throughout the woody strata.

From the forestry point of view, those coppices are characterised by a dominant tree canopy that, in unfavourable edaphic and topographic conditions, becomes arborescent (1–5 m) (Fig. 11.6a). Furthermore, they are mono-specific formations, dense and contemporary, except for specimens from ancient coppice-with-standards, which remain as unique forest elements within these woods. The lower woody strata of the coppices present a certain degree of diversification, fundamentally of bushes, although species such as *Ilex aquifolium* appear to re-colonise these old woods.

These forests are now quite aged, exhibiting little vertical growth, high density, many dead trees and bushes, a high risk of fire, poor fructification, etc. (Serrada Hierro 2008). But although these formations are much degraded, some authors highlight their high ecological value (protection against erosion, carbon sinks, landscape value) (Rotherham 2007).

More recent interventions in these forests attempt to transform the coppices of *Quercus pyrenaica* into seedling forest, thinning out the coppices in order for the forest to become more vigorous, and even to enable silvo-pastoral use. This is the case of several management areas in the forests studied (Horcajo, Braojos, Dehesa Bonita, Dehesa del Alcalde) where a new forest *facies* of semi-open forest is being developed, which initially would take the form of *old growth on stumps* but which, after a few years, would become seedling forest for possible livestock farming.

11.5 Livestock Farming in the Guadarrama Mountains: *Quercus Pyrenaica Dehesas* and Holly Wood With Oak (*Quercus Pyrenaica*, *Quercus Petraea* Subsp *Petraea*)

11.5.1 The Quercus Pyrenaica: From Forest to Dehesa

Before exploitation for fuel transformed much of the *Quercus pyrenaica* forest in these mountains, livestock faming was the principal use (eleventh to thirteenth centuries) (Sáez Pombo 2000; López Estébanez 2003). Livestock herding gave rise to the configuration of a *dehesa*-type structure maintaining the canopy, comprising oak (*Q. pyrenaica* and *Q. petraea* subsp. *petraea*), mountain ash (*Sorbus aucuparia* L., *Sorbus aria* L.) and holly. From this canopy, wood was extracted, branches for livestock and sticks for firewood, under strict bylaws which, among other matters, established how to prune. These livestock farming and pruning systems are still in practice and are an essential part of forest physiognomy, involving the conservation of old specimens. The reduced livestock load has favoured colonisation in the lower woody strata, where one can observe a tendency towards increased specific richness, with the appearance of new species (*Juniperus thurifera* L., *Juniperus communis* L., *I. aquifolium*).

Fig. 11.6 **a** A *Quercus pyrenaica* coppice originally exploited for charcoal at the Dehesa Boyal in Braojos (Madrid). **b** Old specimens of *Quercus pyrenaica* used for wooden beams at the Dehesa Boyal in Horcajo de la Sierra (Madrid). **c** Homogeneous and compact structure of the holly woods in Prádena de la Sierra (Segovia). **d** View of the Cabeza de Hierro or Los Belgas pine forests, Rascafría (Madrid)

11.6 Other Uses of the Oak Forests: Wooden Beams and Pastures

In some sectors of the oak forests we identified scattered stands of *Quercus pyrenaica* which, at certain time in history, were used for the production of wooden beams for house building while their seedling forest structure also facilitated livestock farming under the canopies (Fig. 11.6b). The forest inventories present intense progressive dynamics in the lower wood strata, with specimens of *Taxus baccata* and *Ilex aquifolium* in the arborescent stratum (1–5 m), whereas in the bush and under-shrub strata, we can highlight *Quercus pyrenaica*, with up to 60 % dominance and *Crataegus monogyna* Jacq. Along with *Quercus pyrenaica*, other unique forest species used for this activity were beech (*Fagus sylvatica*) or Spanish juniper (*Juniperus thurifera*).

11.7 The Holly Woods: Recent Expansion of Anthropic Forests

Both on the northern slopes of the Mountains (Prádena de la Sierra) and on the southern ones (Robregordo), estates have been conserved whose dominant formation is the holly woods (*I. aquifolium*). These forests, whose ecological extent reaches the centre of the Iberian Peninsula (Oria de Rueda 2003), appear to have been more abundant in these mountains in past times (López Estébanez 2003, 2006). The uses that shaped these formations are fundamentally browsing for winter feeding for livestock and coppicing, which has generated a structure of very dense stands with covers reaching 100 %. Within these woods, there are old specimens of *Quercus pyrenaica*, *Quercus petraea* subsp. *petraea*, *Sorbus aria* and *Sorbus aucuparia*, remnants of what were likely the original forests. These old deciduous forests gave way to *Ilex aquifolium*, partly due to anthropic action, intended to favour this species which covered the needs of winter livestock feeding (browsing), pastures and refuge within these curious coppices (Fig. 11.6c).

The extremely high density of this holly forest, which re-sprouted from cut stumps, has generated a very simple structure: a tree or arborescent stratum dominated by *Ilex aquifolium*, with some of the above mentioned deciduous species and some holly seedlings in the under-shrub woody stratum, all of which demonstrates the extreme floristic poverty of such a dense *facies*.

11.8 Guadarrama's Historical Pine Forests for Forestry Use

As already noted, historical formations of *Pinus sylvestris* have been conserved, although their use for timber production since the end of the nineteenth century accounts for their current physiognomy to a great extent. This is the case of the pine forest in Navafría (northern slope) or of Cabeza de Hierro (southern slope). The former has belonged to the community of Pedraza since the Middle Ages, and since 1895, a forest management project has regulated the arrangement and structure of tree ages (García López 1994). The latter, which belonged to community of Segovia, and since the seventeenth century, to the El Paular Monastery, was acquired in 1840, during the ecclesiastic *desamortización*, by the *Compañía Belga de Pinares* (Belgian Pine Forest Company). Its exploitation has undergone several stages since then, selective cutting currently being imposed, as a result of its lesser environmental impact, as this pine forest is located in an area of outstanding natural value (Rojo Alboreca et al. 2001).

The present physiognomy of the forests results from their management system, based upon natural regeneration and extraction of the oldest individuals, individually selected; all of which favours a multi-stratum structure in which *Pinus sylvestris* is the dominant species, although in other sectors, *Quercus pyrenaica* and *Ilex aquifolium* become established, favoured by the shade of the pine forest (Fig. 11.6d).

11.9 Conclusions

In Somosierra and Ayllon, as in the rest of Spain, traditional uses remained in force until halfway through the twentieth century. In spite of the fact that changes occurred in the ownership and management of these forests, the type of use remained stable up to 1940. This simple fact leads us to two conclusions: firstly, the succession of types of uses of forests has been very slow in comparison with other European territories (Kirby and Watkins 1998; Tsouvalis 2000; Agnoletti and Anderson 2002; Métailié et al. 2003; Arnould et al. 2003); secondly, all this, together with the slow growth rate of Mediterranean forests, led to the permanence of inherited forest physiognomies and structures which had disappeared from other parts of Europe (Rotherham 2007). We believe that the natural and, above all, cultural values of these forests have not been fully appreciated. Indeed, the technical prescriptions of current managers consider these formations to be dysfunctional and are therefore promoting the transformation of these structures (coppices to seedling forests) to accommodate new uses (livestock farming, reforestation, recreation…).

As a result of our intense efforts with files, fieldwork and lab work, we have been able to confirm that, throughout the long period of traditional exploitation of these forests, their composition has varied, with impoverishment of the tree canopy and modification of the dominant tree formations. The relevance of uses in the different types of forest has given rise to our exhaustive study aimed at characterising these forgotten forestry uses. As we have already stated, charcoal production has constituted an important element shaping the forest landscapes of these mountains, and as proof of the significance of these uses, we have developed an accurate cartography of the charcoal furnaces distributed throughout this area (Fig. 11.3); these present a higher density than in other European mountain territories (Gómez Mendoza et al. 2010).

Will these forest landscapes disappear? Will the absence of uses bring about their extinction? Should managers and technical experts create instruments for the conservation of these cultural forests? Indeed, this type of forests has been disappearing over the last 40 years and along with it any traces of their traditional exploitation. In view of what is occurring in other parts of Europe, where, due to their scarcity, these formations are being increasingly valued, perhaps we should start to pay greater attention to the value and uniqueness of these forests on the Iberian Peninsula.

References

Agnoletti M, Anderson S (eds) (2000) Methods and Approaches in Forest History. CAB International Publishing, Wallingford, Oxon

Agnoletti M, Anderson S (eds) (2002) Forest history: international studies on socioeconomic and forest ecosystem change. CAB International Publishing, Wallingford, Oxon

Arnould P, Marty P, Simon L (2003) Bosques, paisaje, Medio Ambiente. Dos siglos de mercantilización y reconquista forestal en Francia, 1800–2000, In: Amarilla JA, Uriarte Ayo R (eds) Historia y economía del bosque en la Europa del sur (siglos XVIII-XX), Prensas Universitarias de Zaragoza, Zaragoza, pp 89–120

Berkes F, Folke C (1998) Linking social and ecological systems management practices and social mechanisms for building resilience. Cambridge University Press, Cambridge

Bernardos J, Hernando J, Madrazo G, Nieto J (2011) Energy consumption in Madrid, 1561 to c. 1860. In: Massard-Guilbaud G, Mosley S (eds.) Common ground. Integrating the social and environmental in history, Cambridge Scholars Publishing, Cambridge, pp 316–339

Clément V (1993) Frontière, reconquête et mutation des paysages végétaux entre Duero et Système Central du XI° au milieu du XV° siècle. Mélanges de la Casa Velázquez (MCV) XXIX (1):87–126

Clément V (2002) De la marche-frontière au pays-des-bois: forêts, sociétés paysannes et territoires en Vielle-Castille (XI-XX siècle). Casa Velázquez, Madrid

Davasse B (1992) Aspects méthodologiques de l'anthraco-analyse des charbonnières—Historie des forêts de la vallée de soulcem (Pyrénées ariégeoises, France), Protoindustries et histories des forêts, GDR ISARD-CNRS, 3:207–221

Franco Múgica F, García Antón M, Sainz Ollero H (1998) Vegetation dynamics and human impact in the Sierra de Guadarrama, Central System, Spain. The Holocene 8:69–82

García López JM (1994) Fabriqueros el carboneo tradicional en la sierra de Guadarrama. Montes 38:12–16

García Sanz A (1977) Desarrollo y crisis del Antiguo Régimen en Castilla la Vieja. Economía y sociedad en tierras de Segovia 1500–1814. Akal, Madrid

Gómez Mendoza J (1992) Ciencia y política de los montes españoles. ICONA, Madrid

Gómez Mendoza J, Mata Olmo R (1992) Actuaciones forestales públicos desde 1940. Objetivos, criterios y resultados. Agricultura y Sociedad 65:15–64
Gómez Mendoza J, Madrazo García de Lomana G, Saez Pombo E (2009) Dinámica histórica de paisajes forestales y sostenibilidad, in Sextas Jornadas científicas del Parque Natural de Peñalara y del Valle de El Paular. Mayo 2007 5:21–36
Gómez Mendoza J, Gomez Mediavilla G, López Estébanez N, Madrazo García de Lomana G, Sáez Pombo E (2010) Aprovechamientos y dinámicas en los tallares de rebollo de Somosierra-Ayllón. Cuadernos de la Sociedad Española de Ciencias Forestales, 30:247–254
González J (1974) La Extremadura castellana al mediar el siglo XIII. Hispania 127:265–424
Hernando Ortego J (2003) Poder y usos del espacio: la construcción del monte de El Pardo durante el Antiguo Régimen. In: Alcutén Sabio, Goñi Iriarte (eds) La construcción histórica del paisaje agrario en España y Cuba. Catarata, Madrid, pp 131–146
Kirby KJ, Watkins C (eds) (1998) The ecological history of European forests. CAB International Publishing, Wallingford, Oxon
López Estébanez N (2003) Dinámica de la vegetación en relación con los cambios de uso. Universidad Autónoma de Madrid, Tesis doctoral
López Estébanez N (2006) Dinámicas forestales sobre matorrales: *Juniperus thurifera* L., *Pinus sylvestris* L. y *Pinus pinaster* Aiton en el sector oriental del Sistema Central madrileño. Serie Geográfica 13:25–41
López Estébanez N, Sáez Pombo E (2002) Gestión, aprovechamiento y paisaje de las dehesas de Guadarrama y Somosierra (Madrid). Ería 58:230–245
García Madrazo, de Lomana G (2010) La evolución del paisaje forestal en la vertiente segoviana de la Sierra de Guadarrama. Consejería de Educación y Cultura, Valladolid
Manuel Valdés CM, Rojo Alboreca A (1994) Valsaín forest in the XVIII century: an example of forest management in the pre-indutrial era. Investigación Agraria. Sistemas y Recursos Forestales 3:217–229
Martínez García F (1999) *Los bosques de Pinus sylvestris* L. *del Sistema Central español. Distribución, historia, composición florística y tipológica.* Ministerio de Agricultura, Pesca y Alimentación, Madrid
Métailié JP, Bonhôte J, Davasse B, Dubois C, Galop D, Izard V (2003) La construcción del paisaje forestal en los Pirineos orientales, del Neolítico a nuestros días. Un modelo cronológico del bosque en el largo plazo. In: Amarilla JA, Uriarte Ayo R (eds) Historia y economía del bosque en la Europa del sur (siglos XVIII–XX), Prensas Universitarias de Zaragoza, Zaragoza, pp 15–38
Milbourne P, Marsden T, Kitchen L (2008) Scaling post-industrial forestry: the complex implementation of national forestry regimes in the southern valleys of Wales. Antipode 40(4):612–631
Oria de Rueda JA (2003) Las acebedas de España. Aproximación al origen, dinámica y conservación de las mismas. Tesis doctoral, ETSI Montes, Universidad Politécnica de Madrid
Rackham O (1995) Trees and woodland in the British landscape. The complete history of Britain's trees, woods and hedgerows. Weidenfeld and Nicolson, Londres
Rojo Alboreca A, Montero G, Pérez-Antelo A (2001) Cuarenta años de ordenación del monte "Cabeza de Hierro" (Rascafría, Madrid): desarrollo de la ordenación resultados conseguidos. Cuadernos de la Sociedad Española de Ciencias Forestales 11:23–34
Rotherham ID (2007) The implications of perceptions and cultural knowledge loss for the management of wooded landscapes: a UK case-study. For Ecol Manage 249:100–115
Sáez E, Dir (1953) Los fueros de Sepúlveda. Diputación Provincial de Segovia, Segovia
Sáez Pombo E (2000) Montes públicos, territorio y evolución del paisaje de la Sierra Norte de Madrid. Universidad Autónoma de Madrid Comunidad de Madrid y Obra Social Caja Madrid, Madrid
Serrada Hierro R (2008) La conversión de los rebollares en la Comunidad de Madrid. XII Seminario de conservación de la naturaleza, CIAM-Fernando González Bernáldez
Tsouvalis J (2000) A critical Geography of Britain's state forests. Oxford University Press, Oxford

Chapter 12
Of Commoners and Kings

Graham Bathe

12.1 Introduction

This chapter examines the relationship between Royal Forests and commons, exploring the protection afforded to commoners, and examining long-term landscape change associated with commoning. It draws on evidence relating to commons in Southern England, focusing on Savernake Forest in Wiltshire.

12.2 English Commons Since the Norman Conquest

English commons are areas where certain people hold rights to take specified produce from land belonging to somebody else. These rights may include de-pasturing animals, taking sticks, small-wood, fern and heather, cutting peat for fuel, turning out pigs, catching fish and certain wild animals, or extracting minerals. Hence, *by definition*, commons cannot be owned by those with rights, the commoners, setting England apart from many European countries. This arrangement has persisted since at least 1066, when William the Conqueror seized the throne and expropriated all land, establishing himself as feudal overlord. The King ultimately owned every parcel of land. Productive land was distributed amongst Norman lords and held in exchange for military or other services. Within the manorial system, these became the Lords of the Manor. Whilst *all* land was seized, including commons, this did not prevent on-going commoning. Propertied Englishmen were replaced by a Norman elite, but there is little evidence of the eviction of commoners

G. Bathe (✉)
Director, Foundation for Common Land, Byeley in Densome, Woodgreen, Fordingbridge
SP6 2QU Hampshire, UK
e-mail: gmbathe@aol.com

I. D. Rotherham (ed.), *Cultural Severance and the Environment*,
Environmental History 2, DOI: 10.1007/978-94-007-6159-9_12,

or working populace. However, the universality of the Norman system ensured that commoners could never own their commons. They simply had rights on land held by the Lord of the Manor from the King.

12.3 Common Land Before 1066

In early English history the ownership of commons may have been more closely vested in the communities they served, under the purview of a local chief, although the extent of commons prior to the Conquest is difficult to establish. There are allusions to shared pastures within land charters of the eighth to the eleventh centuries. The Laws of Wessex (which predate the kingdom of England) cite penalties for destroying enclosures which could relate to commons, whilst the Domesday Book of 1086 describes large, apparently communal pastures. Like most cultures throughout history, local communities derived their needs from a complex and fluctuating mixture of shared and exclusive resources, involving pastoral, hunted and harvested products. Societies established rules governing the extraction of goods, defining what was legitimate, and generating the notion of defendable 'rights'. Custom and practice was not so much the basis of law, but *was* the law. In due course a legal system based on precedent generated Common Law, and statutes were enacted through Parliament. Whilst some of England's oldest legislation relates to common land, it is certain that commons pre-date both Parliament and legislation.

12.4 The Manorial and Forest Systems

Most commons were managed within the manorial system. Under this, land was divided into a series of more-or-less sustainable, self-governing settlements or manors. The Lord held the manor in exchange for performing services, such as providing military duties (knight-service), specified responsibilities (serjeanty), or rent (socage). Apart from the Lord's demesne land (the home or manor farm), managed for the specific benefit of the Lord, much of the arable, pasture, waste, and sometimes the meadow and woodland, were managed communally.

It was the more productive land that was distributed by Norman Kings to feudal Lords, and where the manorial system continued to develop from its Saxon progenitors. However, this left the King with extensive infertile holdings, incapable of maintaining local communities. A land management regime providing an alternative to manorialism was the Forest system.

12.5 Forest Law

The popular image of Royal Forests— as woodland set aside for hunting—is inaccurate, or at least severely overstated, on both counts. The Forests were specifically areas protected through *Forest Law*, but the designation reveals nothing about the habitats included. In modern times, during attempts to generate income on infertile Crown lands, many Forests were planted with trees, perpetuating the impression that they derived from closed-canopy woodland. It is likely that many Forests were not cloaked in trees, and, for much of their history, could have had open or park-like landscapes. Similarly, whilst the link between hunting and Forest Law is certain, the number and area of Forests grew so substantially in the twelfth century, that a wider function is implied. At its peak there may have been 300 Forests in England and Wales. During the century and a half after the Norman Conquest, the bounds of existing Forests were radically extended. For example, the area of Savernake Forest (i.e. the area subject to Forest Law) expanded some seven-fold to 250 km^2, whilst there is little evidence that the habitat altered at all. At their most extensive, Royal Forests covered over a quarter of the country (Fig. 12.1) (Langton and Jones 2005). No King could possibly visit these in a lifetime, let alone hunt them. Hence a more sophisticated interpretation is needed to explain their purpose.

Fig.12.1 Forests and Chases of England and Wales (after Langton and Jones 2005)

12.6 The Introduction of Forest Law to England

Whilst hunting was enjoyed by the Anglo-Saxon and Danish Kings, who could sport in the great unfarmed tracts of land remaining, the concept of *Royal Forests*, subjected to unique laws, was unknown in England prior to the Conquest. The Normans were probably able to import a mature system from their homelands, with its complex, ritualistic judiciary and administration. This may have had its antecedent in Frankish institutions which secured the protection of game for the exclusive privilege of the King. It is estimated that some 40 % of the Carolingian landscape had been uncultivated, with large areas in Royal hands maintained as hunting reserves (Verhulst 2002). The laws or '*capitularies*' of the Carolingian dynasty relating to the *Forestis*, have close parallels with the later English Assizes of the Forest (Petit-Dutaillis 1915).

Whilst surviving documentation is sparse, eleventh century charters relating to Evereux and Cherbourg confirm that the future King William held Forests in Normandy prior to his invasion of England (Fauroux 1961). Evidence from the early twelfth century, when continental sources are more plentiful, demonstrates that Forests on both sides of the channel were governed under comparable legislative regimes (Petit-Dutaillis 1915). Forest Law was imposed on large swathes of the English countryside in early Norman times. Hunting grounds recognised during the Anglo-Saxon era, such as the Forest of Galtres, formed the basis for certain Royal Forests. In other cases, such as the New Forest established in 1079, extensive heaths and woodland held by the King were newly incorporated within Forest Law. By the time of Domesday in 1086, some twenty-five Forests were already listed.

12.7 The Purpose of the Forest

Whilst the protection of game, especially on the King's own land, constituted its conceptual rationale, Forest Law became equally important for other functions. In particular, it enabled the King to generate revenues on the unproductive land retained in his hands. It strengthened the monarch's authority, displaying prestige and social exclusivity, also conferred on those currently in Royal favour. These functions increased significantly when Forest Law was extended to cover land beyond the King's demesne, ostensibly held by others. The King could demand payments for enclosures and farmland, damaging the vert (trees and shrubs) or wasting woods. Sums charged for depasturing stock, or as a toll for passage, could collectively generate significant income. Where draconian penalties existed, involving loss of life and limb, or maiming, for serious misdemeanours like taking game, these were commuted to large fines inevitably preferred by those convicted. After expanding the area within Forest Law (at its most absurd incorporating the major part of whole counties like Devon and Essex) the King could then seek huge sums for removing the same law ('*disafforestation*'). In the twelfth century Forests accounted for half of the King's revenue (Young 1979). Forests were managed within such a powerful institution that several Kings used them for bargaining, offering to reduce Forest area or severity of the law, in exchange for agreement on national taxation.

From their association with Forests many early Kings of England were despised, a feeling captured in folk tales like Robin Hood, which resonates into modern times. The early Plantagenets in particular manipulated and exploited Forest Law as a tool of wanton self interest. The protection of vert and venison became pretexts for extortion, a practice which extended from the monarch to local officials. No opportunity was lost to raise revenue from the Forests, and extend them, pushing the tolerance of subjects to breaking point. The Forests expanded beyond their original function of managing game. It was the landowning aristocracy who suffered worst. However, commoning may have continued largely untouched through this period, even enjoying a measure of protection not present outside Forests.

12.8 Laws Relating to the Commons

Some of England's oldest legislation concerns commons. The *Statutes of Merton and Westminster* of 1235 and 1285 (the latter of which persisted into the twenty-first century), confirmed the right of '*approvement*'. This enabled Lords of the Manor to inclose and remove from commons any land surplus to commoners' requirements. The Acts reveal certain underlying issues. First, the legislation stated that Lords inclosing '*wastes woods and pastures*' had been hindered by commoners. Hence it confirmed that Lords could manage such land beneficially, settling all doubt about property rights. Second, the Acts assert the rights of commoners, stating emphatically that they must not be prejudiced. Lords were required to save '*sufficient pasture for their tenants and neighbours … with free egress and regress to the same*' and could inclose only the residue. Hence commoners held real protection through law. Third, the legislation enabled land to be set aside for woodland and coppice establishment, at a time when most woodland was grazed, and there were concerns about wood supplies. Many coppices established in late medieval times stem from close to this time. Despite their revealing content, the statutes of Merton and of Westminster would have had no application within Forests, since they had no Lord of the Manor.

Even these Acts included no new provisions, and simply confirmed existing customary procedures. The recognition that commoners held lawful property rights in the thirteenth century was re-affirming through statute something long recognised in practice. Throughout medieval times commons were managed in accordance with custom and practice. Local Courts, focused either on the Manors or Forests, provided a meeting place of commoners and the Lord's or King's representatives. They established rules and byelaws, oversaw rights of tenants, and administered fines on miscreants, either as punishment or sanctioned, chargeable licence to exceed accepted practice. Such laws were often unwritten, although occasionally attempts were made record traditions as a '*custumal*' (customs of the Manor or holding). Despite their local and often unwritten nature, custom and practice carried the weight of law.

Hence commons were managed in accordance with societal customs, recognized and enforced by judgments in courts. Where local decision making was challenged, County or Hundred Courts could determine issues. Local law was increasingly subsumed within *Common Law* from the late twelfth century, founded on precedent, and Parliament could also apply legislation universally across commons where appropriate. Common lands and rights hence exemplify the transition whereby ancient laws derived from custom, were recognised in courts under Common Law, and reflected or modified in national statute. However, custom and practice remained central to the administration of common rights until modern times.

Even within Royal Forests, where it might be anticipated the King's prerogative would override commoners' interests, their rights were protected in law. The *Carta de Foresta* (Charter of the Forests) of 1217 makes explicit reference to commoning. The Charter had been drafted alongside the *Magna Carta*, and issued after King John's death, when his son Henry III was a child. Although withdrawn temporarily when he became of age, it was re-asserted at various times by Henry, Edward I and Edward II in exchange for agreements on taxation. Article 1 of the *Carta de Foresta* states: '*All those woods which belong to the King, which have been afforested* [brought into Forest Law], *shall remain forest, saving the common usage of pasture and other matters, to those who were accustomed to the same*'. The Forests did introduce minor controls on commoners. For example animals had to be removed in the '*fence-month*' when deer give birth. But overall the scale of protection afforded commoning was considerable.

The barrister John Manwood (1615), who was a Justice in Eyre of the New Forest, produced the first book on Forest Law in 1592, which was much expanded and revised as *A Treatise of the Lawes of the Forest*, re-published in 1615 after his death. Drawing heavily on the *Carta de Foresta* and other proclamations, he emphasised the rights of commoners. He states, '*where the King doth afforest his own lands, he doth not by such afforestation, prejudice any man to have common in the same that used, or by right ought to have common therein, but does still reserve the common of herbage, as it was at the Common Law*'. The rights in the *Carta de Foresta* were re-asserted when the *Ordinatio Forestae* was passed in 1305. Manwood claimed that geese, goats, sheep and swine were '*always forbidden*' in Forests, although this is countered by experience elsewhere. For example, sheep were the most abundant commoners' animals mentioned in Savernake Forest courts, and they remain commonable in the New Forest in the twenty-first century. Conversely, the uncertain legitimacy of sheep in the Forest of Dean, led to a standoff between local graziers (called '*badgers*') and Forest officials, which has lasted for centuries.

12.9 The Royal Forest of Savernake

Savernake is one of seven Royal Forests in Wiltshire established by the twelfth century (Fig. 12.2). It is not mentioned in Domesday, and was presumably established between 1086 and 1130 when recorded in a stray surviving Pipe Roll. Savernake exhibits features characteristic of many Royal Forests:

i. It was a large, untenanted Royal holding, outside manorial regimes.
ii. Afforestation enabled beneficial use of uncultivatable land.
iii. It offered recreational facilities for monarchs visiting a nearby castle (Marlborough).
iv. It supplied the castle with timber and commodities, whilst the castle provided a gaol for Forest miscreants.
v. It abounded with game, and was possibly associated with an existing Saxon hunting estate.
vi. It was ideal for introducing Fallow Deer, provisioning larders alongside hunting opportunities.

Fig.12.2 Wiltshire Forestic c1200

12.10 A Landscape of Rights, Privileges and Obligations

Savernake, like other commonable lands, provided different functions for the right-holders and other interests. The former enjoyed agricultural and domestic benefits, especially pasturage for their animals. Lopped or pollarded trees supplied browse and nutrition even outside the growing season. Heath and fern were used for animal bedding and thatching. Coppice poles were used for hurdle making, whilst faggots of furze and thorn, together with deadwood, provided domestic fuel. In contrast, the King derived benefit from the deer and prestige associated with them, constructional timber and sale of wood products, including coppice and bark for tanning. The land-holder's interests suffered from the activities of the borderers (tenants bordering the Forest who had common rights), whilst they, in contrast, suffered deer, which they were not entitled to exclude, competing for browse and damaging crops. Although acorns and beech-mast provided fodder for pigs, any spreading trees could be a nuisance to commoners by shading pasture, whilst they were denied timber. In most cases, borderers were forbidden from taking oak or 'great trees' even on their own holdings.

12.11 Reassertion of Rights

In some cases rights had to be regularly reasserted to be retained. A ceremony at Great Wishford in Wiltshire each Oakapple Day (May 29) reflects this. Villagers are awoken before dawn by a noisy crowd urging people to go into Groveley Forest, and collect oak boughs. These are used to decorate the church and village, and then the villagers carrying their sticks, process to Salisbury Cathedral, where '*knitch*' ladies dance (Fig. 12.3). In the cathedral they read from a 1603 charter asserting rights to collect boughs and their ceremonial traditions in Salisbury, and then they

Fig.12.3 Great Wishford assertion of rights at Salisbury Cathedral

shout '*Groveley, Goveley, Groveley and all Groveley*'. Whilst Great Wishford stories may have been embellished through the years, there are records of other rights lost when not re-asserted. In Savernake in 1337, tenants of Marlborough, Manton, Elcot, Preshute, Wick and Wootton Rivers petitioned the King for restitution of rights, after they were extinguished when commoners failed to come and assert them at a Forest Eyre (PRO SC 8/63/3118). As stated in a legal case of 1607 '*Common is obtained by long sufferance and may also be lost by long negligence*' (Hart 2002).

12.12 Traditional Self-Regulation of the Commons

Within Forests, unlike manorial courts, there was a nationally imposed, sophisticated legal system designed to protect Royal interests and generate income. In its early years, the Forest Eyre, which sat every few years, could impose horrific penalties for taking venison, including execution, amputation, or maiming. In practice, such penalties were nearly always reduced to fines. Whilst the protection of deer was a national affair, the commons continued to be managed locally. In sixteenth-century Savernake, the commons were administered through courts known as Forest Views, each producing a written Certificate of Remembrance. The court records are littered with misdemeanours, and attempts to address them. There are accounts of people with no common rights turning-out stock; exceeding specified numbers; putting out more animals than they could winter on their own holding (the principle of '*levancy and couchancy*'); using incorrect livestock; de-pasturing unauthorised areas, or during unauthorised times including the 'fence month'; occupying coppices closed for regrowth; having pigs un-ringed; animals un-marked; or de-pasturing stock for neighbours. Yet despite voluminous court records, the commons of Savernake persisted for centuries. The records demonstrate checks and balances, with controls on miscreants. Overall, the commons provided an economic mainstay for the community.

12.13 Tragedies of the Commons

Some 40 years after the phrase was first used (Hardin 1968), there is on-going debate whether '*the tragedy of the commons*', i.e. an inbuilt course of destruction when resources are shared, remains a valid concept. There is no evidence of self-imposed tragedy on the Savernake commons. Indeed, in Wiltshire, when commoners were evicted from Forests, there is a more demonstrable '*tragedy of inclosures*', as landless farmers were driven into poverty and desperation. John Aubrey, travelling through Wiltshire in the seventeenth century, remarked following removal of Forest Law: '*… the cry of the poor was … lamentable. I knew several that remembered keeping a cow for 4d per annum. The rule was, however* *many they could winter they might summer, and pigs cost nothing. Now travellers are encumbered with beggars*' (Jackson 1862).

In Savernake, commons survived over 600 years, possibly much longer (since it is easier to map their demise than their origins). There was no tragedy of rural economy. Governance systems were adequate to ensure that commons were managed to the ongoing benefit of commoners. However, commons were nevertheless undergoing massive, gradual change.

Despite its scale, there was a desperate paucity of timber in Tudor Savernake. The local picture mirrored many national and European situations, with particular concerns about the dearth (and hence cost) of naval and shipbuilding supplies. In Savernake, the shortfall affected constructional timber for buildings. The Forest was a mosaic of heath, bracken, furze, grassland and scrub, with deer, sheep, pigs and possibly cattle each numbering several hundred, and smaller numbers of horses. There were large numbers of trees, scattered or in groups. However, many were unsuitable for timber, being young or pioneer species like birch, hawthorn and maple, or felled before maturity for fodder or faggots. There were ancient giant oaks, contorted, hollow and decayed, ill-suited for timber, and other trees pollarded or shrouded (Fig. 12.4). Pollarding involved periodically cutting the trunk well above ground, inducing dense shoot growth beyond the reach of browsing animals. Pollard trunks were typically short. Shrouding (or shredding) involved lopping back to an intact trunk. Both practices produced buds and twigs provided as fodder to livestock and deer, with firewood and short timber. Although these were widespread practices, they were increasingly seen as symptomatic of a profligate system, beneficial to commoners only, depriving estates of quality timber (Petit and Watkins 2003).

Timber shortfalls stimulated legislation to protect woodlands throughout Europe. In England, initial controls focused on Royal Forests. An Act from 1483 extended the period that felled woodlands were enclosed (to exclude animals) to 7 years (Tubbs 1964). Through the *Statute for the Preservation of Woods* in 1544

Fig.12.4 Measuring the cathedral oak, Savernake

timber shortfalls were addressed nationally. The introduction describes the parlous situation: '*The King perceiving great decay of timber and woods, so that unless speedy remedy be provided, there is great likeli-hood of scarcity of timber for houses, ships for the whole community.*' The Act required that coppices retain 12 uncut standard trees per acre, not felled until 10 inches diameter. Young coppices had to be enclosed to exclude grazing animals for 4 years; those over 14 years for six, and older coppices for 7 years. An Act of 1558 extended the closure to 9 years. Once standard trees were a high enough, they were beyond the reach of deer and livestock, and could continue unhindered. These statutes encouraged coppice-with-standards woodland, with great trees forming an intermittent canopy over coppiced hazel and other poles.

12.14 The Private Forest

From the fourteenth to the sixteenth centuries the Forest system broke down, as its revenue-raising significance was eclipsed by national taxation. Land was removed from Forest Law, and many Forests came into private hands. In the same way that the establishment of Forests had not interfered with the rights of commoners, the transfer of Forests into private property was limited in its effect. The commoning communities and their governance systems remained intact. However, the different objectives of landowners and commoners were brought into sharper relief.

In 1553, Savernake was transferred from the King to the Earl of Hertford. The interests of the landowner and borderers were not wholly compatible, generating antagonism between lord and peasant. This antipathy, which persisted for centuries, was described in the House of Commons Journal as '*a perpetual struggle of jarring interests, in which no party can improve his own share without hurting that of others*' (House of Commons 1788). At Savernake opposing interests caused conflict which erupted into threats and brinkmanship.

The paucity of trees on Hertford's extensive estate became apparent when he resolved to build a new mansion in the 1570s. He enquired about his best timber. His steward, Sir John Thynne, gave stark news: '*Concerning your* [sources of] *timber for building, you have none. Your Lordship must now reserve timber in all places*' (Longleat House: Seymour Papers). Despite owning Savernake Forest and a vast estate of thousands of hectares, Hertford was in the absurd position of having to purchase 640 oaks for building his new mansion (WSA 9-26-512).

Hertford resolved to investigate options for restoring his woods. This was unachievable without evicting commoners from part of the Forest, and he sought legal advice concerning "*the lawfulness of his Right to Inclose the Coppices of Savernack*". The instructions speak of "*the lamentation of the people for the great decay and spoils of woods now that much wood ground is converted into pasture as at Southgrove, the Broyle, and Clatford* [Savernake], *with Great Woods cutt downe, at Chute forest, Fortesbury, Stockwood and many other places, and his own Great Wood in Savernack very leere* [empty] *and old, with underwood much decayed*". The Earl commanded his officers to identify causes. They reported "*the Red and Fallow deer are great annoyances and destroy the underwood. The borderers not only oppress the Forest with more cattle than formerly, but many connive to bring cattle from foreign townships* [without rights]. *They also pasture old oxen and cattle without hetching* [tethering?] *and spoil the first spring-growth of wood and coppice, which were formerly preserved*" (WSA 1300-104).

In 1594, Hertford instructed his officers to inclose a portion of the Forest "*which hath ever been coppice ground*", citing Henry VIII's statute for preserving woods. The inclosure was to be ditched and hedged with "*quickplants*". The commoners' capacity to undermine the venture was recognised, and they were assembled and asked for agreement to inclose sufficient ground to allow restoration. Despite some antipathy, Hertford decided to proceed based on the legal assurances he held. His lawyers argued that, as Lord of the soils, he could keep such deer and cattle of his own and accept such other cattle as he pleased. Hertford contended that it would benefit everyone if most of the Forest remained common, but free of deer and his own cattle, whilst planting the remainder.

The area chosen was '*in Circuit two and three quarter miles, and not above one-eighth part of the open ground of the Forest*'. A pale (ditch and fenced bank) was constructed to keep out commoners' animals. The commoners mounted opposition, and prevented the workmen from completing the task. The plan ultimately was abandoned. Hertford's steward wrote concerning the long-running battle with commoners in Savernake, and also in Bentley Wood and other parts of the Estate (WSA 1300-104), bemoaning the '*needless fault finders and repiners*' [complainants]. In a general rant against destruction of Estate woods he bewailed "*these commoners do marvellously murmur and grudge … this Realm will rue the waste of it, yet how lamentable is it to see the perverse disposition of the people*".

Timber production in the Forest remained limited. When the Great Inclosure was again examined in the eighteenth century for landscaping purposes, it was described as '*thin of wood*' (WSA 1300-1928). Only upon the eventual elimination of all commoners through the inclosure movement of the eighteenth and nineteenth centuries was the Forest planted to dense trees.

12.15 Other Examples of Timber Depletion in the Royal Forests

The long term habitat change described for Savernake was not just the result of commoning. However, large numbers of domestic stock, alongside high deer populations, prevented regeneration, whilst existing pollards aged, and oaks were occasionally felled. Timber from Savernake was used in the maintenance of Marlborough and Ludgershall castles. Burdens on timber were shared and larger-scale work satisfied by taking timber from various Forests. Despite the proximity of Savernake, when Salisbury Cathedral was constructed in 1233, it provided just

25 of a consignment of 200 trees, with the bulk coming from Gillingham, Chippenham, and the Forest of Dean (Cal Close Rolls). Timber for the cathedral was even imported from Dublin. Evidence suggests that, despite its size, Savernake timber was in scarce supply. Depredation of the Forests was conspicuous elsewhere, such as Clarendon Forest, where commoners' animals also exerted an influence. In 1301, officers were instructed to raise revenue by selling timber, with £100 coming from Savernake, and comparable amounts from Guildford, Chute, and Clarendon. The King was advised that selling £100 in Clarendon alone '*would be to the utter destruction of that Forest*'. The wardens were instructed to make up the missing £100 from Bere and Pamber (Cal Patent Rolls).

There are also reports of denudation by commoners in the Forest of Dean. The Regard of 1282 reported: '*The wood of Hodenhales is demesne wood of the King and destroyed by the men of St Briavels. Those men claim the right to take it at their will and have always taken it*' (Hart 1987). Over 500 years later, depredation continued: '*At the Hudnalls, poor people cut wood into faggots, and send them by water to Bristol, where it is purchased by the bakers… the oak pollards have been so destroyed, as not to leave a twig behind*'.

12.16 Conclusions

1. The Royal Forests provided an alternative to the manorial system for a large part of England from the eleventh century, imposing Forest Law on a landscape for the protection of vert and venison, and to raise money, whilst protecting existing land-use systems focused on extensive grazing and the gathering of minor products (fuel, bedding, smallwood).
2. Forests were established on less productive lands, where communal management was probably the norm before and after 'afforestation'.
3. The rights of commoners enjoyed good protection under Forest Law, recognised through Royal decrees from the thirteenth century.
4. Some Forest commons persisted for 600 years, possibly much longer. Evidence suggests that there was no 'tragedy of the commons' in socio-economic terms, for the commoners. In contrast, for the poorest members of society, the 'tragedy of inclosure' was everywhere.
5. Despite the resilience of social systems, landscapes may have undergone chronic change. Persistent and burgeoning commoning contributed (alongside high deer populations and unsympathetic land-uses) to a gradual reduction (or elimination) of tree cover. Hence, there might have been a tragedy at an ecological level. In England's unusual commoning regime, where independent property rights are held by commoners and landowners, the interests of each party would have suffered differently as woodland cover declined.

References

Printed Sources

Fauroux M (1961) Recueil des Actes des Ducs de Normandie 911–1066. Societie d'impressions Caron, Caen

Hardin G (1968) Tragedy of the commons. Science 162:1243–1248

House of Commons (1788) Journal 43:561

Hart C (1987) Regard of the Forest of Dean in 1282. Forest of Dean Local History Society, De Archaeologische Pers, Netherlands

Hart C (2002) The Commoners of Dean Forest. Lightmoor, Lydney

Jackson JEJ (ed) (1862) Wiltshire: topographical collections of John Aubrey 1659–1670. Wiltshire Archaeological and Natural History Society, Devizes

Langton J, Jones G (2005) Forests and chases of England and wales, . St John's College Research Centre Oxford, http://info.sjc.ox.ac.uk/forests/Index.html c1500–c1850

Manwood J (1615) A treatise of the lawes of the forest. Society of Stationers, London

Petit S, Watkins C (2003) Pollarding trees: changing attitudes to traditional land management practice in Britain 1600–1900. Rural History 14:157–176

Petit-Dutaillis C (1915) Studies and notes supplementary to Stubb's constitutional history. Sherratt & Hughes, Manchester

Tubbs CR (1964) Early encoppicements of the new forest. Forestry 37(1):95–105

Verhulst A (2002) The Carolingian economy. Cambridge University Press, Cambridge

Young C (1979) Royal forests of medieval England. University of Pennsylvania Press, Philadelphia

Primary Sources—Public Records Office

SC 8/63/3118: *c*.1337, Petition of tenants of Marlborough, Manton, Elcot, Preshute, Wick and Wootton Rivers for restitution of common rights

Primary Sources—Wiltshire and Swindon Archives

WSA 9-26-512: 1575, Sale of Oaks in Stock Woods

WSA 1300-104: *c*.1598, Legal Papers Savernake Forest.

WSA 1300-1928: *c*.1765, Proposals for Landscaping, Savernake Forest

Primary Sources—Longleat House

Seymour Papers, Volume V, fos 56, 57d

Chapter 13
The Cultural Landscape of Royal Hunting Gardens from the Fifteenth to the Eighteenth Century in Białowieża Primeval Forest

Tomasz Samojlik, Ian D. Rotherham and Bogumiła Jędrzejewska

13.1 Introduction

Cultural landscapes are the result of long-lasting interactions of humans with the natural environments (Berglund 1991). Their shapes are an interplay of (1) the type of the regional and local biotic and abiotic conditions (ecosystems, climate, topography), and (2) the specific socioeconomic, cultural, political and historical features of the human society. In Europe, the vast majority of natural habitats in the recent millennia have been forests, which—for all cultures and in all times—provided an immense array of resources and non-material benefits. Differentiated—in time and space—interactions of human societies and cultures with European forests have led to development, persistence and—eventually—demise of a multitude of cultural forest habitats (Muir 2006). In this chapter, we focus on a little known region of Europe: the Grand Duchy of Lithuania and the specific cultural landscape of hunting gardens that developed in its royal forests before the fifteenth century and persisted until the end of the eighteenth century.

Grand Duchy of Lithuania (GDL), which has been one of Europe's "lost kingdoms" (Davies 2011), in the period of its maximum geographic expansion (end of the fifteenth century) included the lands of contemporary Lithuania,

T. Samojlik (✉) · B. Jędrzejewska
Mammal Research Institute, Polish Academy of Sciences, Waszkiewicza 1c,
17-230 Białowieża, Poland
e-mail: samojlik@ibs.bialowieza.pl

B. Jędrzejewska
e-mail: bjedrzej@ibs.bialowieza.pl

I. D. Rotherham
Faculty of Development and Society, Sheffield Hallam University, Pond Street,
Sheffield S1 1WB, UK
e-mail: i.d.rotherham@shu.ac.uk

I. D. Rotherham (ed.), *Cultural Severance and the Environment*,
Environmental History 2, DOI: 10.1007/978-94-007-6159-9_13,

Belarus, Ukraine, parts of North-Eastern Poland and Western Russia, in total around 900,000 km^2 (Błaszczyk 2002). Although its area was cut in the following centuries, it still embraced around 300,000 km^2 before its full annexation by the Russian Empire in 1795. Several features of GDL made its culture and interactions with woodlands unique and interesting. The ruling class (Lithuanians) are a very ancient ethnic group settled in their land since the Iron Age (Sahanowicz 2002). Lithuanians were the last society in Europe to enter Christendom (in 1385) (Davies 1986), which ensured the long-term development of civilisation in relative isolation from influences of neighbouring countries. After 1385, when GDL entered the union with Poland, and the Polish–Lithuanian Commonwealth was born, the legal, societal and cultural traditions of GDL were still respected and its autonomy and differences persisted. Another source of Lithuanian dissimilarity from western European patterns was the country's outstandingly high proportion of forested areas and long-lasting tradition of Grand Duke's ownership over vast areas of land, including forests. The differences embraced, also included the reigning monarchs touring their domain together with their entourage, as they regularly moved between manors built in their forests (Stone 2001). The Grand Dukes' forests served thus as a source of venison and different benefits for the court. The protection of those forests and of big game constituted an important part of the Lithuanian code of laws (Statutes of Lithuania), which was in effect until the mid-nineteenth century. In 1385, with the union between Poland and GDL, the Grand Dukes' woodlands became royal forests. In the sixteenth century, forests belonging to the Grand Duke covered over 30 % of GDL (Sahanowicz 2002). A document from 1559, listed names of thirty-six royal forests (Reviziya pushch 1559), the majority of which were, within few decades, leased under the pressure of economic demand for wood and forest products (Daszkiewicz and Oleksyn 2005). After the reform of 1589, only five forests remained under the kings' dominion as parts of so-called economies of Grodno, Szawle, Brest (with Białowieża Primeval Forest), Olita and Mogilev and served as a direct source of income to the court. A forest in this sense meant not only the wooded area but also a small share of cultivated land surrounding the forests and villages with royal serfs and forests guards.

Białowieża Primeval Forest (BPF) is an example of a forest in which the traditional use of forest resources survived exceptionally long (commercial mass-scale timber production was introduced as late as in 1915). It is also an example of a royal forest of GDL, in which centuries-long anthropogenic impact has contributed to the creation of unique cultural landscapes (Samojlik 2007).

This chapter describes, on the basis of available sources, the origin, main features and decline (after the eighteenth century) of one type of cultural landscape in BPF in the period from the fifteenth to the eighteenth century: the landscape of a royal hunting garden.

13.2 Białowieża Primeval Forest

BPF is located on the contemporary border between Poland and Belarus (52°30′–53° N, 23°30′–24°15′ E) with 635 km^2 on the Polish, and 865 km^2 on Belarussian side (Fig. 13.1). It is the best preserved pristine forest of European lowlands that has been continuously forested for last several thousands of years and has never undergone any major event of large-scale deforestation (Mitchell and Cole 1998). The altitude of BPF ranges from 135 to 202 m a.s.l., and the entire area is predominantly flat (Jędrzejewska and Jędrzejewski 1998). From the point of view of

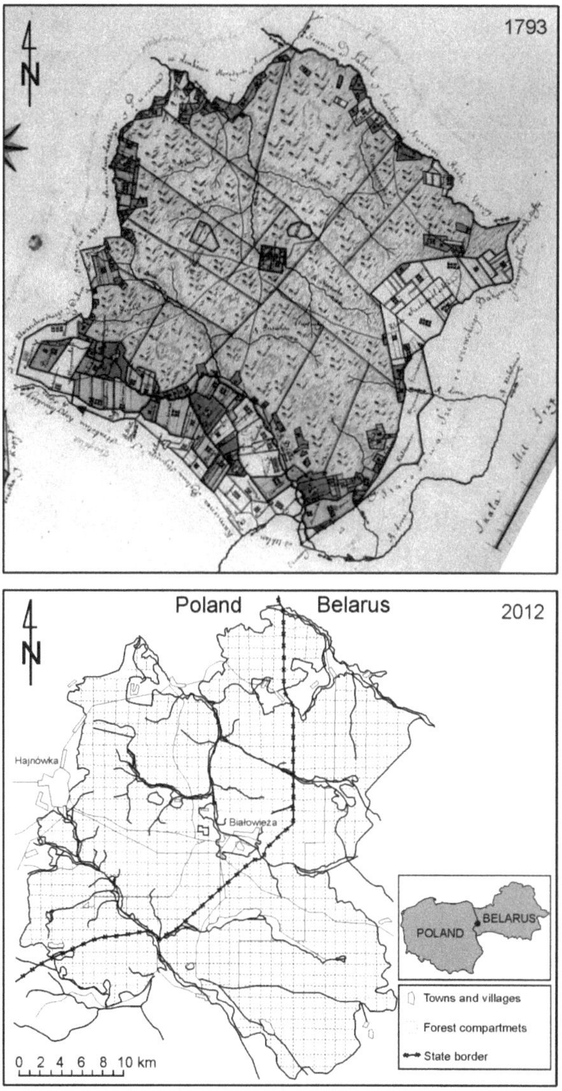

Fig. 13.1 Fragment of "Chart of Mieleczyce Guberniya with Białowieża Forest" from 1793 (Central Archives of Historical Records, Warsaw, Cartographic collection 66-3) and current map of Białowieża Primeval Forest (Polish and Belarussian part). Note the two hunting gardens marked on the map of BPF in 1793

natural studies, it is a unique reference area for many biological and ecological studies of ecological processes in natural conditions (Jędrzejewska and Wójcik 2004; Tomiałojć and Wesołowski 2004). It is also particularly interesting from the point of view of environmental history.

The first discovered remains of human settlements in BPF date back to ancient times (first century BC–fifth century AD; Krasnodębski et al. 2008; unpubl.). Then, after a gap in settlement process, there is a small number of discovered sites dated to early mediaeval times (eighth–eleventh centuries; Krasnodębski et al. 2005, 2011) and connected with Slavonic colonization. Those traces are sparse and do not evidence any large-scale wave of settlement in any of those periods. The early mediaeval settlements in BPF were most probably destroyed in the thirteenth and fourteenth centuries by systematic raids of Teutonic Knights. After several changes of borders and interplays of power between Poland, Lithuania and Ruthenia, BPF finally became the Lithuanian Grand Dukes' domain. In the fourteenth century, after the political union between Poland and Lithuania (Polish-Lithuanian Commonwealth), BPF became a royal forest and served as a hunting reserve of both Lithuanian Grand Dukes and Polish Kings. Royal forest guards, foresters, riflemen and beaters were settled in villages on the border of the forest to prevent any unauthorized access (Samojlik 2005a). The Polish royal period in the history of BPF ended in 1795, when the Russian Empire (in the third partition of Poland) seized this part of the Polish-Lithuanian Commonwealth.

The total area of BPF (including both the forest and arable and cultivated land inside and on outskirts of the forest) at the end of the eighteenth century incorporated 1,468 km^2, out of which 1,120 km^2 were covered with forest (Mikusińska et al. in print).

13.3 Cultural Landscape of a Royal Hunting Garden in BPF

The first and main purpose of Grand-ducal forests of GDL was to serve as game reserves for the court. The hunts by Grand Dukes and—later—kings were well-organised events. Beyond serving a vital role of mass venison providers, they often played a role of special diplomatic meetings (representative events and entertainments) (Samsonowicz 1991). From the thirteenth to the fifteenth centuries, massive royal hunts were known to be part of the preparation of food supplies for the army in times of warfare (Samsonowicz 1991). Therefore, the logistics of the royal hunts were aimed at ensuring a high success rate in harvesting large game in as short time as possible (from a few days to two weeks).

In Lithuanian royal forests, the special service people, *osoczniks*, were settled in villages on the forest peripheries. Endowed with land and waived of the duty to serve in army, the men from *osocznik's* families were obliged to watch and protect the forest and go to work in organising the hunts, wherever called. The *osoka* service included up to 277 men in BPF in 1639 (Ordinatia pusczy 1639;

Hedemann, 1939). Well-organised and intimately familiar with the forest and large game, that service was responsible for selecting the backwoods *(ostęp)* for royal hunts and create hunting gardens (*ogrody do polowań*) there. The Polish word *ogrody* has the same ethymology as the English '*garden*', and originally meant an area surrounded or guarded by fence. Such area was fairly large (5–10 km^2), embraced various forest habitats from wet to dry forest, it had a small stream running through (source of water for animals) and had a relatively good access from a royal hunting mansion (at most 10 km from it). The best areas could serve as hunting gardens repeatedly for as long as 2–3 centuries, yet they were utilised at relatively long time intervals (the minimal frequency of royal hunts in BPF between 1409 and 1784 was calculated at one hunt per 25 years (Samojlik 2006, 2010). According to Połujański (1854), "*there were many hunting gardens both in the contemporary Grodno District and in the entire Lithuania. In each forest a convenient place was selected for an animal passage, enclosed with a fence, so that no animal could escape*". The tall wooden fence of the hunting garden had several wide openings, which enabled large game such as European bison, moose, red and roe deer and wild boar to go in and out freely in long periods between hunts. The forest in the hunting garden was kept sufficiently open to facilitate use by the hunters. This also enabled the grasses and other herbaceous vegetation to flourish and this further attracted ungulates to feed there.

When a royal hunt was announced, the *osoczniks* organised drives of game towards the hunting garden to gather as many large beasts as possible in the enclosed area; they then closed the openings so that the animals were confined. The interior of the hunting garden could be further divided into compartments by wooden fences or by stretching nets across strategic points, in which case the animals were to be captured alive. The latter was more frequent in the centuries when the use of rifles changed the techniques of hunting.

The first records of the regular *osoka* service in Lithuanian royal forests date back to the fourteenth century (Stryjkowski 1582). The first description of a hunting garden was given by the poet Nicolaus Hussovianus in 1523 in his "*Poem on the appearance, ferocity and hunting of bison*": "*Wishing to kill it* [European bison], *we—people of the North—| Adhere to the following, well-established guides: | We fell trees to encircle wide space. | And keep the beast trapped within. | Twelve miles and more counts the perimeter of the fence, | If we keep to the Latin measures, | The guards then surround the place, | So that the quarry will not escape. | Not always is that labour new to us; | Old enclosures wait open in woods, | The beast, which used to freely enter them | For grazing, easier gets entrapped. | In such parks we hunt, equally enclosed. | The fates are on the same scale: ours and the beast's*" (Hussowski 1523).

The poem also reported the accident that happened in 1504 in one of the royal forests. During the hunt by King Alexander I Jagiellonian, the bison struck the wooden bower from which the Queen Helen and her maidservants had been watching the hunt, causing the bower to lean aside and put the queen and other observers at deadly risk. This document certifies that in the early 1500s, the hunting gardens had already had a very long tradition. It also documents the custom of constructing a bower inside the hunting garden, from which guests could observe the hunters on horses chasing game in the large enclosure. Convenient hunting and good observation possibilities were two main reasons to keep the forest more open and clear than in non-fenced backwoods.

With the invention and wide use of firearms, the hunts gradually changed from the traditional form described by Hussovianus to more comfortable and safe one: the King and other participants of the hunt took positions in the bower, while game was driven towards them by *osoczniks* through the compartments of the hunting garden. Such royal hunts in BPF were documented in 1752 by August III Wettin and his court (Brincken 1826) and in 1784 by Stanisław August Poniatowski (Naruszewicz 1784; Samojlik 2005b). In addition, by that time, royal hunts changed their purpose; their practical role of venison provision to the court, army and royal dependants faded, and the role as important entertainment functions began to dominate.

Several sources give a good illustration on how the cultural landscape of a royal hunting garden looked like. These are the maps from 1784 (Fig. 13.2) and 1793 (Fig. 13.1), a drawing from 1784 (Fig. 13.3a) and a brief description by Baliński and Lipiński (1846): "*Royal game preserve, or Grand Kletna, is a spacious and dry backwood overgrown with ancient oaks coiled by ivy. It is enclosed by fences as animals were driven there to be killed by Polish kings from hunting bowers, built especially for this purpose*". A map drawn in 1784 by the King's cartographer shows the location of those two hunting enclosures in BPF (Fig. 13.2). On his way to Białowieża, King Stanisław August Poniatowski, had a chance to observe the construction of one of the hunting gardens: "*in the forest, one mile from Białowieża, we met a group of peasants, guarding the backwood, that is the area fenced for a half mile around to keep various animals: moose, wild boar, bears and European bison*" (Naruszewicz, 1784). Based on the comparison of topographic features of Białowieża village currently and on the eighteenth-century map, we were able to estimate the size of the enclosures at 8 km^2 (Teremiska) and 5 km^2 (Kletna). Furthermore, the map shows the construction features of both enclosures, equipped with inner compartments with internal barriers (Fig. 13.2). In both cases, fences in the last compartment narrowed towards the hunting arbour or bower, constructed most probably on a glade, which allowed hunters gathered in arbour for a clear shot (Fig. 13.3a). Animals were gathered before the King's arrival, and driven by a large group of beaters (*osoczniks*, Fig. 13.3b) towards hunters with rifles (Fig. 13.3c). In order to prevent them from turning back, inner fences were closed by beaters behind the animals being chased.

So far, the Lithuanian royal forests have not been a subject of any wider historical investigation. However, from their general history we can suppose that in the sixteenth and seventeenth centuries, hunting gardens existed in a few such forests, but in the seventeenth and eighteenth centuries, they are known only from BPF. Contemporary place-names *Kletna, Kletnia, Kletno* (enclosed area) are obvious suggestions of the royal hunting gardens from the times before the sixteenth century. In addition to two hunting gardens, Kletna and Teremiska, that are extensively documented with the tradition that Kletna might be in operation since

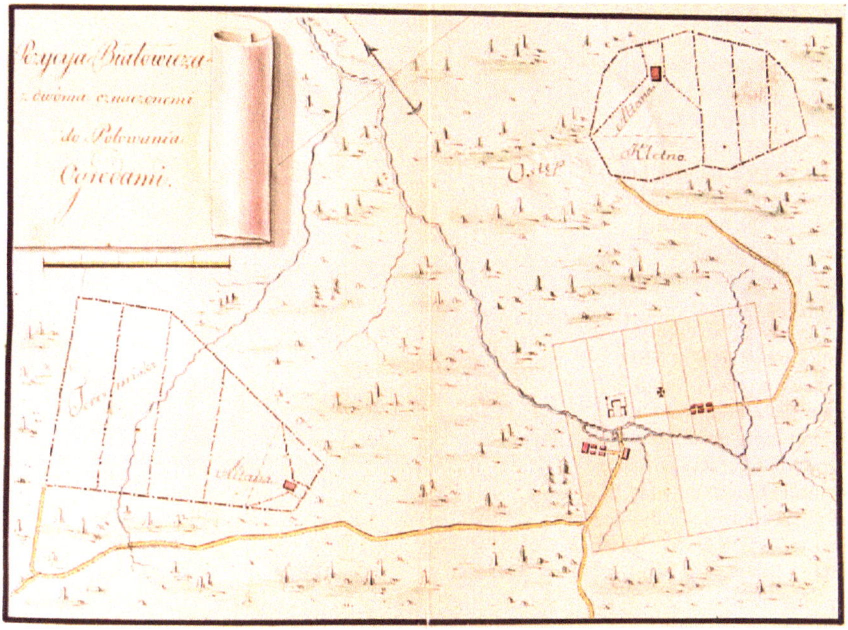

Fig. 13.2 "Location of Białowieża with two marked Hunting Gardens", a hand-drawn map by royal cartographer Michał Połchowski, 1784 (Central Archives of Historical Records, Warsaw, Cartographic collection AK147)

the late sixteenth century (Karcov 1903), there are some suggestions for two other gardens. First, another hunting garden in the Augustowska forest district was used for the hunt of King August III Saxon in 1752: "*After reconnaissance, the foreigners* [August's retinue] *selected Augustowska forest district as a place for hunt, in a distance of one mile* [approx. 7.4 km] *from Białowieża. Next, the preparations began. A battue made of thousand peasants and numerous dogs gathered animals in previously selected part of forest district, the area was hung with sheets and a fence was built. All work was performed with great care*" (Brincken 1826). However, the location of the garden is uncertain. Secondly, a village called Kletno in the eastern (currently Belarussian) part of BPF, known in written documents from 1639 (Hedemann 1939) gives a suggestion of a possible location of a hunting garden before the seventeenth century.

13.4 The Ending of Traditional Utilisation and the Changes in the Hunting Garden Landscape

Although the political end of the Grand Duchy of Lithuania was abrupt (in 1795), the tradition of monarchical and other high-ranked hunts in BPF persisted there under the Russian rule until 1915. Despite several Russian attempts at introducing

Fig. 13.3 **a** Jan Lindsay's painting of a royal hunting arbour in Teremiska hunting garden in Białowieża Primeval Forest, erected in 1784. Apart from a two-storey arbour, the painting shows fragments of inner fences of the enclosure, narrowing towards the glade with arbour (Department of Drawings of Warsaw University Library, Royal Collection P. 878 nr 117). **b** Hunting procession in Białowieża Primeval Forest in 1821, with great number of beaters (*osoczniks*) visible in the background (drawing by Jakub Sokołowski, published in Brincken 1826). **c** "Bison hunt in Białowieża Forest", drawing by Rudolf Żukowski from 1851 (The National Museum in Warsaw, Gr. Pol. 9895) showing a traditional type of hunt in the former royal hunting garden still persisting in the mid nineteenth century

timber extraction in the nineteenth century, the protection of bison and other ungulates remained one of the main goals of BPF's administration. Between 1795 and 1860, the former royal hunting gardens were only occasionally used for hunts organised for high-ranked officials or scientists (to obtain bison for museum collections, e.g. hunts in 1821 or 1851, see Fig. 13.3b and c) (Karcov 1903). In the course of the nineteenth century, Kletna garden fell into abandonment and today there are no visible signs of it in the area. Since 1921, it has been a part of the strict reserve of Białowieża National Park. The only sign of its former existence is a place-name of a backwood (part of forest)—*Wielka* (English: Large) *Kletna.*

The second garden, Teremiska, acquired a new role in the second half of the nineteenth century. Making use of its cultural forest landscape suitable for large-scale official hunts, the Russian rules organised there a massive hunt "according to the local tradition" for Tsar Alexander II and his German relatives in 1860 (Anonymous 1861; Daszkiewicz et al. 2012). The fenced area encompassed 8 km^2 of forest. Inside, a row of shooting posts was placed along the 640 metre-long tract (Fig. 13.4). After the hunt, the garden became a permanent game park (*Zwierzyniec*), where ungulates were kept for show, smaller hunts, and acclimatization before introductions to the wild (Fig. 13.5; Karcov 1903). Meanwhile, the rather open forest regenerated into a dense stand of trees, and recovered its more 'natural'

Fig. 13.4 The view of the Teremiska hunting garden during the Tsar's hunt of Alexander II in 1860—drawing by Michaly Zichy (*Source* Anonymous 1861)

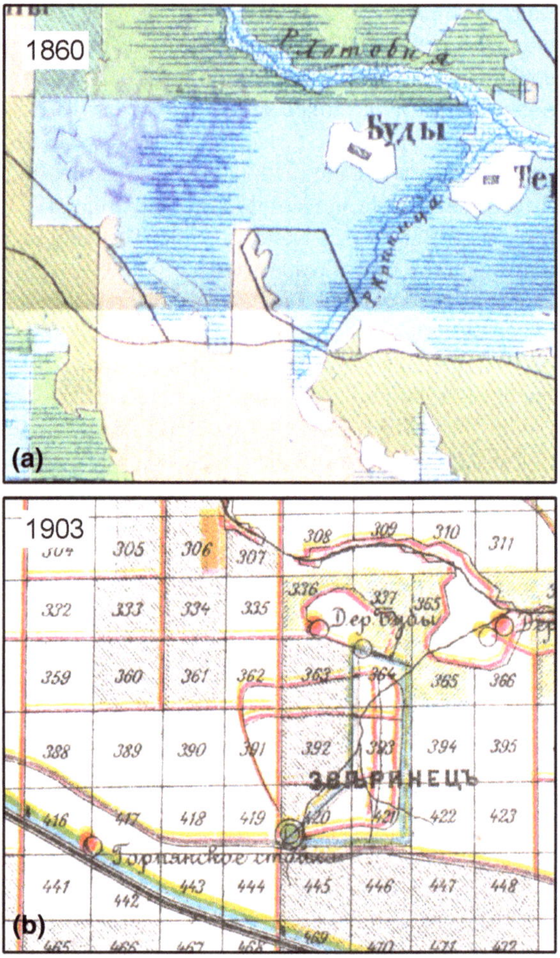

Fig. 13.5 The location of the Russian game park *Zwierzyniec*, created in Białowieża Primeval Forest in the place of the former royal hunting garden Teremiska: **a** On the map from 1860 (*Source* Bobrovskii 1863). **b** On the map from 1903 (*Source* Karcov 1903)

character. In 1888, when BPF was included into the Tsars' personal appanages, the administrator of *Zwierzyniec*, Hans von Auer, was ordered to clear the area and turn it into a park. As he wrote in his memoirs "*it was a terrible strike to me, as the virgin and primeval character of the game park was, in my eyes, its greatest value*" (Auer 1998). *Zwierzyniec* served its function of a game park until World War I ended the Russian rule over the BPF. The war brought also an end to the population of European bison in BPF—the last animal was killed in 1919, and for ten years, the forest was devoid of its key ungulate species. The restitution process started in 1929 by placing back into the enclosure several European bison obtained from zoos (Fig. 13.6a). The Polish managers of BPF located this in a small part of the former Teremiska/*Zwierzyniec* (22 ha in 1929, enlarged to 59 ha in 1931; Fig. 13.6b). This place became an enclosed restitution reserve for European bison (Krasińska and Krasiński 2007). In 1952, first bison was released outside the enclosure, and since then their population grew to 888 in 2010 (473 in the Polish

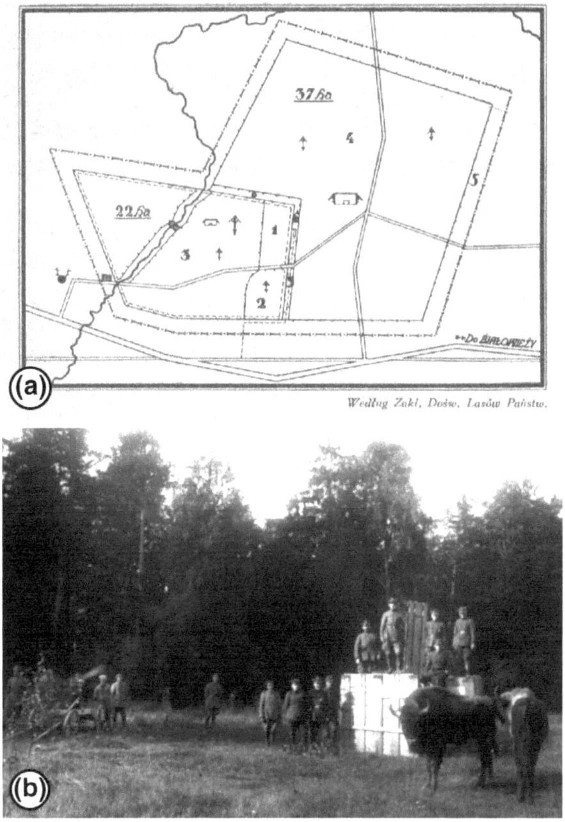

Fig. 13.6 **a** The captive-breeding reserve created in 1929 for the restitution programme of European bison, incorporating part of the former Teremiska royal hunting garden (*Source* Grochmalnicki 1933, after: Krasińska and Krasiński 2007). **b** First European bison being released into the captive-breeding reserve in September 1929. Photograph by J.J. Karpiński (after: Krasińska and Krasiński 2007)

and 415 in the Belarussian part of the BPF; Raczyński 2010). Nowadays, the enclosure with a small number bison (up to 35 animals) is maintained as a reserve of animals with known pedigree (important as founders of new populations) and the last-resort reserve in case of epizootic disease (Krasińska and Krasiński 2007). The grazing impact of captive animals keeps the fenced part of the forest—part of the former royal hunting garden—open, with single old trees as remnants of past densely wooded landscape.

The history of cultural landscapes of royal hunting gardens in BPF is common for all royal forests of the Grand Duchy of Lithuania, and at the same time unique for the forests of Białowieża. Hunting gardens were most probably widely used as a hunting method in the majority of royal forests of GDL, enabling for a relatively safe and secure organisation of *venatio magna*, grand royal hunts. Cultural landscapes similar to the ones in BPF were thus probably widespread in Lithuanian forests in the fifteenth and sixteenth centuries, but then gradually declined throughout the seventeenth century. In the eighteenth century, royal hunting gardens were evidenced only in BFP, where they persisted until the beginning of the nineteenth century. What seems unique in the case of BPF is the fact that the area originally designated as a hunting garden in the eighteenth century was in the following centuries adapted to serving new purposes. This area was a Russian game reserve (where animals were kept constantly, not gathered only before hunts) in the period 1860–1915, and has been the Polish breeding reserve of European bison (since 1929). In this way, the modern cultural landscape has replaced the royal hunting garden landscape from the deep past.

This case study and the findings of the research, further illustrate and emphasise the need to understand the past in order to inform our knowledge of contemporary landscapes. The hunting gardens mix complex interactions between people and nature over long time-periods. Furthermore, they demonstrate a depth of cultural utilisation in these historic landscapes with sequential displacement of uses and ultimately a break or severance from their past traditions. To manage and conserve these areas today it is important that their historic origins and uses are better understood.

References

Anonymous (1861) Okhota v Belovezhskoi Pushche (A Hunting in Belovezha Forest). The Imperial Academy of Sciences, Sankt Petersburg, 71 p (in Russian)

Auer H von (1998) Unter Wisenten im Urwaldrevier. Bialowice um 1900. Landbuch Verlag, Hannover, 151 p

Baliński M, Lipiński T (1846) Starożytna Polska pod względem historycznym, jeograficznym i statystycznym opisana (The ancient Poland in historical, geographical and statistical description). S. Orgelbrand, Warszawa (in Polish)

Berglund BE (ed) (1991) The cultural landscape during 6000 years in southern Sweden: the Ystad project. Ecological Bulletin 41, 495 p

Błaszczyk G (2002) Litwa na przełomie średniowiecza i nowożytności 1492–1569 (Lithuania at the turn of Middle Ages and modern period 1492–1569). Wydawnictwo Poznańskie, Poznań, 336 p (in Polish)

Bobrovskii PO (1863) Materialy dla geografii i statistiki Rossii, sobrannye oficerami generalnago shtaba. Grodnenskaya Gubernya. Chast Pervaya (Materials for the geography and statistics of Russia, collected by officers of the general staff. Grodno District. Part I). Department of the General Staff, St. Petersburg, pp 404–459 (in Russian)

Brincken J (1826) Mémoire descriptif sur la Forêt Impériale de Białowieża, en Lithuanie. Glücksberg, Warsaw, 127 p

Daszkiewicz P, Oleksyn J (2005) Introdukcja "sosny ryskiej" w osiemnasto- i dziewiętnasto-wiecznej Francji (Introduction of the "Riga pine" in the 18th and 19th century France). Rocznik Dendrologiczny 45:7–40 (in Polish with English summary)

Daszkiewicz P, Samojlik T, Jędrzejewska B (2012) Puszcza Białowieska w pracach przyrodników i podróżników 1831–1863 (Białowieża Primeval Forest in the works of naturalists and travellers 1831–1863). Wydawnictwo Naukowe Semper, Warszawa, 208 p (in Polish)

Davies N (1986) Heart of Europe. A short history of Poland. Oxford University Press, Oxford, 511 p

Davies N (2011) The Vanished Kingdoms: The History of Half-Forgotten Europe. Allen Lane, 848 p

Grochmalnicki J (1933) Results of ten years' bison-breeding in Poland. Polski Oddział Międzynarodowego Towarzystwa Ochrony Żubra, Poznań, 24 p (in Polish with English summary)

Hedemann O (1939) L'histoire de la Forêt de Białowieża (jusqu'a 1798), Institut de Recherches des Forêts Domaniales, Warszawa, 312 p (in Polish with French summary)

Hussowski M (1523) Carmen de statura, feritate ac venatione Bisontis (Poem on the appearance, ferocity and hunting of bison). Published by: Białowieski Park Narodowy, Białowieża, 1994, 103 p (in Latin and Polish)

Jędrzejewska B, Jędrzejewski W (1998) Predation in Vertebrate Communities. The Białowieża Primeval Forest as a Case Study. Springer-Verlag, Berlin-Heidelberg, 450 p

Jędrzejewska B, Wójcik JM (eds) (2004) Essays on mammals of Białowieża Forest. Mammal Research Institute, Polish Academy of Sciences, Białowieża, 214 p

Karcov G (1903) Belovezhskaya Pushcha. Ee istoricheskii ocherk, sovremennoe okhotniche khozaistvo i vysochaishe okhoty v Pushche (Belovezha Primeval Forest. Historical description, contemporary game management, and monarchial hunts in the Forest). S. Marks, Sankt Petersburg, 414 p (in Russian)

Krasińska M, Krasiński ZA (2007) European Bison: The Nature Monograph. Mammal Research Institute, Polish Academy of Sciences, Białowieża, 317 p

Krasnodębski D, Samojlik T, Olczak H, Jędrzejewska B (2005) Early medieval cemetery in the Zamczysko Range, Białowieża Primeval Forest. Sprawozdania Archeologiczne 57:555–583

Krasnodębski D, Dulinicz M, Samojlik T, Olczak H, Jędrzejewska B (2008) A cremation cemetery of the Wielbark culture in Kletna Range (Białowieża National Park, Podlasie province). Wiadomości Archeologiczne 60:361–376 (in Polish)

Krasnodębski D, Olczak H, Samojlik T (2011) Early mediaeval cemeteries in Białowieża Primeval Forest. In: Cygan S, Glinianowski M, Kotowicz PN (eds) In silvis, campis… et urbe. Średniowieczny obrządek pogrzebowy na pograniczu polsko-ruskim (In silvis, campis... et urbe. Mediaeval funeral rite in the Polish-Ruthenian borderland). Collectio Archaeologica Ressoviensis 14:145–174 (in Polish with English summary)

Mikusińska A, Zawadzka B, Samojlik T, Jędrzejewska B, Mikusiński G (in print) Quantifying landscape change during the last two centuries in Białowieża Primeval Forest. Applied Vegetation Science, doi: 10.1111/j.1654-109X.2012.01220.x

Mitchell FG, Cole E (1998) Reconstruction of long-term successional dynamics of temperate woodland in Białowieża Forest, Poland. J Ecol 86:1042–1059

Muir R (2006) Ancient Trees, Living Landscapes. The History Press, Stroud, 256 p

Naruszewicz A (1784) Diariusz podróży najjaśniejszego Stanisława Augusta króla polskiego na sejm grodzieński [The diary of the travel of His Majesty Stanisław August, the King of Poland, to the Seym in Grodno], Warszawa (in Polish)

Ordinatia puszczy (1639): Ordinatia puszczy J. K. Mści leśnictwa Białowieżskiego y Kamienieckiego Roku 1639 miesiąca Octobra [Ordinatia of the King's Białowieża and Kamieniec forests, in October 1639]. Published by: Vilenskaya Arkheogeograficheskaya Kommissieya, Vilnius, 1870 (in Polish and Russian)

Połujański A (1854) Opisanie lasów Królestwa Polskiego i guberni zachodnich Cesarstwa Rosyjskiego pod względem historycznym, statystycznym i gospodarczym (The historical, statistical and economic description of forests of the Kingdom of Poland and western districts of the Russian Empire). Vol. 2. Józef Unger, Warszawa, pp 176–206 (in Polish)

Raczyński J (2010) European Bison Pedigree Book. Białowieża National Park, Białowieża, 92 p

Reviziya pushch (1559): Reviziya pushch i perekhodov zverinykh v byvshem Velikom Knyazhestve Litovskom [...] v 1559 godu (Inspection of forests and animal passages in the former Grand Duchy of Lithuania [...] in 1559). Published by: Vilenskaya Arkheogeograficheskaya Kommissieya, Vilnius, 1867 (in Russian)

Sahanowicz H (2002) Historia Białorusi od czasów najdawniejszych do końca XVIII wieku (The history of Belarus from the oldest times until the end of the 18th century). Instytut Europy Środkowo-Wschodniej, Lublin, 411 p (in Polish)

Samojlik T (ed) (2005a) Conservation and Hunting. Białowieża Forest in the Time of Kings. Mammal Research Institute, Polish Academy of Sciences, Białowieża, 90 p

Samojlik T (2005b) Stanisław August Poniatowski in the Białowieża Forest (30th August–2nd September 1784). Kwartalnik Historii Kultury Materialnej 53(1):35–52 (in Polish with English summary)

Samojlik T (2006) Hunts and stays of Kings of Poland and Grand Dukes of Lithuania in the Białowieża Primeval Forest in the 15–16th century. Kwartalnik Historii Kultury Materialnej 54(3–4):293–305 (in Polish with English summary)

Samojlik T (2007) Antropogenne przemiany środowiska przyrodniczego Puszczy Białowieskiej do końca XVIII wieku (Anthropogenic changes of the environment of Białowieża Primeval Forest until the end of the 18th century). PhD thesis, Mammal Research Institute, Polish Academy of Sciences, Białowieża, 181 p (in Polish)

Samojlik T (2010) Traditional utilisation of Białowieża Primeval Forest (Poland) in the 15–18th centuries. Landscape Archaeol and Ecol 8:150–164

Samsonowicz A (1991) Łowiectwo w Polsce Piastów i Jagiellonów (Hunting in Poland of the Piasts and Jagiellons). Studia i Materiały z Historii Kultury Materialnej 62, Wrocław, 410 p (in Polish with English summary)

Stone D (2001) The Polish-Lithuanian State, 1386–1795. History of East Central Europe. Vol. 4. University of Washington Press, Seattle, 379 p

Stryjkowski M (1582) Kronika polska, litewska, żmódzka i wszystkiéj Rusi Macieja Stryjkowskiego (Maciej Stryjkowski's chronicle of Poland, Lithuania, Żmudź and all Russia). Published by: Wydawnictwa Artystyczne i Filmowe, Warsaw, 1985, 572 p (in Polish)

Tomiałojć L, Wesołowski T (2004) Diversity of the Białowieża Forest avifauna in space and time. J Ornithol 145:81–92

Chapter 14
The End of Common Uses and Traditional Management in a Central European Wood

Péter Szabó

14.1 Introduction

Common uses and traditional management were once important elements of woodland management in Central Europe (Johann et al. 2011). Among them, we can list for example pasturing, haymaking, coppicing, pannage, bee-keeping and wild fruit collecting. At different times and for different reasons, by the second half of the twentieth century virtually all of these uses disappeared. In this chapter, I examine a lowland woodland in the Czech Republic to illustrate some of the issues connected to this long process.

Hodonínská Dúbrava (the name literally means 'the oakwood in Hodonín [a nearby town]) is a sub-continental oak (*Aceri tatarici-Quercion*) and hornbeam (*Carpinion*) woodland. It is situated on the right bank of the river Morava (Fig. 14.1) at the Czech-Austrian-Slovak border. The site is almost flat, gently sloping towards the southwest, with the lowest and highest points at 164 m and 242 m. The climate is relatively warm and dry with 9 °C of average annual temperature and 550 mm of precipitation (Tolasz et al. 2007). The substrate is blown sand of 150–200 cm thickness with underlying clayey sediments (Novák and Pelíšek 1943). The prevailing soil type is the arenic variant of dystric cambisol (AOPK 2002–2010), a sandy soil poor in nutrients. A peculiar oakwood vegetation type (*Carici fritschii-Quercetum roboris*) is probably endemic for this place and its close surroundings (Chytrý 1997; Roleček 2004). Due to low ground water levels, other vegetation types, such as alluvial forests (*Alnion incanae*) and alder carrs (*Alnion glutinosae*) also occur. The fine vegetation mosaic is probably the result of a mixture of calcareous rock and acidic blown sands, fine micro-topographic

P. Szabó (✉)
Department of Vegetation Ecology, Institute of Botany of the Academy of Sciences of the Czech Republic, Lidická 25/27 60200 Brno, Czech Republic
e-mail: peter.szabo@ibot.cas.cz

I. D. Rotherham (ed.), *Cultural Severance and the Environment*, Environmental History 2, DOI: 10.1007/978-94-007-6159-9_14,

Fig. 14.1 The location of Hodonínská Dúbrava

differences, and the vicinity of the Morava River producing numerous seasonally or permanently wet sites (Roleček 2007).

Written sources about the history of Hodonínská Dúbrava are available from the mid-fourteenth to the twentieth centuries. All sources are presently kept at the Moravian Archives (MZA) in Brno. The following kinds of sources were used:

(1) *Charters* are records of legal transactions. They are usually concerned with the granting of rights or property. The first charters dealing directly with Dúbrava are from the mid-fourteenth century, while the last ones are from the eighteenth century.
(2) *Urbaria* are conscriptions of all incomes from an estate. The main concern of *urbaria* was the conscription of tenants and the dues they had to pay for using the land they farmed. Four *urbaria* survive that provide relevant information about the Hodonín estate. They date from 1600 (MZA F5 kniha 1a, analyzed in Chocholáč 1994), ca. 1654 (incomplete: MZA F5 kniha 1), 1691 (MZA F5 kniha 3) and 1805 (MZA F5 kniha 4).
(3) *Estate conscriptions* describe the value of landed property in a given historical moment. A detailed conscription survives from 1692, when the Hodonín estate was sold to count Johann Adam von Lichtenstein (MZA F5 kniha 5).
(4) *Account books* were kept by woodland owners to register incomes generated by the cutting and selling of underwood and timber. For Dúbrava unfortunately most of the woodland accounts books were lost; the few that remain cover the years 1765–1772 (MZA F5 karton 538).
(5) *Forestry management plans* (FMP) are detailed surveys containing information on the name, size and position of each woodlot (Bürgi 1999), as well as on tree composition and forest structure, supplemented by current and planned management. The earliest surviving FMP of Dúbrava dates from 1851 (MZA F5 kniha 232). Further FMPs were prepared in 1864 (MZA F5 kniha 233–236, 238–240) 1906 (MZA F5 kniha 242–245), 1925 (MZA F263 kniha 1), 1936 (MZA F263 kniha 4–5) and 1952 (MZA F302 kniha 4).
(6) *Forestry documents* were produced by the local forestry administration. Such documents include for example various surveys of woodland areas, discussions on types of management and detailed diaries of yearly activities.

(7) *Large-scale maps* (ca. 1: 30,000 and larger) are generally available in the Czech Lands from the eighteenth century onwards. The best known examples are the three nationwide surveys prepared by the Austrian Army in 1764–1783, 1836–1852 (both 1: 28,800) and 1876–1880 (1: 25,000) (available online at http://oldmaps.geolab.cz). A much more detailed (1: 2,880) set of maps were drawn in 1824–1843 as part of the so-called Stable Cadastre (Bičík et al. 2001). In addition to these overall surveys, many large-scale maps depict individual estates or parts thereof. For the study of woodland history, the most useful are maps produced as parts of forestry management plans. Three sets of maps survive from Hodonín: 1851 (MZA F5 mapa 54–59), 1884 (MZA F5 mapa 60–64, note that the FMP itself is lost) and 1906 (MZA F5 mapa 65–70).

14.2 Common Versus Communal

To be able to examine common uses, first the meaning of the term ‘common’ must be clarified in the given context. As a start, we should note that the kind of absolute property we are used to in the twenty-first century was very rare in earlier periods. Another significant difference from current legal customs is that in the Middle Ages and Early Modern Period property rights were defined in a utilitarian manner that is hard to translate into tight legal terminology. The situation is further complicated by the fact that the terms devised (partly already by the contemporaries) to describe property relations work acceptably in the case of simple land-use types (such as arable) but often fail to grasp the complexities of woodland use. The most basic distinction existed between ownership and usufruct. Ownership meant “the right to alienate the property” (Warde 2006, p. 105). Usufruct, on the other hand, entailed the right to particular resources as well as the right to manage the land to get these resources. The clearest example of this double structure is arable: the fields of the average village were owned by the overlord, but the usufruct belonged to the tenants who paid a yearly rent for the right to farm the land. Usufruct was not simply an economic concept. Tenant lands and dues were hereditary and it was not easy for the overlord to change the existing conditions. The term ‘common’ has a different meaning with respect to ownership and usufruct. ‘Common’ (or communal) ownership means that a certain piece of land was owned by a community (such as a town, a village or a chapter) rather than by an individual (such as a count or a bishop). In usufruct, ‘common’ land is where certain individuals or communities other than the owner had rights to a set of produce. Sometimes they had to pay for these rights, but they were often legally or traditionally privileged not to do so.

In Dúbrava, the available sources are very clear about the fact that ownership was strictly private. The Wood belonged to the overlord of the Hodonín estate from the Middle Ages until the twentieth century; it was not ‘common’ in this sense. ‘Common’ woods as such existed, because the 1600 and 1691 *urbaria* (MZA F5 kniha 1a and 3) both make a difference between regular (private) and communal woods. It is especially instructive to compare Dúbrava with the forest neighbouring it on the northeast, which was donated in its entirety to the town of Bzenec and thus became a communal wood already in 1214 (Boček 1839, pp. 75–76). By contrast, in the sense of usufruct Dúbrava was a 'common' wood. The first detailed source on the management of Dúbrava (the 1600 AD *urbarium*—MZA F5 kniha 1a) describes a complex system in which copious common usage rights mostly in the form of wood-pasture and woodland meadows played a significant part. Pasture included the all-year-round presence of certain domestic animals as well as pannage: the fattening of domestic pigs on acorns. The latter was also closely regulated; the amount of pigs allowed depended on the quantity of the acorn crop in any given year. Woodland meadows were tenanted rather like arable land. In addition, firewood was cut, strawberries and oak galls were collected, there were managed ponds, beehives and even arable fields within the Wood. Except for a few carefully described cases, the tenants had to pay for every woodland produce. This was markedly different from communal woods on the estate, which the communities could use as they pleased and for free. Some earlier references imply that this system with was not new in 1600. By 1350, at the latest (the document at our disposal claims to include the copy of a charter from 1228, but the authenticity of the latter is debated), the town of Hodonín had the right to collect dry wood, cut grass and pasture their animals in Dúbrava (Boček 1839, pp. 204–205; Brandl 1874, p. 16). In 1370, the village of Šardice received the right to timber and underwood (except for oak) in parts of the Wood (Brandl 1878, pp. 116–118). In 1531, the same village claimed that their rights included pasture as well (except for the territory of the woodland meadows). The overlord apparently agreed and the conditions were reformulated in detail because a new pond was established that flooded those parts of Dúbrava that had been used by the villagers (MZA F5 karton 3 inv. č. 29).

This system of common uses and traditional management functioned intact for centuries. However, during the past two hundred years all elements were gradually abandoned and at the other end of this long process the entire system disappeared. Since the mid-twentieth century Dúbrava has been a closed high-forest (including extensive *Pinus* plantations) with no common rights or traditional management.

14.3 Wood-Pasture and Woodland Meadows

The system observed in 1600 survived until the mid-eighteenth century. Foreshadowing major changes to happen a few decades later, in 1752 the town of Hodonín renegotiated its pasture rights with the estate overlord (MZA F5 karton 6 inv. č. 76). The initiative came from the overlord, who considered that extensive pasturing in the Wood was detrimental to tree growth. This almost coincided with the appearance of the first general 'Forest Code' (in effect rules for woodland management) issued by the Habsburg central government for Moravia in 1754. Whilst not prohibiting pasturing in woodlands, this code certainly tried to limit it

(Nožička 1957). New wood-pastures were delimited in Dúbrava and animals were strictly banned from areas outside these pastures. At the same time, meadows were abolished in the 'woodland' parts, and an existing meadow next to the Wood was given over to Hodonín in exchange. In short, the overlord wanted to separate different land-uses. When in 1762 the imperial Habsburg family bought the Wood, they continued with this process on a grander scale. In the late 1780s, they reduced the area of the Wood by ca. 30 %, which was given over completely to the neighbouring villages as pasture—this was apparently based on mutual consent between the estate and the villages. The remaining two thirds of Dúbrava were enclosed by a large woodbank (Szabó 2010a) (Fig. 14.2) and turned into a coppice (MZA F5 kniha 232). The purpose of the woodbank was to prevent animals from entering the Wood, and—although the foresters complained that the villagers would drive their livestock into the Wood at night—it provided enough protection to radically reduce pasturing, which gradually disappeared.

Interestingly, the original plan had been slightly different. Instead of the coppice, the forestry administration had wanted to create a gigantic game preserve (effectively a private wood-pasture), which would have been surrounded by a hedge/fence rather than a bank. The idea was based on the (at that time) higher profitability of game keeping as opposed to the money to be made from trying to introduce productive forestry into a traditional wood-pasture landscape. This project failed for several reasons. First, even though most of the trees in the new pasture parts were felled to provide material for the hedge/fence, this was not enough and funds to buy more were insufficient. Second, the price of wood began to rise, and soon it seemed financially reasonable to try out forestry instead of game keeping. In the end the game preserve was entirely given up and large-scale coppicing was introduced.

The only local settlement that was apparently unhappy with the new system was Dubňany, on the northern edge of the Wood. They entered into a long and complicated dispute with the local representatives of the Habsburg family in the first half of the nineteenth century (MZA F5 karton 38 inv. č. 1029). Their basic argument was that the pasture areas allocated to them in the late 1780s were

Fig. 14.2 The enclosure woodbank around the wood, constructed in the late 1780s. April 2009

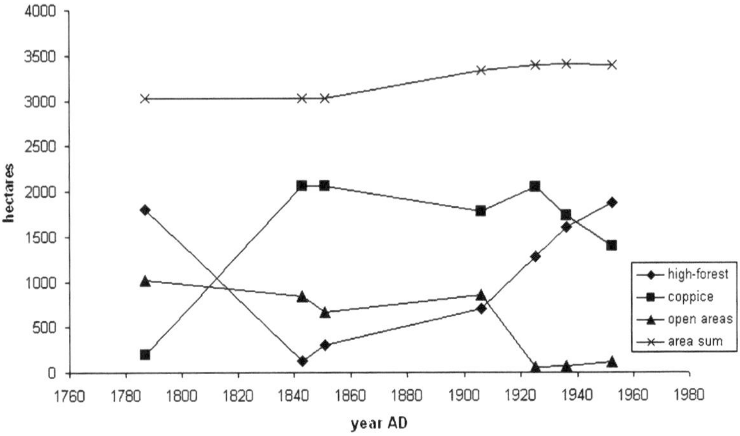

Fig. 14.3 Coppices, high-forests and open areas in Hodonínská Dúbrava from the late 1780s until the twentieth century. (*Sources* FMPs 1851 – MZA F5 kniha 232; 1864 – MZA F5 kniha 233–236, 238–240; 1906–MZA F5 kniha 242–245; 1925 – MZA F263 kniha 1; 1936: MZA F263 kniha 4–5; 1952 – MZA F302 kniha 4. Data for the late 1780s are estimates)

insufficient for their animals. When the imperial administration gave the logical answer that all neighbouring settlements had the same treatment and no one else complained, the villagers of Dubňany tried to argue that the pasture areas were in fact their *property* and they should therefore be allowed to *use* some more pasture. In other words, they tried to misuse the concepts of ownership and usufruct to their advantage. Even though they presented various documents to back their claim, this was of course legal nonsense and was promptly dismissed by the imperial administration. The affair closed the history of common wood-pasture in Dúbrava. Further and increasingly effective central legislation allowed for less and less pasturing in forests, and at present wood-pasture is entirely prohibited by law. Woodland meadows also ceased to exist; today we do not even recognise their former locations.

The disappearance of wood-pasture and woodland meadows posed a considerable challenge for the forestry administration. Under traditional management, Dúbrava was a semi-open landscape. It was obvious that open patches would have to be afforested. However, many of them—because of their poor soil or poor drainage—were close to unsuitable for modern forestry. It was not until the 1920s, that foresters succeeded in getting rid of open areas through pine plantations and a complex network of drainage channels (Fig. 14.3).

14.4 Coppice

Coppicing has an unusual history in this Wood. It was part of the pre-1780s system and generated some income from wood-sales, but it was not dominant (MZA F5 karton 538 inv. č. 1569–1579). As can be gleaned from relatively late documents, coppice and wood-pasture functioned side by side. Trees were always the overlord's property, and the tenants had to take extra care not to allow their animals into freshly cut areas, although we do not know if this was done with temporary fences or by close herding. In the post-1780s system, coppicing was going to have a major role. Contemporary forestry management plans inform us that by 1851 latest, apart from ca. 300 ha of conifer plantations, the entire Wood was coppiced on a 28-year and later 36-year rotation. According to the foresters, the coppices were a mixed success. The stools left after cutting the very old oaks in the wood-pastures were often reluctant to sprout, and aspen (which suckers rather than sprouts from the stool) grew better in lower-lying areas than on higher ground (MZA F5 kniha 232). In 1864, the freshly installed forestry administration heavily condemned the coppices, which "hardly produced half of the former yield [*Bodenertrag*]" (MZA F5 kniha 233). Instead of the coppices, they proposed high-forests and pine plantations. Their advice was followed, but the transformation process was slow and coppices remained a majority use until the 1950s (Fig. 14.3).

The question is whether we can accept the foresters' reports at face value and conclude that coppicing was not a viable management option in Dúbrava. Closer examination reveals that most of the foresters' arguments were less than sound. First of all, let me stress again that the foresters who condemned coppicing were newly appointed and came with an intellectual heritage that was against coppicing in general. As for particulars, it is often claimed that oak does not coppice very well, but experience has proved the contrary (Crockford and Savill 1991). However, when they are old, coppicing standard trees for the first time is problematic. It is certainly not very surprising that cutting the veteran oaks (which were, according to the foresters 300–400 years old) on the former wood-pastures did not produce spectacular results—not least because, despite efforts to the contrary, grazing did not immediately stop in the Wood.

14.5 Conclusions: Larger Context and Local Colour

Traditional management and common uses disappeared from Hodonínská Dúbrava gradually from the mid-eighteenth century to the mid-twentieth century. That multiple woodland uses and common rights were despised was a universal phenomenon in eighteenth- and nineteenth-century Europe (e.g. FritzbØger 2004). This was governed by the ideology of the Enlightenment: "the belief that all the world's problems can be solved by a combination of science (or what is presented as being science) and government. Land uses are to be categorised and simplified; common land and other multiple land uses are bad" (Rackham 2006, p. 446). The other basic influence was that of economic liberalism, which stressed that "the purpose of forests was to maximize profit for landowners" (Puettmann, Coates and Messier 2009, p. 6). Considering that the Wood was the property of one of the foremost rulers of eighteenth-century Europe (the Habsburgs), its history can be considered a quintessential representation of a general European phenomenon.

What happened in Dúbrava was the woodland version of the process known as 'enclosure' (Neeson 1993). Coppicing, on the other hand, had a somewhat surprising upswing before its abandonment in the mid-nineteenth century. This was in fact due to a coincidence. Modern forestry appeared in southern Moravia around 1800, and foresters originally did not oppose coppicing as long as it was regulated according to their standards (Szabó 2010b). This is somewhat in contrast to other regions in the Czech Republic and elsewhere in Europe, where high-forests were preferred from the beginnings of modern forestry (Puettmann, Coates and Messier 2009). Already in 1864, however, the freshly installed forestry administration had a very different attitude from their predecessors, whom they accused of "not knowing what they were doing" (MZA F5 kniha 233). In summary, before the end of the eighteenth century Hodonínská Dúbrava was a traditional wood-pasture with extensive common usage rights. The demise of these common rights and the gradual emergence and final dominance of plantation forestry happened in a way that was usual in many other European woods, with the exception of a short coppicing intermezzo in the early nineteenth century. It is rather interesting to notice that as opposed to many examples from other European regions, around Dúbrava the locals showed little resistance to enclosure. The reasons behind this are most probably threefold. First, the villages were involved in the decision-making process. Second, they seem to have agreed with the general principles—as we have seen above, only one village complained and even that was a side issue that did not concern underlying principles. Third, and arguably most important in any debate, the villages would have had to face the mighty Habsburg family, and they apparently managed to estimate their chances in such a dispute in a realistic manner.

Acknowledgments This chapter was written with the help of grant IAA600050812 and long-term research development project no. RVO 67985939, both from the Academy of Sciences of the Czech Republic.

References

AOPK CR [Agency for Nature and Landscape Conservation of the Czech Republic] (2002–2010) Soil maps 1:50000, digital edition.http://www.nature.cz/monitoring-pud. Accessed 16 May 2011

Bičík I, Jeleček L, Štěpánek V (2001) Land-use changes and their social driving forces in Czechia in the 19th and 20th centuries. Land Use Policy 18:65–73

Boček A (ed) (1839) Codex Diplomaticus et Epistolaris Moraviae, tom. 2. Olomouc

Brandl V (ed) (1874) Codex Diplomaticus et Epistolaris Moraviae, tom. 8. Verlag des Mährischen Landes-Ausschusses, Brno

Brandl V (ed) (1878) Codex Diplomaticus et Epistolaris Moraviae, tom. 10. Verlag des Mährischen Landes-Ausschusses, Brno

Bürgi M (1999) A case study of forest change in the Swiss lowlands. Landscape Ecol 14:567–575

Chytrý M (1997) Thermophilous oak forests in the Czech Republic: syntaxonomical revision of the Quercetalia pubescenti-petraeae. Folia Geobotanica 32:221–258

Chocholáč B (1994) Hodonínské panství počátkem 17 Století. Časopis Matice Moravské 113:59–70

Crockford KJ, Savill PS (1991) Preliminary yield tables for oak coppice. Forestry 64:29–50

FritzbØger B (2004) A windfall for the magnates: The development of woodland ownership in Denmark c, 1150–1830. University Press of Southern Denmark, Odense

Johann E, Agnoletti M, Bölöni J, Erol SY, Holl K, Kusmin J, García Latorre J, García Latorre J, Molnár Zs, Rochel X, Rotherham ID, Saratsi E, Smith M, Tarang L, van Benthem M, van Laar M (2011) Europe. In: Parrotta JA, Trosper RL (eds) Traditional forest-related knowledge: Sustaining communities, ecosystems and biocultural diversity. Springer, Dordrecht, Heidelberg, London and New York, pp 203–250

MZA – Moravský zemský archiv (Moravian Archives)

Neeson JM (1993) Commoners: Common right, enclosure and social change in England, 1700–1820. Cambridge University Press, Cambridge

Novák V, Pelíšek J (1943) Stručná charakteristika půd na přesypových pískách v lesní oblasti Dubrava u Hodonína. Lesnická Práce 8:225–235

Nožička J (1957) Přehled vývoje našich lesů. Státní zemědělské nakladatelství, Praha

Puettmann KJ, Coates KD, Messier C (2009) A critique of silviculture: Managing for complexity. Island Press, Washington, Covelo and London

Rackham O (2006) Woodlands. Collins, London

Roleček J (2004) Subcontinental oak forests of Carici fritschii-Quercetum roboris association in Záhorská nížina Lowland (Slovakia). Bulletin Slovenskej botanickej spoločnosti 26:163–176

Roleček J (2007) Vegetace subkontinentálních doubrav ve střední a východní Evropě. Masaryk University, Brno (Unpublished PhD dissertation)

Szabó P (2010a) Ancient woodland boundaries in Europe. J Hist Geogr 36:205–214

Szabó P (2010b) Driving forces of stability and change in woodland structure: a case study from the Czech lowlands. For Ecol Manage 259:650–656

Tolasz R, Míková T, Valeriánová T, Voženílek V (eds) (2007) Climate atlas of Czechia. Czech Hydrometeorological Institute and Palacký University, Olomouc

Warde P (2006) Ecology, economy and state formation in early modern Germany. Cambridge University Press, Cambridge

Part III
The History and Use of Landscape Commons

Chapter 15
'A very fair field indeed...': An Archaeology of the Common Lands of English Towns

Mark Bowden and Nicky Smith

15.1 Introduction

Commons are often seen as existing in a specifically rural milieu but historically towns and cities in England, and elsewhere, were provided with common pasture lands. These were used primarily for grazing the draft animals of those townspeople engaged in trade, but they also provided pasture for farm animals in an economy where the rural and the urban were inextricably mixed and where, in many cases, towns had grown out of or been developed upon existing villages. Many towns and cities retain at least some of their commons: examples include historic cities such as Lincoln and York; great industrial cities such as Newcastle-upon-Tyne; small towns such as Hungerford and Stockbridge; places which have almost lost their urban status, such as Minchinhampton and Corfe; and conversely, places which gained urban status at a late stage, such as Brighton or Plumstead. The common lands attached to these places can therefore move from a rural to an urban milieu or vice versa.

The commoners had many rights, as on any other common: to graze animals (pasture); to feed pigs on fallen acorns and beech mast (pannage); to collect wood and other materials for building, fuel and crafts (estovers); to dig for minerals (common in the soil); to cut peat or turves for fuel (turbary); and to catch fish in ponds, streams and rivers (pescary). Beyond these main rights there were others peculiar to particular areas or commons, for example: the right to dry laundry and beat carpets or the right to set up tenter frames to dry dyed cloth; the right, or indeed the obligation at some periods, to practice archery. It should go without saying that the rights existing on any particular common were constrained by the underlying geology, geomorphology, existing vegetation cover and its direct

M. Bowden (✉) · N. Smith
English Heritage, Kemble Drive SN2 2GZ Swindon, UK
e-mail: Mark.Bowden@english-heritage.org.uk

I. D. Rotherham (ed.), *Cultural Severance and the Environment*, Environmental History 2, DOI: 10.1007/978-94-007-6159-9_15,

management; gravel could not be dug from commons where there was no gravel and pannage for pigs could not be found where there was no woodland. Town commons also developed as places of recreation and entertainment, as extensions of domestic and industrial space. They were a locus for public building and an arena for military, religious and political activities; the latter often subversive or rebellious.

Although there have been valuable studies devoted to aspects of the history and archaeology of some individual town commons, they have been largely disregarded as a unique class of historical landscape by historians and archaeologists. Honourable exceptions include F.W. Maitland, who, as early as the 1890s, was urging urban historians that they must deal with fields and pastures (1897) and more recent commentators such as French (2000, 2003), Giles (1950) and Hammond (1931). The Royal Commission on the Historical Monuments of England (RCHME) and English Heritage (EH) carried out archaeological surveys of the commons at Minchinhampton (Smith 2002), Corfe (Fletcher 2003) and Newcastle Town Moor (Lofthouse 1995).

The remaining urban commons are under threat and are not adequately protected, despite recognition of their wildlife and recreational value. In 2002, English Heritage embarked upon a project to study town commons in England, to match its existing initiatives in other aspects of the urban scene—EH had highlighted, in *Power of Place* (HERSG 2000) and other policy documents, issues of urban conservation and regeneration—a matter of high level government strategy. The aim of the project was to investigate, through taking a representative sample, the archaeological content and Historic Environment value of urban commons in England, and to prompt appropriate conservation strategies for them. The objectives were to research and survey a selection of urban commons in England, to make available the results of that work in the most appropriate ways to the widest constituency, and if possible to promote local community conservation initiatives.

A few town commons were surveyed in detail and more cursory work was undertaken in over fifty places. The result is the first overview of the archaeology of town commons (Bowden et al. 2009).

15.2 Archaeological Remains on Town Commons

Common land generally is a rich resource because its relatively benign traditional land-use preserves the physical evidence of past activities, including prehistoric and Roman remains as well as the traces of common use itself. Only 3 % of the English land surface is common but 11 % of Scheduled Ancient Monuments are on common land. Petersfield Heath Common has a well preserved Bronze Age barrow cemetery; Beverley Westwood has rare Iron Age barrows and an enclosure which is probably of early medieval date; Minchinhampton Common has what is probably the earliest monument on any town common—a Neolithic tomb—as well as traces of prehistoric field systems and the famous, probably late Iron Age,

Bulwarks; Newcastle Town Moor preserves the earthworks of a late prehistoric settlement within a mile and a half of the modern city centre. Bootham Stray, York, has a Roman marching camp, partly well preserved on the edge of the common pasture and partly flattened by the ploughing of the adjacent half-year lands.

A wide swathe of common land once existed at the confluence of the Rivers Stour and Avon, in Christchurch, Dorset. Surviving remnants of this include Sopley Common, Ogber, Coward's Marsh and Town Common, while a residential area immediately to the south-east is suggestively named 'Jumpers Common'. The open space formed by Town Common and St Catherine's Hill is a Site of Special Scientific Interest (SSSI) and part of the wider Dorset Heaths Special Area of Conservation (SAC) (Smith et al. 2012, p. 7). Part is still subject to common grazing rights held by tenants or occupiers of residential property in the old borough of Christchurch, as delimited on a map of 1870. In addition to its ecological importance, the area supports an array of archaeological features, most of which are Scheduled Ancient Monuments (SAMs). Industrial-scale sand and gravel quarrying during the first half of the twentieth century has mutilated a number of the area's Bronze Age round barrows, yet nine of the eleven recorded on the OS 1st edition 25-inch maps (1872) survive in some form and are designated as SAMs; they form two linear cemeteries, one on either side of the prominent north–south ridge that forms the backbone of Town Common. Other features are better preserved, but poorly understood. These include an enclosure, roughly oval, defined by a low, broad bank (10 m wide) with a slight external ditch. It has no obvious entrance and its earthworks appear excessively broad for their height and the relatively small area (0.1 ha) they enclose. The feature defies easy explanation. Also poorly understood is a very unusual double-banked square enclosure, sub-divided by a further bank (Fig. 15.1). On the basis of documentary evidence and excavation finds this has been identified as the site of a medieval chapel of St Catherine, which belonged to Christchurch Priory (Peers 1912, p. 85). Although there seems no obvious reason to doubt the veracity of the identification, enclosures such as this were not normally constructed to house medieval chapels. Exceptional circumstances may have occurred to prompt enclosure of this particular chapel; alternatively, it could have been built within an earlier work. A slight bank and ditch leading from its north-eastern corner abuts a substantial cross-ridge bank, aligned on a bowl barrow. Further earthworks consist of linear banks and ditches, some of which are probably woodland boundaries, and one feature named the 'Hen Ditch'.

A new ten-year management plan will safeguard these remains for the future. It recognises the historical and archaeological importance of the area and stresses that there is still much to learn about its rich and fascinating past (Smith et al. 2012, p. 56). Importantly, it also recognises that relatively recent features such as reservoirs and telecommunications masts form part of the story of the area and link it with the town and the surrounding populace (ibid, 57).

Egremont, Cumbria, is a further example, where prehistoric barrows, cairns and a stone circle were recorded on the common in the eighteenth century (Hutchinson

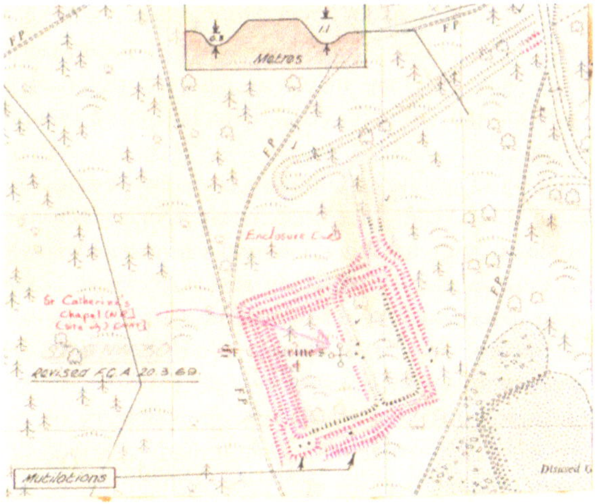

Fig. 15.1 Antiquity model prepared for OS map revision by Fred Aldsworth in 1969, showing the rectangular enclosure and related earthworks on Town Common, St Catherine's Hill, Christchurch; the drawing includes a profile across the enigmatic and undated cross-ridge bank; original scale 1:2500; north to top. (© Crown copyright. NMR.)

1794, p. 25). Here, however, the common has subsequently been enclosed and these remains destroyed, though the approximate site of the stone circle, on the southern end of a spur to the west of the town, can still be located (Burl 1976, pp. 69–342) close to a quarry named Ringingstone on the 1st edition OS 25-inch map (1895). The area of former common pasture is marked on the same map by small regular square enclosure fields with new or re-aligned roads, while the former common arable fields of the town still exhibit a reverse-S pattern.

Crucially, in some cases this remarkable preservation reveals how earlier landscape organisation has influenced the layout of present urban space. Surviving traces of fields on Hungerford Common that are almost certainly prehistoric or Romano-British in origin, share an alignment with the planned medieval town. It seems likely that these ancient fields influenced the layout of the town (Bowden et al. 2009, pp. 12–13; Fig. 15.2).

The uses of urban commons, as noted above, included grazing, which leaves no archaeological trace, and the getting of timber, furze, herbs, and so on, which is also mostly archaeologically invisible. Mineral extraction has, by contrast, left extensive remains on many commons; coal mines on Newcastle Town Moor, massive quarries and their extraction route ways on Beverley Westwood, Hungerford Common and Corfe Common, to name but a few.

Sports have also left their mark in the shape of the nearly ubiquitous race courses, playing fields, golf courses and even the remains of a bull-baiting pit. New evidence for such use continues to be discovered. Recent work combining aerial photographic and lidar (light detecting and ranging) analysis has revealed earthwork traces of a forgotten racecourse at Alston, Cumbria (Fig. 15.3). It roughly oval circuit can be traced intermittently as parallel linear ditches 11 m apart. Specific mention of the race course in documents has so far eluded researchers but, from its absence on the earliest OS maps of the area, the fact that it is partly overlain by late eighteenth- or early nineteenth-century properties and that

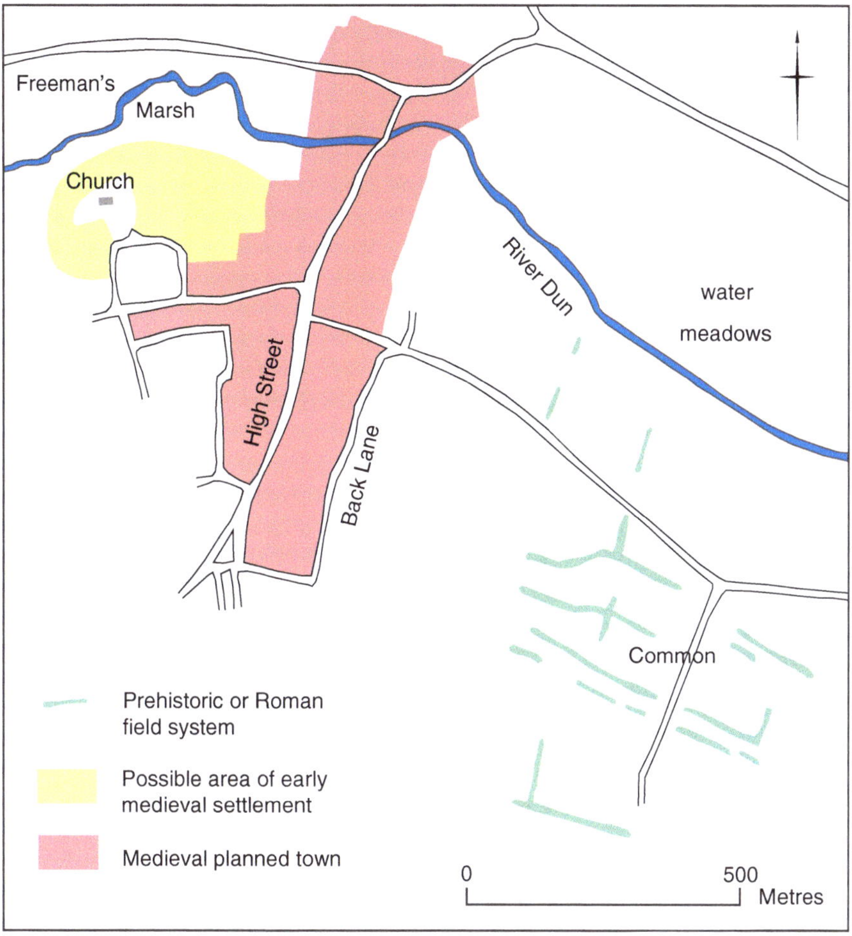

Fig. 15.2 Traces of late prehistoric or Romano-British fields survive on Hungerford Common; the main streets of the medieval new town seem to follow their alignment, emphasising the time-depth visible on so many surviving commons (Bowden et al. 2009, fig 2.3. © English Heritage.)

it appears to respect a mid-eighteenth-century toll road. Its period of use has been narrowed down to a few decades in the mid to late eighteenth century (Oakey 2010).

A long history of military activity on many town commons has also left a wide variety of remains, including rifle butts, practice trenches, airfields, Second World War defences and Cold War structures. Greenham Common, Newbury, has become the most notorious example, hosting military activity for hundreds of years before being requisitioned as a Royal Air Force airfield in 1941, and serving as the base for American (US) bombers and nuclear armed cruise missiles during the Cold War.

Fig. 15.3 Lidar image of part of Alston Moor with surviving elements of the eighteenth-century race course picked out in green. The remains were discovered during the course of English Heritage's 'miner/farmer landscapes of the north pennines' project. (© English Heritage.)

A large proportion of the common lands of English towns have been built over but even in these cases the shape of the common and its arrangement can often be observed in the resulting street layout; Liverpool's Townfield was engulfed by urban expansion in the eighteenth century but map evidence shows that the furlong layout can still be traced in the street plan, demonstrating that it had not been consolidated and enclosed piecemeal before that date (Elliott 1973, p. 89, Fig. 2.6). Liverpool's once extensive commons have now been reduced to two traffic islands.

15.3 Governance and Continuing Use

Town commons were administered by officials elected from among the local population but the variation in governance of commons across the country was considerable. Common rights could be vested in all householders in a town or they could be restricted to individual freemen, particular properties, burgesses, or even to senior members of the town's corporation. In some places common rights could be shared, conveyed, leased or sub-let and in some towns different commons had different rights attached to them. Not everyone who had common rights exercised them but on the other hand, because of leasing, commons were often used by people who had no rights. These complexities lead to illogicalities and apparent contradictions that have been noted by a number of writers (Moore 1945, pp. 12–14; Short and Winter 1999, p. 615; Thompson 1991, p. 159).

The main threat to town commons, historically, has been enclosure. This began at a very early date and was met by strong, though often futile, resistance at places such as Berkhamsted, Norwich, Atherstone and Newbury. Enclosure was perhaps always inevitable as the original reasons for the commons' existence and their purposes waned. In many cases, the only persisting use of commons was as a recreational space for the townspeople. As this had no perceived 'value' for those in power, it was ignored and over-ridden. Even where economic value existed, in the right to graze a cow or to gather materials for domestic crafts or small-scale industries, these rights could too easily be extinguished and the benefits appropriated by the ruling elite (Bowden et al. 2009, pp. 8–10). Perhaps it was a perception of town commons as places of rebellion and resistance that spurred on the enclosers. The commoners of Mousehold Heath, Norwich, had a long history of resistance to enclosure through the eighteenth and nineteenth centuries. However, the Heath's reputation for subversion goes back much further; in 1381 it was a camping ground for rebels during the Peasants' Revolt and in 1549 it was the base from which Robert Kett's rebels besieged Norwich in their fight against enclosures. However, it was the more recent history of the Heath and its use for uncontrolled activities, both economic and recreational, that drove the movement for enclosure and emparkment. Crucially, it was its economic importance to the people of Pockthorpe and other surrounding settlements that fuelled their remarkable resistance to enclosure (MacMaster 1990).

Some town commons were probably saved by the many other activities that they attracted. Some of these were acceptable to the establishment—fairs and agricultural shows (for example, Newcastle Town Moor's 'Hoppings', originally a temperance festival, and its Royal Agricultural Show). Sports such as golf and cricket were also generally acceptable. Military training assumed a major significance on some town commons, such as Woolwich; and even weapons testing, from Congreve rockets on Wimbledon Common during the Napoleonic Wars to the first tank, 'Little Willy', on Lincoln South Common during the First World War. Military use has been followed, poignantly, by commemorative uses, as in the case of the prominent war memorial on Berkhamsted Common, commemorating the thousands of men who were trained there during 1914–1918. Relatively modern military structures are also present, for example concrete vestiges of an underground Royal Observer Corps monitoring post on Town Common, Christchurch, designed to confirm and report hostile aircraft and nuclear attacks between 1959 and 1991 (*Defence of Britain Project* 1995–2005, UID 2829).

Other activities that took place on town commons were of a more contradictory nature. These included political and religious meetings; famously, for instance, the 'Grand Reform Dinner' of 1832 on Brandon Hill, Bristol. This began as a respectable affair but ended riotously after it was gate-crashed by opponents (Poole 1999). Similarly, Kennington Common, London, was where George Whitefield preached 'the power of god' to crowds of *c.* 20,000 people in 1738, but in 1848 it became the venue for the 'Great Chartist Meeting', prompting government fears of revolution and the evacuation of Queen Victoria to the Isle of Wight for safety. Horse racing, a popular pastime on town commons, was often patronised by the nobility and the royal family, yet it was banned on some because of the disreputable characters it attracted. In the early years of the twentieth century, new attractions appeared on town commons in the form of flying displays and competitions, augmenting the circuses, fairs and Wild West shows that had been a part of the calendar for many years.

John Moore's description of the town common of his fictitious 'Elmbury', actually Tewkesbury, emphasises the more *ad hoc* on-going social activities:

> '... Elmbury's own field, called the Ham ...was something of a legal curiosity, and mixed up in its title-deeds were some of the principles of feudalism, capitalism ... and communism...every man, woman and child in Elmbury had the right to walk and play in the field, which gave them a good possessive feeling about it. It was always 'our Ham'. In winter we shot snipe there, and sometimes hares, without let or hindrance. In May, when buttercups gilded it, and the grass was as high as your waist, the courting couples used its cover for their amorous games...But in June the lovers' hiding places were laid bare, and those same lovers, probably, were toiling and sweating on the wagons, bringing in the hay. Then, while the quick-growing aftermath painted the field green again, and the ochreous sheep or the white-faced Hereford cattle were turned out to graze on it—then the Ham became more than ever Elmbury's playground. Cricket-pitches, on which the ball broke unpredictably, made brown scars on the turf. From the banks of the river jutted out numberless fishing rods ...
>
> Meanwhile along the towpath, on summer evenings and Sunday afternoons paraded ... shopkeepers and their wives ... mothers wheeling their babies out for an airing; boys and girls 'walking out' ...it was part of the life as well as the landscape of Elmbury ... It was a very fair field indeed ...' (Moore 1945, pp. 12–14).

Such innocent pleasures were not supported by the establishment and indeed in many places were rigorously stamped out, leading to the emparkment and gentrification of many former commons.

15.4 Recent Losses and Response

Despite many successes for the Commons Preservation Society towards the end of the nineteenth century (Bowden et al. 2009, p. 10) urban commons remained under threat throughout the twentieth century. So Mousehold Heath for example, still needs its Defenders. At Bungay, while there was a long-running legal dispute over the ownership of Outney Common, the integrity of the common itself has not been seriously threatened, except by a proposal for housing development in 1938 which, fortunately, came to nothing (Reeve 1996, pp. 98–102, 111–139). Small parts of the common have nevertheless been lost to peripheral development, and it is this 'nibbling' which is so often destructive.

In many cases, 'nibbling' has been justified by the need to provide space for buildings such as hospitals, prisons, schools and colleges. Town commons also accommodate public utilities such as reservoirs, water towers, sewage works, refuse tips and urban infrastructure such as pipelines, cables and telecommunications masts (Fig. 15.4). They provide convenient empty corridors for railway tracks, by-passes and ring-roads (Fig. 15.5). While such features form part of the recent history of most urban commons, they also initiate the destruction of many—altering their character and breaking them up into convenient parcels for further development.

Fig. 15.4 Traditional uses of town commons often continue alongside the later infrastructure which most support. Here, on King's Meads, Hertford, horses graze near the branch line which connects the town to the national rail network. (Photo: Nicky Smith)

Fig. 15.5 Town commons provide convenient open corridors to accommodate new road and rail routes. The Kingsmead viaduct, seen here, was built in 1976 to carry the new A10 trunk road across Hertford's town common. (Photo: Nicky Smith)

The fact that many town commons are on marginal land, unsuitable for building and undesirable for cultivation—heath, bog, river floodplain, salt marsh and coastal strip—has afforded them an element of protection for many centuries. However, this is diminishing with ever more effective drainage technology, new flood alleviation schemes, a growing shortage of building land and the movement of industry and housing out of town centres. Threats to the remaining town commons in England include housing encroachment, industrial development, mineral extraction, flood alleviation schemes and park and ride schemes. The latter nearly destroyed the West Common at Lincoln, but strong local opposition, especially from those who graze their horses on the common and are actively involved in its management, stopped the proposed scheme.

The recognition of town commons as a valid historical entity and a valued part of the modern urban environment is a vital first step towards successful informed conservation. This is what our project aimed to assist. An important consideration for the future is maintaining the character of town commons as a different sort of urban open space, distinct from parks and public gardens. The fact that they are no longer, generally, working as agricultural commons should not mean that they are treated as urban parks. A local, 'bottom up' approach to the management of these spaces would enable townspeople to enjoy these commons as active participants, developing their interest in both their natural and historical aspects.

Acknowledgments We are grateful to our many colleagues who worked on the English Heritage Urban Commons project and in this case particularly to Dave Field. We also acknowledge the help of other colleagues, Peter Addison and Hugh Beamish, for supplying information on the commons of Christchurch. We would like to take the opportunity to apologise to Neil MacMaster for our unaccountable lapse in mis-attributing his work on Mousehold Heath in our previous publication; we have corrected the error here.

References

Bowden MCB, Brown G, Smith NA (2009) An archaeology of town commons in England: 'a very fair field indeed'. English Heritage, Swindon

Burl A (1976) The stone circles of Britain, Ireland and Brittany. Yale University Press, New Haven

Defence of Britain Project (1995–2005) (http://ads.ahds.ac.uk/catalogue/specColl/dob/index.cfm)

Elliott G (1973) Field systems of northwest England. In: Baker ARH, Butlin RA (eds) Studies of Field Systems in the British Isles. Cambridge University Press, pp 41–92

Fletcher M (2003) Corfe Common, Purbeck, Dorset. (Archaeological investigation report series AI/18/2005), English Heritage, Swindon

French H (2000) Urban agriculture, commons and commoners in the seventeenth and eighteenth centuries: the case of Sudbury. Suffolk. Agr Hist Rev 48(2):171–199

French H (2003) Urban common rights, enclosure and the market: Clitheroe Town Moors 1764–1802. Ag Hist Rev 51(1):40–68

Giles PM (1950) The enclosure of common lands in Stockport. Trans Lancs Cheshire Antiq Soc 62:73–110

Hammond B (1931) Two towns' enclosures. Econ Hist Rev 2:258–266

HERSG (Historic Environment Review Steering Group) (2000) Power of Place: the future of the historic environment. English Heritage, London

Hutchinson W (1794) A history of the county of Cumberland. Law and Son, Carlisle

Lofthouse CA (1995) Town Moor, Newcastle. (unpublished archaeological survey report), RCHME, Newcastle

MacMaster N (1990) The battle for Mousehold Heath 1857–1884: "Popular Politics" and the Victorian public park. Past Present 127(i):117–154

Maitland FW (1897) Township and borough: with an appendix of notes relating to the history of Cambridge. (re-printed 1964), Cambridge University Press, Cambridge

Moore J (1945) Portrait of Elmbury. Collins, London

Oakey M (2010) Photo finish for England's highest racecourse. Research News, vol 15. English Heritage, Swindon, pp 3–5

Peers CR (1912) Christchurch: architectural descriptions of the town, castle and priory. Vict Cty Hist Hamps, Constable Lond 5:83–109

Poole S (1999) ''Till our liberties be secure'': popular sovereignty and public space in Bristol, 1750–1850. Urban History 26:40–54

Reeve T (1996) A common privilege: the story of Outney Common. Bungay, Morrow, Bungay

Short C, Winter M (1999) The problem of common land: towards stakeholder governance. J. Environ Plan& Manag 42(5):613–630

Smith NA (2002) Minchinhampton Common: an archaeological survey of the earthwork remains. (Archaeological Investigation Report Series AI/12/2002), English Heritage, Swindon

Smith JE, Parfitt A, Minter R and Christchurch Countryside Service (2012) St Catherine's Hill & Town Common Management Plan, 2012–2022: Consultation Draft. Prepared for St Catherine's Hill & Town Common Management Plan Steering Group, Christchurch Borough Council, Christchurch

Thompson EP (1991) Customs in common. Merlin Press, London

Chapter 16
From Pasture Woodland, via Deer Park and Common, to Cultural Severance: A Case Study of the Commons of Ashampstead, Berkshire

Dick Greenaway

16.1 The Landscape Setting

The Commons lie on the dip slope of the Berkshire Downs in West Berkshire. The setting is bounded on the north by The Ridgeway along the crest of the Berkshire Downs, on the east by the River Thames at the Goring Gap, on the south by the channel of the River Pang and on the west by the western watershed of the Pang Valley (see Fig. 16.1).

Topographically the area lies on the interfluves between the River Thames on the east and the valley of the River Pang on the west. A narrow steep-sided dry valley runs roughly northwest to southeast through the area and creates two plateau areas (see Fig. 16.2).

The solid geology is chalk but the soils vary widely from strongly calcareous to the most acid Eocene clays, gravels and sands (Soil Survey of England 1967). The soils are everywhere very permeable and as a result there are few streams or surface waters other than the River Pang. The Commons themselves lie along the southern edge of the parish of Ashampstead which lies astride the central dry valley and extends on to the plateaux on either side.

16.2 Occupation in the Landscape

Scattered finds of implements from the Paleolithic, Mesolithic and Neolithic periods indicate occupation from earliest times. The Pang Valley has two Neolithic long barrows at its northern extreme, is rich in Bronze Age round barrows, has two

D. Greenaway (✉)
The Cottage Ashampstead Common, Pangbourne,
Reading West Berkshire RG8 8QT, UK
e-mail: rg.greenaway@btinternet.com

I. D. Rotherham (ed.), *Cultural Severance and the Environment*,
Environmental History 2, DOI: 10.1007/978-94-007-6159-9_16,

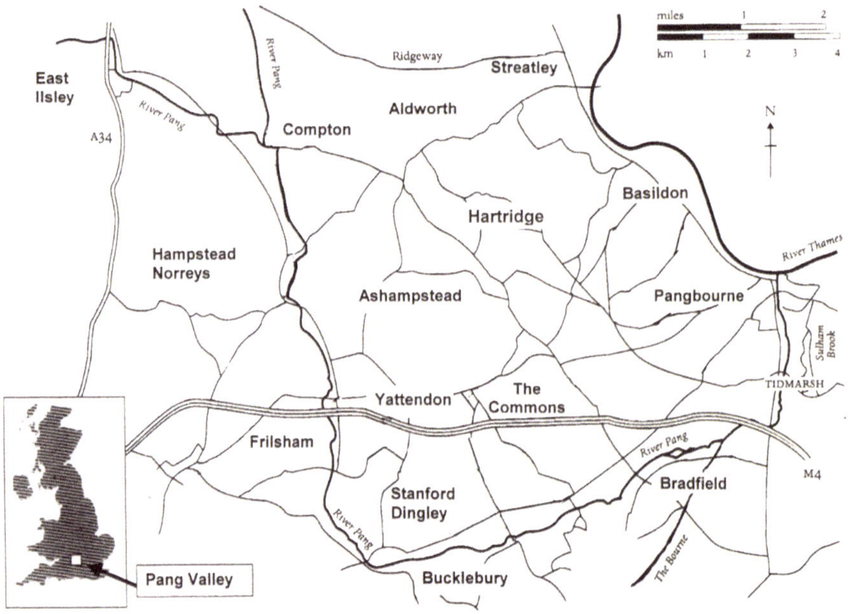

Fig. 16.1 The Pang Valley

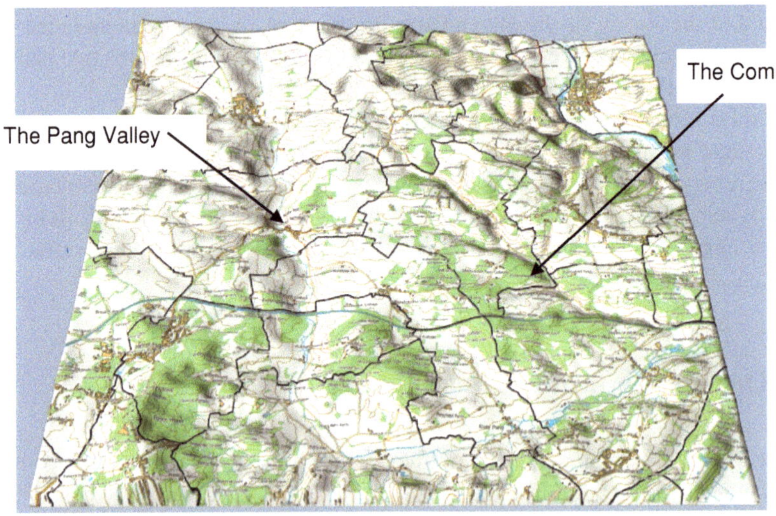

Fig. 16.2 The topography

Iron Age hill forts and several Roman sites. The Goring Gap section of the Thames valley has evidence of occupation in all periods. The setting would seem to typify Tom Williamson's description of

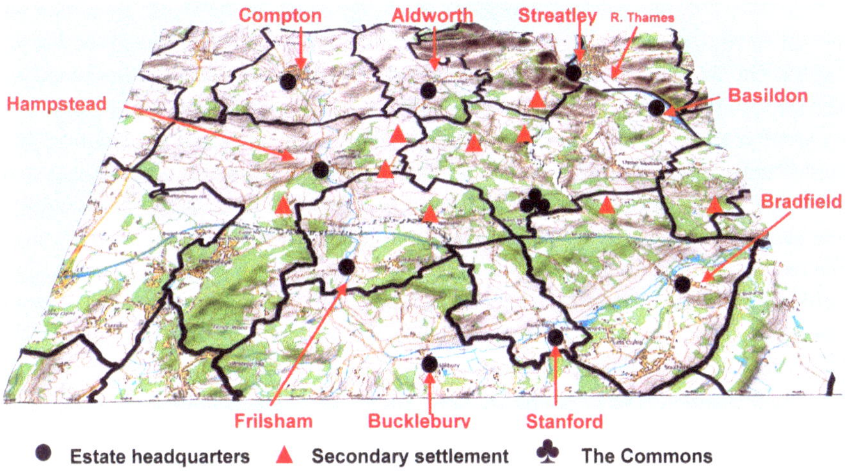

Fig. 16.3 Late Saxon estates based on modern parish boundaries

> *.. high interfluves lying between the principle valleys ...... occupied by intractable clays or poor, acidic sands, and often .... lacked dependable water supplies. When population densities were low these uplands tended to be occupied by woods and pastures...... In most cases, as populations rose they came to be permanently occupied, but even then they generally remained dependant—tenurially, economically and socially—on settlements in the neighbouring valleys* (Williamson 2008).

The Pang valley and the interfluve to the west of the dry valley, together with the interfluve between the dry valley and the River Thames, were divided into a number of Saxon estates based on settlements in the Thames and Pang valley bottoms. They extended up on to the interfluves on either side thus sharing the full range of resources of water, hay meadows, arable land, dry pastures and woodland (see Fig. 16.3).

16.3 Woodland History

We can only speculate about the character of the larger areas of woodland. The opinion that has prevailed for a long time has been that it was Pasture Woodland rather than dense high forest. The available evidence is sparse.

There are few reliable early charters for the area as a whole. Grundy (1925) quotes a charter of King Eadwig (955–959) which grants to Abingdon Abbey a wood named *Hafochrycg* (Hawkridge) on the eastern edge of the Pang Valley to provide oaks for the repair of their roof. A second charter, of which at least the boundary description is thought to be genuine, covers an exchange of land between King Alfred (870–899) and the Bishop of Winchester. The land described lies on the western side of the Thames stretching up on to the interfluve from the river.

The first charter implies that the wood contained mature oak trees and the second contains references to a *game enclosure* and to *the Hanging Wood of the Calves*, but also contains references to a *slade* and to a *hamstaede*—homestead and possibly estate centre (Ekwall 1960)—implying open farm land.

Apart from the charters it is necessary to rely on place name evidence. It is immediately obvious from this that, at least in the immediate pre-Conquest period, there was very little woodland, although Margaret Gelling locates the Forest of Barroc—from which Berkshire derives its name—immediately to the west of the Pang valley in the parishes of Chieveley, Ilsley and Fawley (Gelling M and Cole A 1974). Apart from Streatley on the eastern edge of the area there is not a single place name with *leah* (a clearing)—and only one with *feld* (a field or an open space within sight of woodland) (see Fig. 16.1).

When we come to the Domesday record (Morris J 1979)the lack of woodland is equally obvious. Of the twelve manors in the area covering nearly 14,000 hectares (33,438 acres) the total pig value is 401, with a note that the manor of Yattendon has 'wood for fences'; indicating the existence of some coppices. The largest concentrations are at Bradfield (100 pigs), Bucklebury (100 pigs) and Basildon (120). These are all south-eastern manors based in valley bottoms but running up on to extensive areas of acid infertile soils on the interfluves. The pattern was still apparent in 1761 when the first large scale map was made (Rocque 1762). At that date the area of woodland was approximately nine per cent and again concentrated in the southeast. The manors of Basildon and Bradfield include the area later covered by the Commons.

Finally, to add to the problem, there are numerous later field names indicating the clearance of woodland, 'Stubbles', 'Breach Field' etc. It would seem that we need to reassess our ideas of the eleventh-century landscape.

At this point I would like to consider the impact of the Statute of Merton and its subsequent amendments (Ruffhead 1786). It is generally accepted that medieval legislation was intended as much to raise revenue for the king as to govern the country, but this statute, passed in 1235, makes no mention of money or licences. It is purely permissive. It implicitly recognises Lords of Manors' ownership of manorial waste and states that '… *many great men of England (which have enfeoffed Knights, and their freeholders of small Tenements in their great Manors) have complained that they cannot make their profit of the residue of their Manors, as of Wastes, Woods and pastures* …' and goes on to allow them to enclose portions of the waste provided they leave enough for their tenants and do not block access to their tenants holdings.

Medieval kings frequently played off the lower orders against the baronage and this statute would have been very unpopular. Considerable pressure would have been needed to achieve it. I suggest that 'the great men of England' could see the rapidly growing population eating into the woodland resource with their ploughs and animals and took action to preserve this source of so many of the essentials of life. As a result many areas were enclosed with banks and ditches and then managed intensively for woodland products. Evidence of this in the Pang Valley is

Fig. 16.4 800 year old ash stool on a coppice boundary bank

provided by numerous coppices with massive 800 year old ash coppice stools on their boundary banks (see Fig. 16.4).

The Statute also made it easier for lords to create deer parks which provided other kinds of woodland products in addition to being private hunting grounds. There was a noticeable increase in park creation in the thirteenth century as has been demonstrated for Hertfordshire by Anne Rowe (2007). I do not think it is an exaggeration to claim that the countryside, particularly of southern England, would be a very different place had this statute not been granted.

16.4 The History of the Commons

As stated above the Commons lie along the southern boundary of the parish of Ashampstead (see Fig. 16.5). They cover approximately ninety hectares (200 acres) of the plateau to the west of the central dry valley ending on the east with a steep slope into the dry valley. The soils are mainly infertile and acid with Berkhamstead series on the north, Winchester series on the south and Frilsham series on the west. Small areas of chalky Icknield soils exist on the eastern slopes. These are the same soils as in the surrounding arable fields. The site was not chosen because it was especially infertile and of little use for anything else.

Apart from a thin scatter of Mesolithic, Neolithic and Bronze Age flint implements the first evidence of occupation is provided by a few eleventh-century pottery shards. In the twelfth and thirteenth centuries a substantial pottery industry existed on the western edge. This was exploiting clay from the side of the minor dry valley separating the two soil areas and was using kilns dug into the steep slope. The kilns were heated by wood from the area. Pottery from this site has been found in Oxford, Reading and Newbury (Mepham and Heaton 1995).

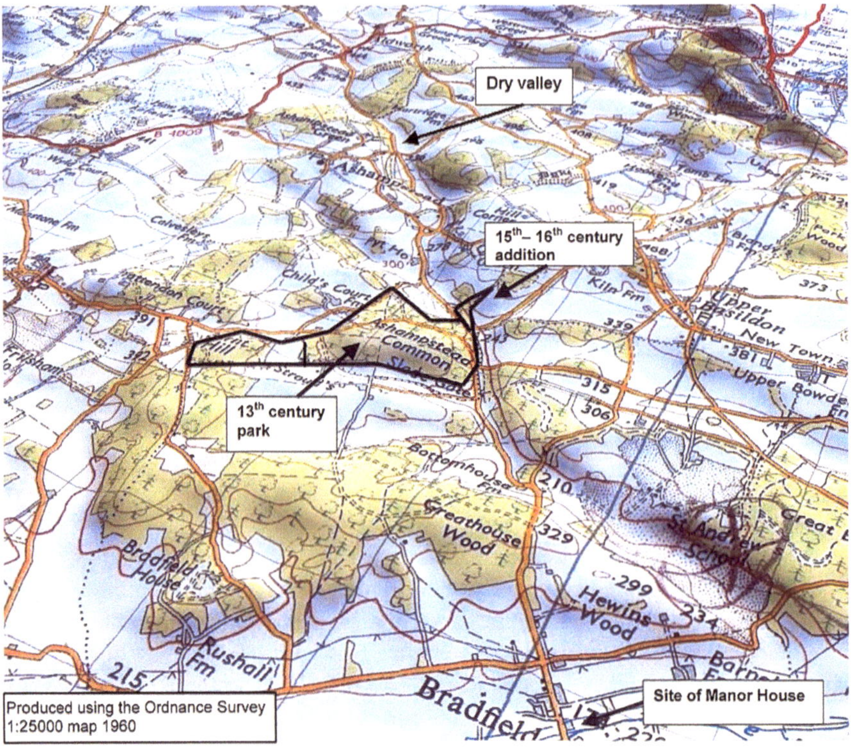

Fig. 16.5 The setting of the Commons and their relation to the manorial centre

16.5 The Deer Park

At some date early in the thirteenth century an area of common wood pasture was enclosed by a bank with a ditch on the inside to form a deer park but it seems that the commoners retained their rights of common. It may seem strange that a park should contain common rights and revert to being a common when ceasing to be a park. However, there is another documented example at Binswood (Hampshire County Council undated) and Stamper (1988) gives other examples.

Documentary evidence for the park's construction is hard to find. The shape might be thought unusual (see Fig. 16.1) but its construction and furnishing are so typical that I am convinced that it is a park and simply occupied an available area of land. There is a rather unconvincing alternative western boundary which would reduce the size of the park but give it a more regular shape. In the best preserved area along the southern side, the pale is built in short irregular lengths. Oliver Rackham (1997) considers this to be a characteristic of early gentry parks where the pale avoided major trees (see Fig. 16.2). Also, along the south-western and western lengths the pale coincides with the parish boundary.

The Pang Valley lay just outside Windsor Forest from which Forest Law was removed in 1227 (Hatherly and Cantor 1979). In 1253, Roger de Somery obtained a charter from Henry III granting him Free Warren on his lands outside the Forest, but he had Free Warren at his Manor of Bradfield before this date (Victoria County History - Berkshire (1923)). In 1240, in return for agreement not to oppose the enclosure of the wood, he had granted the right of free chase in it to William Englefield (Hatherly and Cantor 1979). This enclosure may mark the construction of the park at Ashampstead Commons. In 1291, a park at Bradfield is mentioned when it was assigned with the manor to Agnes de Somery (Cal. Close Rolls 1288–96). Bradfield lies on the Pang just three kilometres SSE of the Commons and is linked to them by a very ancient route connecting the Pang Valley to the high chalk in the north (Fig. 16.5). The route's antiquity is demonstrated by massive lynchets that have formed along its sides and by it being part of the boundary of the Domesday manor of Hartridge. In 1245, as stated above, the Prior of Poughly Abbey in West Berkshire brought a charge in Reading. *Assize of nuisance to declare whether Ralph of Aston, John of Bagpuize, Simon le Serjeant, Robert Coleman, Walter Bernard, Roger Pupyn, Geoffrey of Querner and Henry Grete, have unjustly erected a certain fence in Asseden to the nuisance of the Prior of Poughly's free tenement in Werdham, whereon he complains that by that fence he is prevented from going to his common of pasture in a certain wood where he always used to have common. .....* (Selden Soc.a. 1973)

It would seem that the fence complained of was the park pale. The list of jurors contains at least three men with local names—Ralf de la Beche, John de la Hawe, Gilbert of Everington—making it probable that the case concerned the Pang Valley area. A final piece of admittedly circumstantial evidence is that the manor of Bradfield eventually descended to John Langford who died in 1509. He enclosed four acres of 'moor', part of his demesne land, and added them to his park (Royal Historical Society 1897a, b). A very obvious extension of four acres exists on the eastern side of the Commons. From the above it would seem that the deer park constructed along the southern edge of Ashampstead parish was built shortly before 1245 by the de Somery family of Bradfield.

The de Somerys could not have picked a more appropriate piece of land for their park. Stephan Mileson has analysed the characteristics of deer park topography (Mileson 2009): *Existing natural features in parkland could be used to aid in the hunt. Suitable slopes within parks would have helped guide deer towards a place where archers were stationed, as well as making the hunting visible from higher ground. In fact the presence of gradients or valleys may even have played a part in determining the choice of a site for a park: the great majority of parks contained areas of sloping terrain.*

The area chosen is the southern end of a flat topped plateau the edges of which slope down steeply on the east and south. A minor dry valley cuts into the western slope of the main dry valley. A road runs along this to the west and is overlooked by the plateau. Further to the west another steep sided straight dry valley provides a perfect site for a coursing run and divides the eastern area from a flatter western area. However, there is no evidence that a coursing run was created (see Fig. 16.6).

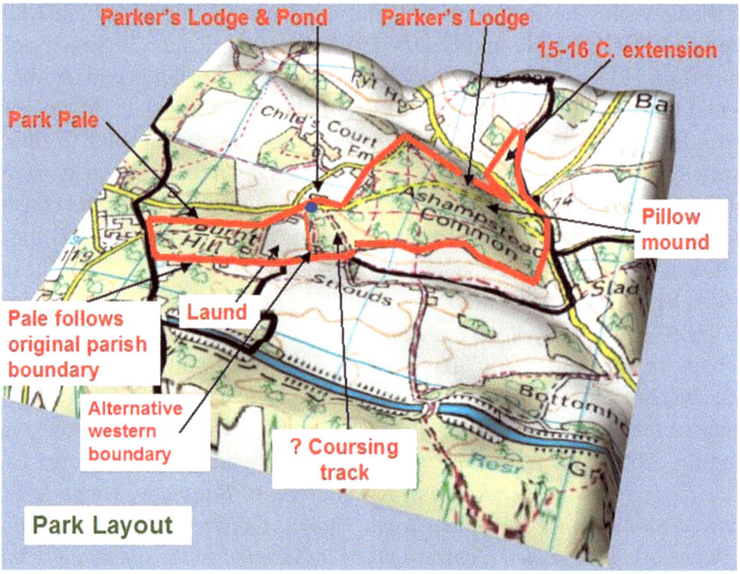

Fig. 16.6 The park layout

Fig. 16.7 The design of the pale and freeboard

The park was constructed by building a substantial bank around the area with a ditch, on the inside (see Fig. 16.7). A fence of the type shown in Fig. 16.8, or a hedge, would have topped the bank. A track, the freeboard, was created along the outside of the bank to give access to the pale for repairs. This still exists along the southern edge. It is bounded on the outside by a low bank which probably supported a hedge (see Figs. 16.9, 16.10, 16.11).

The area of the park is virtually devoid of surface water and so a pond was dug at the northern end of the dividing valley. Oliver Rackham mentions similar ponds being dug in Ongar Park and Monk's Park (Rackham 1997). A parker's lodge was

Fig. 16.8 Profile through the southern pale and freeboard

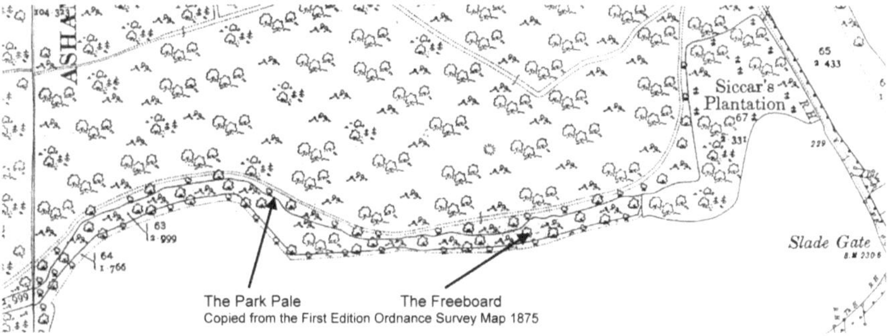

Fig. 16.9 The Park Pale

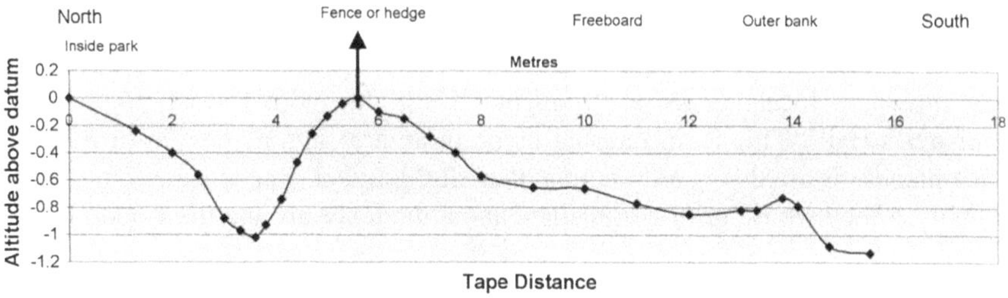

Fig. 16.10 An example of a park pale fence from Moccas Park

built on the slope overlooking the pond. Its position would allow the parker to monitor the deer using the pond, and would also allow a watch to be kept on people using the road entering the park from the west. This was an important function since poaching was a serious problem in all parks. A second lodge with a similarly extensive view was built on the plateau edge at the eastern end of the site overlooking both the ancient road running north from Bradfield and also the road

Fig. 16.11 Outer bank

crossing the area up the dry valley. Both these are typical. A *laund*—an open grassy area—was provided south of the pond and may have contained a small standing from which deer were shot. At least one pillow mound was built for rabbits. All these were the standard equipment of a deer park. It is curious to note that the eastern parker's lodge, which survived until the late nineteenth century, was known to local people as 'Maggot's Hall'. Its more formal title was 'St Margaret's Hall'. This is an unusual name in an area almost devoid of religious place-names. It is also worth noting that St Margaret was the patron saint of Poughly Abbey whose prior complained about the fence around the wood in 1245.

16.6 The Ecological Evidence

A detailed study of the trees and ground flora within the park has provided valuable evidence for its date and for land uses within it. For example an ancient yew (*Taxus baccata* L) growing on the boundary bank (Fig. 16.12) has been independently dated as approximately 800 years old.

Since it is growing *on* the bank, the bank must have been there in the thirteenth century and the yew may have started life as part of a hedge. Any fence in

Fig. 16.12 800-year old yew on the pale bank

proximity to a source of seed will become a hedge within a very few years unless carefully and regularly weeded and it is likely that the yew became established from seed deposited in droppings from birds perched on the fence.

Distribution maps of the ground flora, particularly the areas of dense bluebells (*Hyacinthoides non-scriptus* L) and dense dog's mercury (*Mercurialis perennis* L), have highlighted areas of undisturbed soils and areas of cultivation. The ecology behind this is peculiar to the local soils. Bearing in mind that the soils are predominantly acid, in undisturbed woodland bluebells and dog's mercury will coexist in reasonably equal proportions. However, when woodland is cleared for cultivation the acid soils have to be chalked or limed to raise their pH to a level which will allow cereals to grow. Local soils can have a pH as low as 4.5 whereas a pH of 7 is needed for cereals. Raising the pH was achieved by spreading chalk dug from below the acid soils thickly on the cleared land. If the area subsequently reverted to woodland the dog's mercury, being a lime loving plant, would quickly invade the area and form a dense mat. Bluebells with their very slow rate of spread would not be able to compete. The plots show this very clearly (Figs. 16.13, 16.14).

The area around the eastern parker's lodge has no bluebells but does contain cowslips. This indicates an area of paddocks or fields. Similarly, the sides of the

Fig. 16.13 A clear boundary between bluebell and dog's mercury

dry valley from which clay was quarried for the pottery, have no bluebells and are densely carpeted with dog's mercury. An area of dog's mercury around another quarry shows that it was enclosed within a field. The areas of dense bluebell probably mark areas remote from settlement which were rarely visited or areas of coppice where the lack of soil disturbance and the cropping regime encouraged the development of a dense mat of the plant.

Looking at later land use, there are a number of pollarded oaks and sweet chestnuts dating from the seventeenth and eighteenth centuries. Their existence not only shows that the area was providing timber, probably for firewood, but also that it was being grazed.

16.7 Use Made by the Community in the Historic Period

One of the few manorial documents so far found is a record of a Manor Court held in 1885 setting out the rights of common and presenting the findings of an inspection of the manor boundary (Greenaway 1998). The list of common rights is so informative that it is worth quoting them in full.

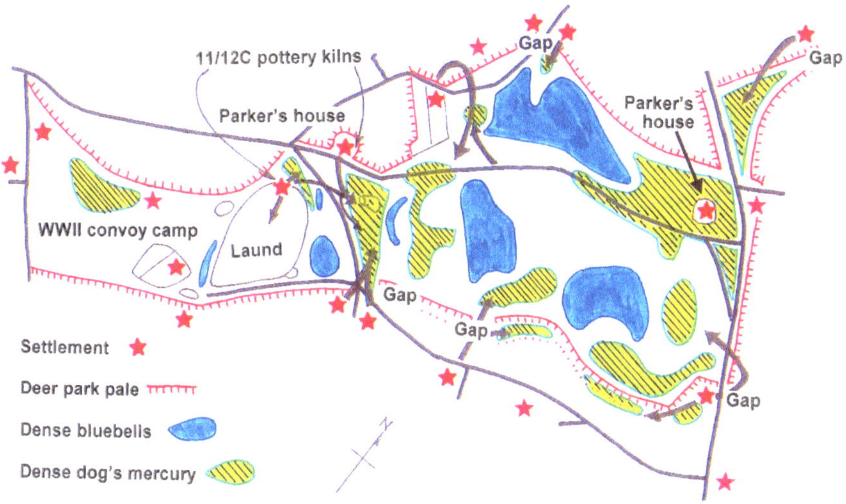

Fig. 16.14 The distribution of dense bluebell and dense dog's mercury in relation to park features and settlements

16.7.1 The Court Leet, Court Baron and Court of Survey of Roberts(sic) John Hopkins

Lord of the said Manor there held on Tuesday 9th day of June 1885 before me William Frank Blandy Gentleman steward there.

Names of the jury. John Butler (Foreman), James Trin, Robert Palmer, James Townsend, William Nullis, James Harrington, Josiah Woodley, George Haines, Joseph Rivers, Thomas Street. Presentments of the Jury.

We present that all the longtopped trees, and pollards growing on the Commons and Wastes within the said Manor are the property of the Lord of the Manor.

We Present that the inhabitants of the said Manor have the right to cut furze and bushes growing on the Common within the said Manor, but not elsewhere, and that no other person whatsoever has a right to cut furze off, or from the said Common but the Parishioners, and that the same furze and bushes are to be used within the said Manor and not elsewhere.

We present that all persons that shall feed any cattle on the Waste or Commons of the Manor without having a right of Common shall forfeit ten shillings for every horse or cow, and five shillings for every score [20] of sheep.

We present that no person shall turn more cattle into the commons or commonable places than he can winter on the lands in respect of which he claims his common.

We present that if any Hogs shall be found digging in any part of the parish, and are not ringed, the same shall be impounded by the Hayward, who shall have the same well ringed and receive of the owners thereof, one shilling for every Hog.

<u>We present</u> *that the Hayward shall impound all sheep that are not well marked with one, or more, letters of the owners name and shall receive at the rate of two pence for every sheep for any number not exceeding one score, and one shilling a score afterwards, and so in proportion for any greater or less number than a score, and that he shall take for impounding every horse sixpence, for every cow four pence, for every hog four pence, for every sheep under a score two pence, and for any greater number one shilling a score.*

<u>We present</u> *All persons who shall permit their cattle without a follower to take care of them, to feed in any of the lanes within this Manor, and direct that if cattle shall be found so feeding without a follower they shall be impounded by the Hayward.*

<u>We present</u> [Here follows a description of perambulating the boundaries of the Manor and of encroachments noted].

<u>The Jury present</u> *that the right to wash sheep in Burnt Hill Common Pond having been mooted on behalf of the owner of Hartridge, it has not been practised since the last Court and ought not to be allowed.*

Signed by the Jurymen above.

The formality of this document and its careful use of archaic terminology make it plain that there was no dispute about the area being common land. Indeed, the ancient regulations were followed at least until the early twentieth century (Watney 1906). They make it clear that intensive use was regularly made of the Commons and Manorial Waste, not only small scale use by smallholders for gathering firewood and for grazing the family cow, but also for commercial use with large numbers of animals involved.

The map evidence shows the result of this use. The 1762 map (Rocque 1762). Figure 16.15 shows a very open landscape with few scattered trees. By 1824, the situation had changed little. Greenwood's map (Greenwood C & J 1824) shows a very open park-like area (Fig. 16.16), the symbols used being quite different to those used to depict woods. The Tithe Award map of 1845 similarly shows a landscape of bushes and scattered trees (Fig. 16.17). However, by the time the First Edition of the Ordnance Survey 6 inches: 1 mile map (1875) the surveyors used a woodland symbol to depict the vegetation. This has been questioned. One resident, Mr A.S.J. Hancock whose memory stretched back to 1890, wrote an impassioned letter to the local parish magazine in 1966 when tree planting was being discussed (Hancock ASJ 1966). He made it clear that a large swathe along the NW edge had been open grassland used by residents living along the common boundary to graze cows, goats, pigs, poultry and bees. He calls it 'The Milk and Honey Land'. He describes how animals had to be tethered and how he became a cow keeper on the commons when he returned from the army in 1919. The numbers of people using the commons in this way steadily declined and Mr Hancock describes how ….. for the last quarter of this century the saplings, thorns and bramble bushes has (sic) been growing fast and covering up the grass acres which used to feed the tethered animals.

The Rocque Map of 1761, shows a scatter of small settlements all around the periphery of the commons with some actually inside. These settlements were

Fig. 16.15 1762

Fig. 16.16 1824

cultivating the surrounding arable land and also making use of the resources of the commons. The effects of this on the flora can be detected. Detailed plots of the areas of densest bluebells and densest dog's mercury when linked to the available breaches through the park pale indicate which settlements were exploiting which areas of the common (see Fig. 16.14). By the time of the Tithe Award map (1845) and then the Ordnance Survey First Edition (1875), many of these settlements had disappeared. Again the results of this can be seen in the flora. For instance, there is a noticeable component of trees which began life in the mid-nineteenth century as grazing pressure slackened. The late nineteenth century decline in use can be seen along the north-western side in the extensive growth of sycamore (Fig. 16.18).

In the late nineteenth century the landowner tried to give the area a more parkland appearance by planting a large number of exotic trees. The principal tracks became conifer lined avenues and groups of wellingtonia (*Sequoia gigantean (Lindl.*) and even Monkey Puzzle trees (*Araucaria araucana (Mol.*) were planted. The Commons were the community's playground. A clear area at the western end was used to play cricket (Fig. 16.19) and Methodist camp meetings were held where itinerant ministers preached to large congregations.

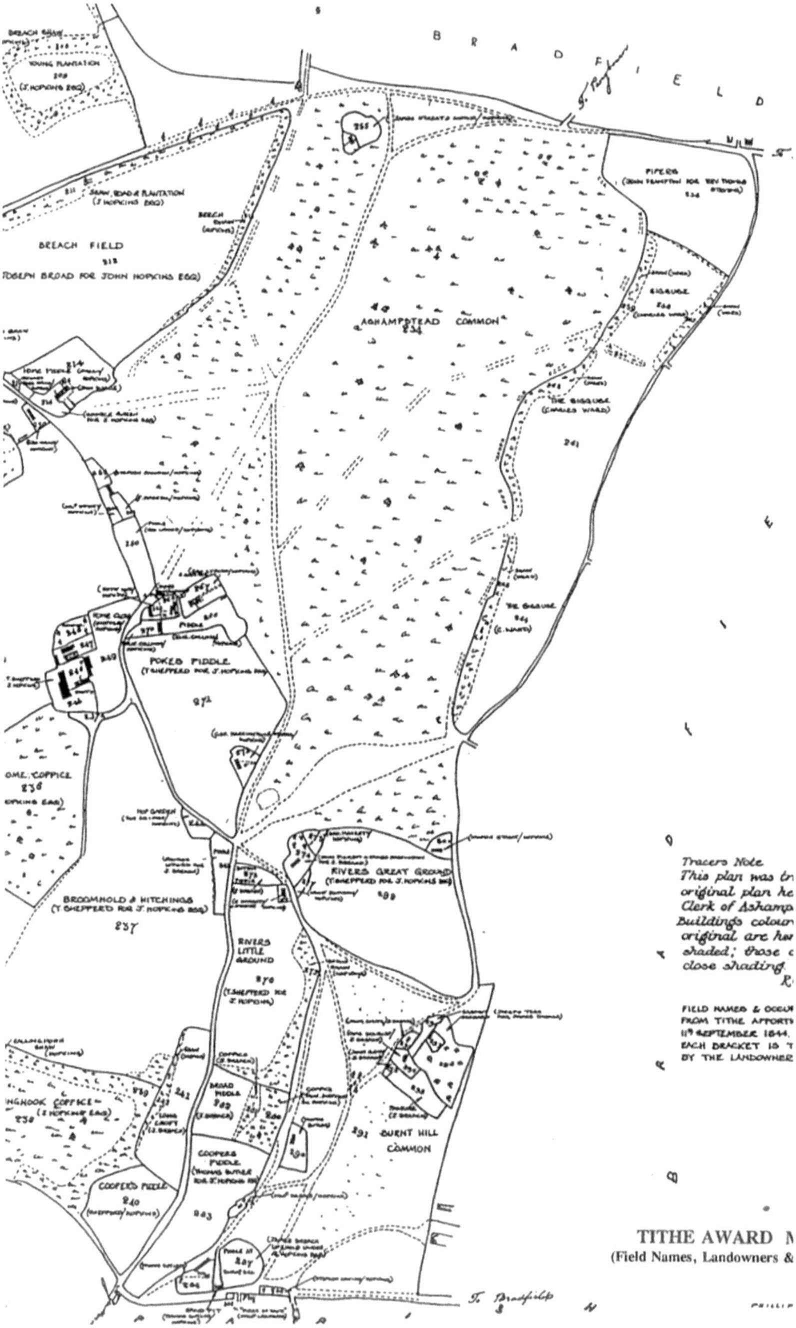

Fig. 16.17 1845

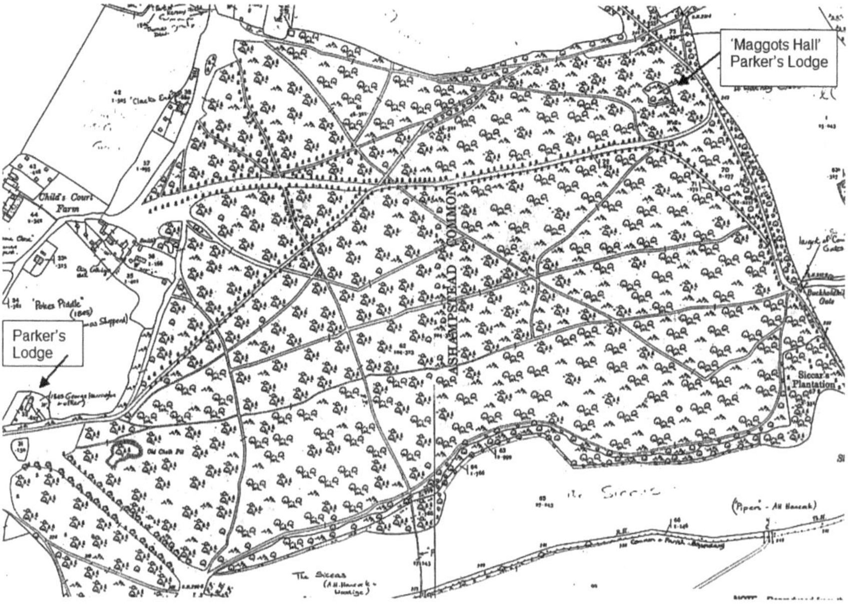

Fig. 16.18 1875

Fig. 16.19 The cricket team in 1902

Until 1889, parish councils were responsible for the up-keep of roads in their parishes. Materials were quarried from local commons and roadside wastes to reduce the charge on local taxation. The resulting large but shallow pits exist on many commons and can be dated to this period by the size of trees growing in them. Other, deeper, pits provided chalk to put on the surrounding arable fields and small shallow quarries provided clay for a local medieval pottery.

During World War 2, being within a day's drive of the south coast ports, the by now wooded commons were used to conceal troop movements. Although the troops lived under canvas, semi-permanent buildings were erected to service them and some foundations of these still remain. They provide an interesting archaeological study as they disappear under the leaf mould. After only sixty years some are already two centimetres down. A larger complex of huts and concrete roads was built at the western limit of the commons. After the war this was used to house displaced persons. It was demolished in the 1950s and is already a thriving oak wood.

16.8 What of the Future?

In the past, our commons were working and productive areas, essential to the life of the community. How do we return them to this position without destroying them as an equally essential source of tranquillity, mental and physical stimulus and sheer enjoyment in an increasingly demanding world? Currently the Commons are providing firewood in varying quantities with questionable attempts at planning for sustainability. This is opening the canopy and thus benefiting the flora and understory. However, increasing deer grazing is having a deleterious effect on both of these. The soils, mainly hard gravel, would not allow pig grazing as in the Wyre Forest, and there are insufficient local people prepared to monitor grazing cattle, so those traditional uses are not viable. There can only be a limited number of Field

Fig. 16.20 600 tonnes of firewood

Study Centres no matter how fine the site, so that is not an option. Volunteers are already working with the landowner's blessing to conserve and enhance the more important features—such as the veteran trees and flower rich glades—and this is maintaining the sense of the Common's importance in the local population and deterring grosser exploitation. This seems the best we can do until some future crisis focuses minds more clearly (Fig. 16.20).

References

Cal Close Rolls (1288–96) p 208

Ekwall E (1960) English Place-names. Oxford University Press, Oxford

Gelling M, Cole A (1974) The Place-names of Berkshire. Part III. English Place-name Society, Nottingham, p 837

Greenaway RG (1998) The Commons of Ashampstead Parish. Ashampstead Parish Council, Ashampstead

Greenwood C. & J. (1824) *Map of Barkshire* [Sic]

Grundy GB (1925) Berkshire Charters. Berkshire Archaeol J 29:204, 205, 210, 211

Hancock ASJ (1966) Letter to the Broadsheet. Published in Greenaway, R.G.(1998) The Commons of Ashampstead Parish. Ashampstead Parish Council, Ashampstead

Hatherly J, Cantor L (1979) The Medieval Parks of Berkshire. The Berkshire Archaeol J 70:74, 100, 190, 397

Hampshire County Council (undated) Pasture Woodlands—commons and forests. Hampshire County Council

Mepham L, Heaton MJ (1995) A Medieval Pottery Kiln at Ashampstead, Berkshire. Medieval Ceram 19:29–43

Mileson S (2009) Parks in Medieval England. Oxford University Press, Oxford, p 33

Morris J (1979) History from the Sources. Domesday Book. Berkshire, Phillimore, Andover p60

Rackham O (1997) The History of the Countryside. Dent, London, pp 98, 126

Rocque J (1762) A Map of the County of Berks by the late John Rocque Topographer to his Majesty

Rowe A (2007) The Distribution of Parks in Hertfordshire. In: Liddiard R (ed) The Medieval Park. Windgather Press, Macclesfield, p 131

Royal Historical Society (1897a) Domesday of Inclosures. I p 149–150

Royal Historical Society (1897b) Domesday of Inclosures. I p 110, 137, 138

Ruffhead O (1786) The Statutes at large from Magna Carta to the 25th year of the Reign of King George III (Cap IV). Charles Runnington, London

Selden Society (1973) Roll & Writ Vol. XC Berkshire Eyre Civil Pleas. p28

Soil Survey of England Sheet 268 'Reading' (1967)

Stamper P (1988) Woods and parks. In: Astill G, Grant A (eds.) The Countryside of Medieval England. Oxford University Press, Oxford, p 145 and quoting VCH East Yorkshire (1976) VI, 3

Victoria County History—Berkshire (1923) iii, p 397

Watney Dr H (1906) Unpublished letter to Ashampstead Parish Council

Williamson T (2008) Sutton Hoo and its Landscape. Windgather Press, Macclesfield, p 95

Chapter 17
Changing Cornish Commons

Peter Herring

Abbreviations

CRoW Act	Countryside and Rights of Way Act 2000
HLC	Historic landscape characterisation
MBA	Middle Bronze Age

17.1 Introduction

Taking a four thousand year view when considering changes in use, extents and perceptions of commons in Cornwall helps us place 'the end of tradition' and 'cultural and ecological severance' in long-term context. The histories of the surviving and well-known commons of Bodmin Moor and the west Cornwall downs (Lizard, Goonhilly and West Penwith), and of those forgotten commons in central and northern Cornwall that were either largely or completely enclosed in the last two or three hundred years demonstrate both continuity (of practice and ecology) and change. This chapter considers not only the loss of commons (to either change of use or privatisation and enclosure), but also the effects on communities, economies and landscape of their creation and extension.

Historic landscape characterisation (HLC) establishes likely medieval and later prehistoric extents of rough ground in Cornwall (Fig. 17.1), while more particularist landscape archaeology and landscape history demonstrates how that rough ground might have been used and perceived at various times. From the Middle Bronze Age (MBA) through to the later medieval period, and in some places into the post-medieval and early modern periods, the grazing and other resources

P. Herring (✉)
English Heritage, Penhale, Portloe, Truso PR2 5PS Cornwall, UK
e-mail: peter.herring@english-heritage.org.uk

I. D. Rotherham (ed.), *Cultural Severance and the Environment*,
Environmental History 2, DOI: 10.1007/978-94-007-6159-9_17,

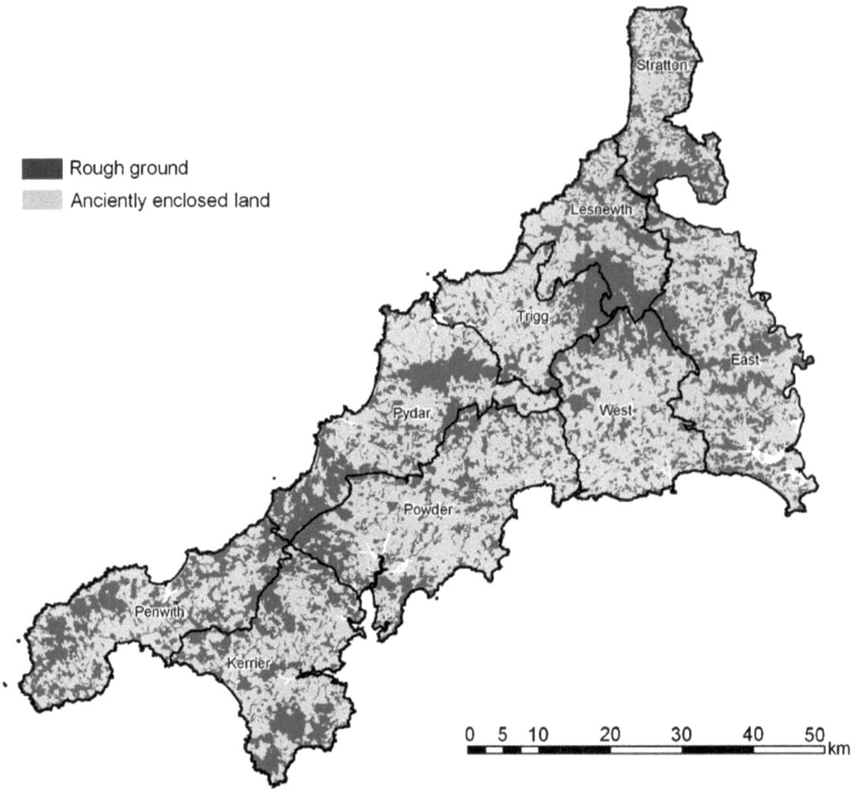

Fig. 17.1 An interpreted version of Cornwall's HLC showing the likely extent of rough ground (*darker shade*; mainly commons), and lowland farmland, the anciently enclosed land (*lighter shade*) in the later prehistoric and early medieval periods. The Cornish hundreds, which might have prehistoric antecedents, apparently carefully divided access to both, and to the coast (Cornwall Council, with permission)

obtained from rough ground was typically shared between members of small cooperating hamlets, and in the larger areas, by several or many groups of these (Herring 2004, 2006b, 2008a).

Prehistoric and medieval agricultural arrangements on Cornwall's largest granite upland, Bodmin Moor, can be reconstructed with confidence due partly to abandoned settlements and associated field and boundary systems surviving above-ground as stony banks with occasional upright slabs (Johnson and Rose 1994). Similar remains survive on the granite hills of West Penwith and to a lesser extent on the Lizard downlands (Dudley 2011).

On Craddock Moor, St Cleer, a probably MBA (MBA, *c*1500 BC) round house settlement has broad lanes passing through accreted curvilinear fields to open ground surrounding the field system, ground that was apparently used in common by all the households here, and also by others in separate settlements nearby (Fig. 17.2).

Fig. 17.2 Probably MBA settlement and field system on Craddock Moor, St Cleer. Lanes lead from round houses through fields to open ground beyond, apparently shared with other settlements and so operated as a form of common (From Herring 2008a, Fig. 12; aerial photograph by Steve Hartgroves, Cornwall Council; copyright reserved)

Perhaps slightly later, but through close similarity to Dartmoor's reaves, possibly also of the broad MBA period, are coaxial field systems such as those on Bodmin Moor's north-eastern edge. Lanes ran through parallel-walled fields to issue livestock onto large areas of open land, as at East Moor and on Carne Downs (Johnson and Rose 1994). On East Moor one coaxial system is attached to another and both appear to have shared an area of open ground to their west and south. This was apparently used as a common by the several households whose ruined homes survive in the field systems which were themselves commons, at least created by the whole of a community made up of several neighbourhood groups of households. Owned livestock were probably driven up and down those lanes in the spring and autumn to make use of seasonally available herbage. Here on East Moor and there at Craddock Moor were examples of the 'commons' and 'commoners' of prehistoric Cornwall. Other coaxial field systems edging areas of probable common land have been recorded in west Cornwall (Herring 2008a).

Coaxial field systems appear to have been created by communities determined to rationalise access to the two main components of their mixed farming system: the land within the coaxial field systems being potentially arable and the land beyond the terminal banks being open grazing land almost certainly used in common. The coherence of the Dartmoor reave systems, with several separate ones ringing each of the areas of unenclosed or open higher ground of north, south and east Dartmoor, suggest that the communities were themselves working in concert, with a still higher level of society, a sort of local authority, overseeing creation of the system and perhaps managing access to the extensive commons that were parts of it (Fleming 1988).

Other apparently MBA settlements on Bodmin Moor, like Brockabarrow Common, had either no enclosures or very few and lay in large areas of open ground in the heart of the Moor (Fig. 17.3). These field-less settlements, several with scores of ruined houses, might have been either specialist pastoral settlements or summer homes of transhumants, those who accompanied livestock to summer grazings from permanent homes elsewhere on the Moor or in lowland Cornwall

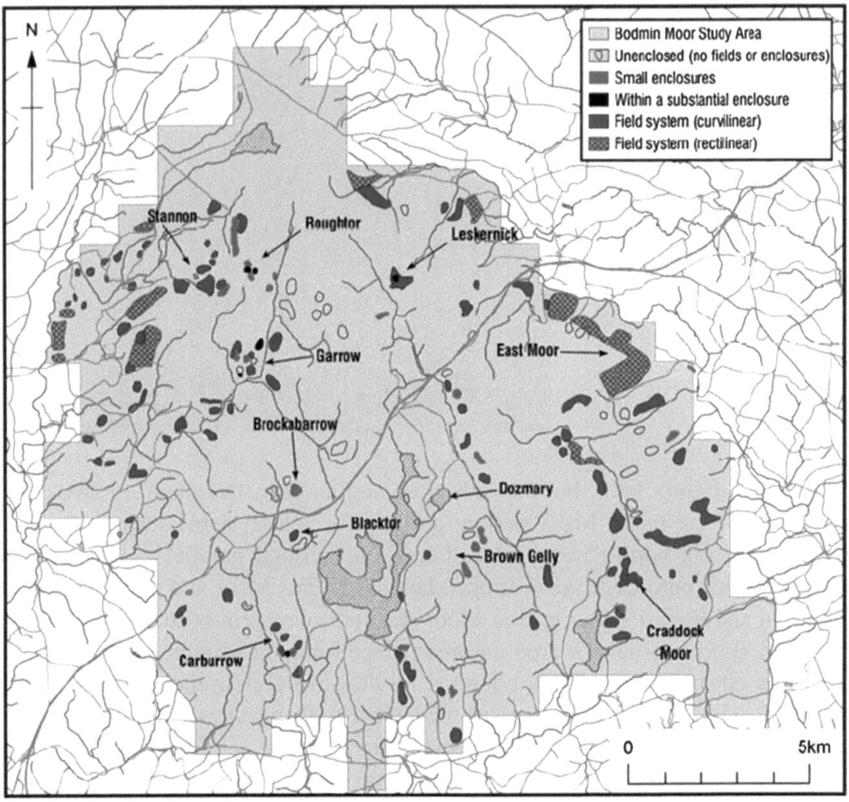

Fig. 17.3 Prehistoric settlements and fields on Bodmin Moor. On excavation most are demonstrated to be MBA. Most of the unenclosed land appears, from landscape archaeology, to have been unenclosed rough ground, effectively extensive commons (From Herring 2008a, Fig. 13)

(Herring 2008a). Their inhabitants were probably commoners again, their houses standing in many square miles of open land shared with many others (*ibid*, 81).

That prehistoric (and medieval) Cornish rough ground was open is suggested by the growing body of palaeo-ecological evidence (mainly pollen analysis). Land use systems created and supported semi-natural communities of open ground on Bodmin Moor by the early second millennium BC, around 500 years before the earliest known permanent dwellings. Heather dominated the vegetation overlain by Early Bronze Age cairns at Colliford (Maltby and Caseldine 1984, pp. 94–111). Those visiting the great hills in north-west Bodmin Moor also experienced an open landscape at that time, with trees largely confined to valleys (Gearey and Charman 1996; Gearey et al. 2000). Careful positioning and design of contemporary monuments like stone rows and stone circles also often depended on having long distance views to particular hills, notably Roughtor, being clear of intervening woodland or scrub (Herring 2008b). The palaeo-ecological evidence does, however, also suggest that this open acid grassland vegetation was varied, including, heath, scrub and scatterings of trees, and that patches of each moved around the commons as grazing levels, fires and other factors affected it.

Keeping the Moor's downs clear would have required numbers of livestock in the tens of thousands: scales of activity, investment and problems, in terms of managing sustainable access to a limited resource, that surprise those who expect early prehistoric farming to have been small-scale and localised (Herring 2008a, 2011; and see Pryor 2006). In addition to the implications for understanding contemporary demography, economy, administration and society, appreciation of this encourages us to imagine how prehistoric people would have felt about the transformations they themselves had wrought on a familiar world. Mile after mile of open land with wide and long views and in which they could strike out in any direction had replaced a largely closed and tree-dominated world, even if one accepts the model of earlier prehistoric tree cover being more like wood pasture rather than dense forest (see Vera 2000). We are accustomed to the British uplands being clear, but for those who created this world in which hills ran uninterrupted into each other, with yet others visible beyond, this open-ness and freedom may be expected to have been something to come to terms with. The ways that many monuments were designed in response to these qualities, and the ways they were experienced, suggest that communities were indeed dealing with the implications for their sense of being, their sense of place, and their identity of creating commons (Herring 2008b).

Over long periods, Cornish communities built and used gathering places, such as the fourth millennium BC tor enclosures and quoits. Archaeological interest tends to focus on the rituals that drew people to them and the various exchange and social activities that probably took place once there. But gatherings also provided opportunities for communities to discuss and organise access to shared or 'common' resources (Herring 2008a). Early second millennium BC gatherings at stone circles may have also included resolution of issues relating to the then newly created commons. Roughly contemporary stone rows on Bodmin Moor (and Dartmoor and Exmoor) tend to edge higher ground, standing close to valleys that were probably still wooded. We may imagine groups emerging with their livestock from those wooded valleys in the late spring and following their stone rows towards their accustomed grazing lands (Herring, 2008b).

Ceremonies performed at row, circle or standing stone might have included initiation of young adults into the inherited communal geographies of the adult world (Bender et al. 2007); these probably included knowledge of grazing and related regimes. Most rows had remarkably low stones, just a few inches high, only discovered by archaeologists in the 1980s when grazing levels on Bodmin Moor kept vegetation very low. This design feature appears deliberate. At Leskernick the stone row builders were able to find and erect large stones, over 2 m high, at the row's western terminal, yet chose to use stones less than 0.2 m high for most of the 320 m long line. For such monuments to have functioned in the Early Bronze Age vegetation levels would have been equally low and grazing levels equally high. The tininess of the stones may even represent celebrations of intensive grazing by domesticated livestock, of establishment and maintenance of commons,

Fig. 17.4 The Leskernick Early Bronze Age stone row, with fallen large stones at its terminal (*foreground*). The largest of the tiny stones in the rest of its length, picked out by the people, is that under the right foot of the long-haired third person. It is likely that the row was built within an area of open ground created and maintained by large numbers of grazing animals (Photograph by Graeme Kirkham, copyright reserved)

and of a community's ability to create this new way of being in the world (Herring 2008b) (Fig. 17.4).

As we have seen, landscape archaeology and palaeo-ecology suggest that the use of large commons continued through the second millennium BC. Nevertheless, archaeological evidence also suggests that permanent settlement on the uplands of Bodmin Moor and West Penwith (and Dartmoor) were apparently abandoned in later prehistory, around the turn of the first millennium BC. Sometimes this has been attributed to climate change, with people retreating from worsening weather, but the evidence is equivocal (Gearey et al. 2000). Light can be thrown on this issue by on-going examination of the lowlands (around 60 % of Cornwall's area), which seem to have been largely enclosed and settled by at least the early first millennium BC. Human and domesticated livestock populations appear therefore to have been much larger than previously supposed.

An alternative model to explain the abandonment of those upland settlements suggests a rationalisation around 1000 BC of the balance between home farmland and common grazing. The margins of the large commons were extended at the same time that other commons, such as the smaller ones found on cliff tops were reduced through continued extension of fields, especially on top of earlier coaxial fields (as at Bosigran, Zennor) or onto cliffs as at Maen Cliff in Sennen. Substantial lynchets being created by cultivation at the foot of all fields in West Penwith's later prehistoric 'brickwork' field systems shows they were more intensively utilised than the 'arable' elements of the earlier coaxial field systems. Increasing pressures on both arable and pastoral elements of the mixed farming system appear to have stimulated a reorganisation around 1000 BC, designed to increase the pool of common grazing, to intensify use of the arable component, and to increase overall agricultural productivity (Herring 2008a).

The apparent universality of abandonment of the uplands suggests that there still existed in Cornish society, as well as on Dartmoor, an administrative body exerting authority over all or most farming communities. As the rationalisation's chief beneficiaries were the farming communities themselves, it may be suggested that this local authority had again been established by the farmers themselves and perhaps included their representatives, a sort of council of respected leaders from the several hamlets (Herring 2008a).

If this interpretation is accepted, interesting situations would have developed three thousand years ago as local authorities oversaw removal of upland settlements and the absorption of privately held cultivated land into the extended commons. At the same time they permitted or encouraged extensions of private enclosed land onto cliff-top commons. People in Cornwall experienced two contrasting changes to their world, similar in some ways to those experienced in the eighteenth century AD in the Scottish Highland clearances and the English enclosure of commons. Correlation of extents of later prehistoric rough ground with the distribution of Cornish Iron Age hillforts shows many of these, now increasingly seen as meeting places for communities, probably compliant to or contributing to the local authorities, placed on the edges of commons and perhaps playing a role in overseeing their use (Dudley 2011, Fig. 94). Medieval pounds for containing trespassing livestock may have had late prehistoric antecedents in the secure spaces between hillfort ramparts.

The enlarged commons would have continued to be used for summer grazing, probably involving the transhumants whose single-person huts can be found within the shells of abandoned earlier houses and ring cairns (e.g. Bender et al. 2007, p. 171; Jones 2004–2005, pp. 36–40; Herring 2011).

Where later prehistoric settlements were constructed of granite; and where subsequent agricultural use has not removed all remains we can see that hamlets (of round houses or courtyard houses) were fairly evenly spaced. This is most visible in west Cornish coastal parishes like Zennor and suggests that land was fully exploited. Areas of rough ground, sometimes enclosed by stock-proof boundaries, lay beyond arable land organised into fairly regular field systems with hamlets at their heart. In Zennor, it appears that the early medieval pattern of narrow ribbon-shaped *tre*-lands (see below), running from cliff top to hilltop (Herring 2006a), were directly developed from prehistoric arrangements. A prehistoric footprint of blocks of land capable of supporting a hamlet may underlie much of early medieval Cornwall although archaeological research indicates that positions of settlements sometimes shifted within these. Patterns of fields certainly changed, from brick-work to strip-shaped, after the middle centuries of the first millennium AD, but there seems to have been an essential continuity in tenure, with numbers of hamlets, each organising use of its tiny 'estate', being grouped together in what are more usefully seen as 'multiple estates' rather than as manors (Herring 2011).

Many Cornish farming hamlets of the first millennium AD had names with the *tre* prefix, meaning 'farming estate'. The tenurial arrangements represented by *tre* names, of which there are at least 1,300 surviving, 'seem to belong to the native system of land-tenure which preceded the manorial system current under the Anglo-Saxons and Normans, and therefore belong to the period of native administration' (Padel 1985, p. 224). This 'native' system may have been inherited from later prehistoric arrangements with cooperating groups of households practicing a form of mixed farming in small territories often naturally defined (small hills or valley sides) and delineated by natural features, like streams or ridges. In form and practice they were probably similar to the ancient Irish 'townlands'; to avoid confusion with those they are named *tre*-lands here (Aalen 1997; Herring 2011).

Within *tre*-lands (not all of which had *tre* names) there were usually blocks of rough ground shared between the households and serving as hamlet-level commons. In West Penwith's ribbon-shaped *tre*-lands such rough ground was found on clifftop, downland and steeply sloping valley sides (Dudley 2011, Fig. 43). The *tre* estates and constituent households also probably had rights on the larger commons, such as Bodmin Moor and the Lizard Downs.

Summer transhumant use of those larger uplands (and cliffs and lowland marshes) continued through the early medieval period, as demonstrated by place-names (especially *hendre* and *havos*, the winter/home and summer dwellings respectively) and by archaeological remains of single-person summer huts (Herring 2011). Policing of rights to extensive pastures, probably via drifts in which all summering livestock were rounded up and checked, with trespassing livestock distrained in pounds were apparently organised by the large Cornish hundreds (see Fig. 17.1) that may have perpetuated the local authorities modelled above for later prehistory (*ibid*).

It seems likely that control over rights and access to commons would have been in the hands of those who held or managed the multiple estates, probably most visible now as the parishes that apparently had an economic basis, judging from how they neatly subdivided extensive resources, like the Bodmin Moor and Lizard uplands (Herring 2004, Figs. 4.4 and 4.5). These were themselves subservient in terms of hierarchy of land administration to the hundreds.

Transhumance in which members of households accompanied their livestock to summer grazings, perhaps initiated in the MBA, seems to have ceased around AD 1000. It was replaced by extension of settlement and associated 'private' rough pasture or *tre*-land commons onto the larger commons and the employment of fewer herds, some of them professional, to watch over livestock that were no longer milked on the downs and moors (Herring 2011).

We have then, around a thousand years ago, another set of substantial changes to both extents and use of Cornish commons; another set of disjunctions or severances and perhaps also another set of responses to them. Permanent cessation of transhumance, a fundamental part of Cornish farming and social organisation for over two thousand years, must have seen a severe dislocation, especially for the young women who were probably the main participants (Herring 2011).

Privatisation and enclosure of both communally worked arable and pasture took place considerably earlier in Cornwall than in many parts of Britain. Much involved hedging in and improving land within those *tre*-land commons (Herring 2004) and by the fifteenth or sixteenth centuries many hamlets had shrunk to single farms, with the *tre*-land's rough ground falling into the hands of a single farmer. Elsewhere the previously intermixed land within hamlets' *tre*-lands was divided between two or more households and so privatised by a different route. The rough pastures of Brown Willy's *tre*-land were reorganised into three large blocks associated with secondary settlements probably by the late thirteenth century (Herring 2006b). Privatisation of *tre*-land commons was often more gradual and in west Cornwall included piecemeal intake of hamlet common as privately held rough ground known as crofts (Herring 2010).

The larger blocks of manorial common were also reduced through colonisation of uplands by new farms or hamlets, themselves often with hamlet commons secured within curving ring fences, as at Louden Hill, Brown Willy and Garrow (Johnson and Rose 1994; Herring 2006a). At Louden a 1288 law case reveals that the new settlers there did not automatically obtain rights on the extensive Hamatethy Common onto which they had placed their new hamlet's ring fence and so made do with the rough ground contained by their ring fence (Johnson and Rose 1994).

Despite the creation of scores of new hamlets and farms on Bodmin Moor (*ibid*, map 2) and on the sheltered edges of the other Cornish uplands, HLC (Cornwall County Council 1996), and a range of landscape history sources, confirm that there were still large areas of open rough ground in Cornwall (mostly used in common) at the end of the medieval period. The first accurate map of Cornwall prepared by Thomas Martyn in 1748 illustrates their extent then quite clearly (Herring 2004, Figs. 4.1–4.3).

In the next two centuries, partly stimulated by increased economic activity accompanying tin, copper and lead mining, china-clay working and slate and granite quarrying, many commons were lost to enclosure, mostly apparently through local agreement (or imposition); very little was subjected to formal Parliamentary process.

We know virtually nothing of the responses of ordinary early modern commoners to losing their rough grazing and fuel grounds. It seems that few would have received compensation for the absolute losses from the pool of common resources that establishment of new farms and smallholdings represented. Many whole commons were enclosed in this period, obliging farmers to reorganise not only their summer grazing, but as this usually entailed keeping livestock on the home farm, to also reallocate arable and pasture in the farm's fields. A knock-on effect was therefore intensification of exploitation of the home farm, and a tightening of the ley (grass) element of the convertible husbandry regime (Herring

2006b). This all coincided with the urgings of agricultural reformers who saw commons as wasteful of resources and convertible or ley husbandry as unnecessarily primitive (e.g. Worgan 1811), both ideas reflecting lack of understanding of Cornish conditions and the sustainability of the long-established regime that involved both commons and ley husbandry.

Establishing intakes from commons continued well into the twentieth century. St Breock Downs was largely fenced and broken in for new farms in the immediate post Second World War period.

This rapid survey of commons' long-term history demonstrates their importance to Cornish agriculture, which to a large degree has always revolved around pasture, from meadows to rough ground, the latter mostly lying in some form of common. Seasonal use of the rough grazing freed home fields for hay and grain, making it a crucial element of the farming landscape. The blocks of Cornish rough ground also helped subdivide the country into several sub-regions, contributing much to the look and feel of Cornwall for several millennia and to the peculiar senses of Cornish and sub-Cornish place and identity.

The commons also developed peculiarly semi-natural vegetation communities whose components fluctuated through time and space, creating shifting mosaics of heath, furze, bracken and scrub, all set on a ground of acid grassland. While continuity in basic land use can be traced back four thousand years for most surviving commons, the superficial land cover, aside from that underlying acid grassland has probably been amongst the most fluid of all of Cornwall's artificial habitats.

The usage and abandonment of the remaining commons during the later twentieth century and the first decade of the twenty-first vary considerably and interestingly. Bodmin Moor's grazing levels have remained high, leaving large parts open for easy movement of livestock, farmers and those who enjoy the Moor's wide spaces. While the acid grassland dominated vegetation is a direct descendant of that which palaeo-ecological studies traces back to the Early Bronze Age, ecologists now value the Moor and guide management of it not entirely for what it is and has been, but for what it might perhaps become if grazing is reduced and other management adjusted. They aim to increase biodiversity, and the ecological and cultural services it provides society, through encouraging development of heathy scrub by diverting commoners from long established practice, in terms of volume of grazing and its timings within the farming year.

In contrast, in west Cornwall, the late nineteenth and early twentieth century specialisation of agriculture (dairying, market gardening etc.) and the consequent concentration of effort on the better in-bye land has left most commons, and indeed most rough ground of whatever legal status, neglected for several decades. This is long enough for some people who enjoy these scrubbed over downlands to develop a belief that the secondary vegetation communities, mostly less than a century old, are wild, and relics, like the area's famous megalithic remains, of prehistoric times. By most indices, including those of biodiversity and agricultural value, vegetation communities here have become impoverished through domination in places by furze and bracken; this is often at the expense of much more varied lower grasses, heathers, and herbs. Elsewhere, archaeological remains (including those created by or for prehistoric, medieval and early modern grazing) have been smothered by the scrub. Attempts by agencies in partnership with local farmers to reintroduce the grazing that prevailed here for millennia in order to obtain amenity, landscape, ecological and archaeological benefits meet resistance from those who perceive commons as open and wild (Kirkham 2011).

What we are witnessing, perhaps, are shifts in the agencies of change on Cornwall's commons. For around four thousand years, agricultural communities—whether the farmers themselves as commoners or their lords, when they either led or enabled the several waves of enclosure—appear to have held sway. Now, in the last century or so, there has been a claiming of the commons by a much wider society, partly through democratically supported designation (Sites of Special Scientific Interest, Areas of Outstanding Natural Beauty, Scheduled Monuments etc.), based on the widely accepted ecological, landscape and archaeological values of rough ground. Government, agencies and societies acting on the basis of these designations have actively conserved and managed commons. The experience of resistance to the well-meant management regimes instigated by such agencies may be a precursor of a more general trend in which commons, especially post the 2000 Countryside and Rights of Way (CROW) Act, become arenas of contestation again. This may no longer be over the loss of valued or vital economic rights as when commons were diminished through enclosure, or when rights on them were more closely defined. Now that multiple sets of values are entwined in a much more complex way, the registered rights of farmers as commoners are increasingly entangled in other strands of interest and action, those of ecologists, archaeologists, ramblers and other users. The way forward for areas that may once have appeared among the simplest places in Cornwall to manage once again depends on careful and considerate negotiations between all those with rights and interests.

Interestingly, it may be that this apparent broadening of the range of stakeholders with an interest in the Cornish commons (commoners, landowners, specialists, tourists, artists, walkers, riders, explorers, and many others) is itself deceptive. Cornwall's upland and coastal commons were probably always open lands to which all society had a form of *de facto* right of access. The 2000 CROW Act may have formalised as legally secure those rights, behaviours and desires that had pertained for generations. Those who now enjoy the exhilaratingly wide spaces of Bodmin Moor often do so in ways not too dissimilar from how the first, stone-row building commoners may have appreciated them (above). And, in turn, those who established Cornwall's commons, around 4,000 years ago, may have also appreciated them as landscape as much as land or territory. If our interpretation of the monuments they created is at all accurate, they seem to have seen the commons as areas, 'as perceived by people, whose character is the result of the action and interaction of natural and/or human factors', as landscape is defined in the European Landscape Convention (Council of Europe 2000) (Fig. 17.5).

Clearly, as commons are subject to plural perceptions and values, all agents and agencies active on them need to be aware they are still working within such broadly

Fig. 17.5 Grazing created and now maintains the open acid grassland of Cardinham Common and its mosaic of furze, ferns, heath and thorns

defined landscape, albeit a peculiar landscape with rights and responsibilities legally and customarily attached to it. As British biodiversity is culturally determined and culturally maintained, or changed, then strategies for its management will be most sustainable if they are built on thorough historic understanding on the one hand and if they successfully accommodate contemporary cultural needs on the other. Communities finding common cause and using common concepts and language will ensure that the opportunities contestation offers for designing agreed (or common) solutions are not missed. Extending or building on tradition then becomes realistic rather than romantically nostalgic.

Acknowledgments Thanks to Bryn Tapper, Steve Hartgroves and Graeme Kirkham of Cornwall Council's Historic Environment Service for images.

References

Aalen FHA (1997) The Irish rural landscape: synthesis of habitat and history. In: Aalen FHA, Whelan K, Stout M (eds.) Atlas of the Irish rural landscape. Cork University Press, Cork, pp 4–30

Bender B, Hamilton S, Tilley C (2007) Stone worlds narrative and reflexivity in landscape archaeology. Left Coast Press, Walnut Creek

Cornwall County Council (1996) Cornwall Landscape Assessment 1994. Cornwall County Council, Truro

Council of Europe (2000) European Landscape Convention, European Treaty Series, No.176. Council of Europe, Florence

Dudley P (2011) Goon, Hal, Cliff and Croft: west Cornwall's rough ground. Cornwall Council, Truro

Fleming A (1988) The Dartmoor Reaves. Investigating prehistoric land divisions. Batsford, London

Gearey B, Charman D (1996) Rough Tor, Bodmin Moor: testing some archaeological hypotheses with landscape palaeoecology. In: Charman DJ, Newnham RM, Croot DG (eds) Devon and east Cornwall field guide. Quaternary Research Association, London, pp 101–119

Gearey BR, Charman DJ, Kent M (2000) Palaeoecological evidence for the prehistoric settlement of Bodmin Moor, Cornwall, south-west England: part II—land-use changes from the Neolithic to the present. J Archaeol Sci 27:493–508
Herring P (2004) Cornish uplands: medieval, post-medieval and modern extents. In: Whyte ID, Winchester AJL (eds) Society, landscape and environment in upland Britain. Society for Landscape Studies, supplementary series 2, pp 37–52
Herring P (2006a) Cornish strip fields. In: Turner S (ed) Medieval Devon and Cornwall. Windgather Press, Macclesfield, pp 44–77
Herring P (2006b) Medieval fields at Brown Willy, Bodmin Moor. In: Turner S (ed) Medieval Devon and Cornwall. Windgather Press, Macclesfield, pp 78–103
Herring P (2008a) Commons, fields and communities in prehistoric Cornwall. In: Chadwick A (ed) Recent approaches to the archaeology of land allotment. British Archaeological Reports (International Series), Oxford, pp 70–95
Herring P (2008b) Stepping onto the commons: south-western stone rows. In: Rainbird P (ed) Monuments in the landscape. Tempus, Stroud, pp 79–88
Herring P (2010) The Cornish landscape in the Lanhydrock Atlas. In: Holden P, Herring P, Padel OJ (eds) The Lanhydrock Atlas. Cornish Editions, Fowey, pp 27–40
Herring P (2011) Shadows of ghosts: early medieval transhumants in Cornwall. In: Turner S, Sylvester R (eds) Life in medieval landscapes; people and places in the Middle Ages; papers in memory of H.S.A. Fox. Windgather Press, Macclesfield, pp 89–105
Johnson N, Rose P (1994) Bodmin Moor: an archaeological survey. The human landscape to c. 1800. vol 1, English Heritage, London
Jones AM (2004–2005) Settlement and ceremony. Archaeological investigations at Stannon Down, St Breward, Cornwall. Cornish Archaeol 43(4):1–140
Kirkham G (2011) Managing the historic environment on west Cornwall's rough ground. Cornwall Council, Truro
Maltby E, Caseldine CJ (1984) Environmental reconstruction at Colliford. In: Griffith F (ed) Archaeological investigations at Colliford reservoir, Bodmin Moor, 1977–78. Cornish Archaeol 23:92–117
Padel OJ (1985) Cornish place-name elements. English Place Names Society, Nottingham
Pryor F (2006) Farmers in prehistoric Britain, 2nd edn. Tempus, Stroud
Vera FWM (2000) Grazing ecology and forest history. CABI Publishing, Hungerford
Worgan GB (1811) General view of the agriculture of the county of Cornwall. The Board of Agriculture and Internal Improvement, B. McMillan, London

Chapter 18
The Commons of the Ancient Parish of Sheffield

David Hey

18.1 Introduction

In considering the common lands of a particular district we need first of all to understand the administrative arrangements and to define the boundaries. These can be complex. I have chosen the ancient parish of Sheffield, the whole of which became a borough in 1843 and the city 50 years later, not because it had any authority over the commons but simply to describe the variety of experience within it. Like the other enormous parishes on the edge of the Pennines, it contained numerous small commons and greens and extensive stretches of moorland. Medieval Sheffield formed part of the huge lordship of Hallamshire, a Northumbrian shire that had been taken over intact by the Normans and whose name survives to this day. Sheffield was originally a chapel-of-ease of the parish of Ecclesfield, which covered the whole of the 71,526 acres of Hallamshire and whose *eccles* name suggests that the Northumbrian shire had even earlier origins. Within this chapelry the Norman lords created a market town around their castle and they provided their prominent knights with sub-manors at Darnall, Ecclesall, Owlerton, and Shirecliff. Manorial records survive only for Ecclesall and for the chief manor of Hallamshire, based on Sheffield castle. Other lands were given to the burgesses of Sheffield in 1296. (Hey 2010).

D. Hey (✉)
12 Ashford Road, Dronfield Woodhouse S18 8RQ Dronfield, UK
e-mail: dg.hey@btinternet.com

I. D. Rotherham (ed.), *Cultural Severance and the Environment*,
Environmental History 2, DOI: 10.1007/978-94-007-6159-9_18,

18.2 The Parish of Sheffield

The chapelry became the parish of Sheffield in the twelfth century. It covered 22,370 acres, stretching from the present Meadowhall Shopping Centre by the M1 motorway across the town and over the moors to the county boundary. It was far too large to be used for civil purposes, such as overseeing the poor or the highways, or for the practical arrangements for day-to-day farming. Like other Pennine parishes, it was therefore divided into townships, which were sometimes referred to by the Viking word *bierlow* or in medieval documents as *vills*. Until the creation of the borough in 1843, Sheffield had six townships: the small urban township of Sheffield, with the lord's enormous deer park extending over the skyline, and the six rural townships of Attercliffe-cum-Darnall, Brightside Bierlow, Ecclesall Bierlow, Upper Hallam, and Nether Hallam (Fig. 18.1).

Each of the medieval institutions mentioned above had a say in the ways in which the local commons, wastes and greens were organised on a day-to-day basis. Farming systems within the parish of Sheffield varied considerably between the six townships. The contrast between the two that formed the eastern section of Sheffield parish illustrates this point. South of the River Don, the township of Attercliffe-cum-Darnall contained two settlements, both of which were farmed as classic Midland open-field villages, each with three fields divided into strips and a common beyond; hardly a tree was to be seen. Attercliffe seems to have been a medieval planned village arranged around a triangular green—known unromantically as Goose Turd Green in John Harrison's survey of 1637(Ronksley 1637)—or strung out along the Sheffield-Rotherham road past Beighton or Oaks Green (both of which names were derived from local families) and the township's

Fig. 18.1 Sheffield parish in 1795. This simplified version of William Fairbank's map of the parish of Sheffield in 1795 depicts the 6 townships and their commons and woods. The whole of the parish became the borough of Sheffield in 1843 and the city 50 years later

meadows by the river. Darnall had a separate 3-field system, with a common and 2 greens covering 5 acres between them to the east. The place-name means 'secluded nook of land on the boundary', (Smith 1961) and references to the 'Old Town' on the parish boundary in the 1637 survey (Ronksley 1637) suggest that the village was re-planned on either side of the street (Fig. 18.2).

On crossing the river Don to the north, we come across an entirely different farming landscape amongst the low hills of Brightside Bierlow. The contrast with the neighbouring township of Attercliffe-cum-Darnall, south of the river is immediately apparent. Here were no open-fields and no nucleated settlements. Instead, we find scattered farmsteads and cottages, numerous closes, pastures and meadows held in severalty, and plenty of coppiced woods, including one that had been converted from a medieval deer park. These woods were managed largely to supply charcoal for the local iron industry. Brightside Bierlow's only common was Pitsmoor, where iron ore had once been mined. However, it had three small greens at the hamlets of Brightside and Grimesthorpe, and the small Assembly or 'Sembly' Green on the opposite bank of the River Don to Sheffield Castle. Here a muster of manorial tenants held their land by a special tenure that had once taken place each year (Ronskley 1637). Brightside Bierlow had more in common with the western townships of Sheffield parish than with its southern neighbour (Fig. 18.3).

By the eighteenth century, the urban township of Sheffield was largely built over. New roads, houses and industrial buildings preserved the former strip boundaries of the 'townfield'—the usual northern term for a small open-field—which went over the hill to Shalesmoor, the only common in the township, which lay on the boundary with the township of Nether Hallam. This common was known anciently as Sheremore, a name that was derived from 'shared'. (Smith 1961) The only greens in the township were Castle Green, bordering the medieval castle and the market place, and Balm Green, high on the western border, where the medieval Barker's Pool provided water for domestic use and to cleanse the streets by a system of sluices and channels leading down to the River Don. Today, one of Sheffield's central shopping streets is known as 'The Moor', but this 17-acre common did not belong to the burgesses, freeholders and tenants of Sheffield township. It was the common attached to Little Sheffield, a hamlet half a mile or so to the south of the town within Ecclesall Bierlow, and it was marked as 'Little Sheffield Moor' on Ralph Gosling's map of 1736.[1]

When we turn to the western townships we come to more rugged terrain and more complicated arrangements. The inhabitants of Little Sheffield also had common rights on Sharrow Moor, a name meaning 'the hill where the shared land was'. (Smith 1961) This 42-acre common was used by the inhabitants of all the nearby hamlets and scattered farmsteads of Ecclesall Bierlow, and a school was built on it in 1668 at the joint expense of all the householders. To the east of the common, the strips of an open 'townfield' descended the steep hill from Sharrow

[1] Sheffield Central Library Local Studies Department: map collection.

Fig. 18.2 Attercliffe and Brightside.The contrasting field patterns, commons and woods of the two townships on either side of the River Don is evident from this section of Fairbank's 1795 map

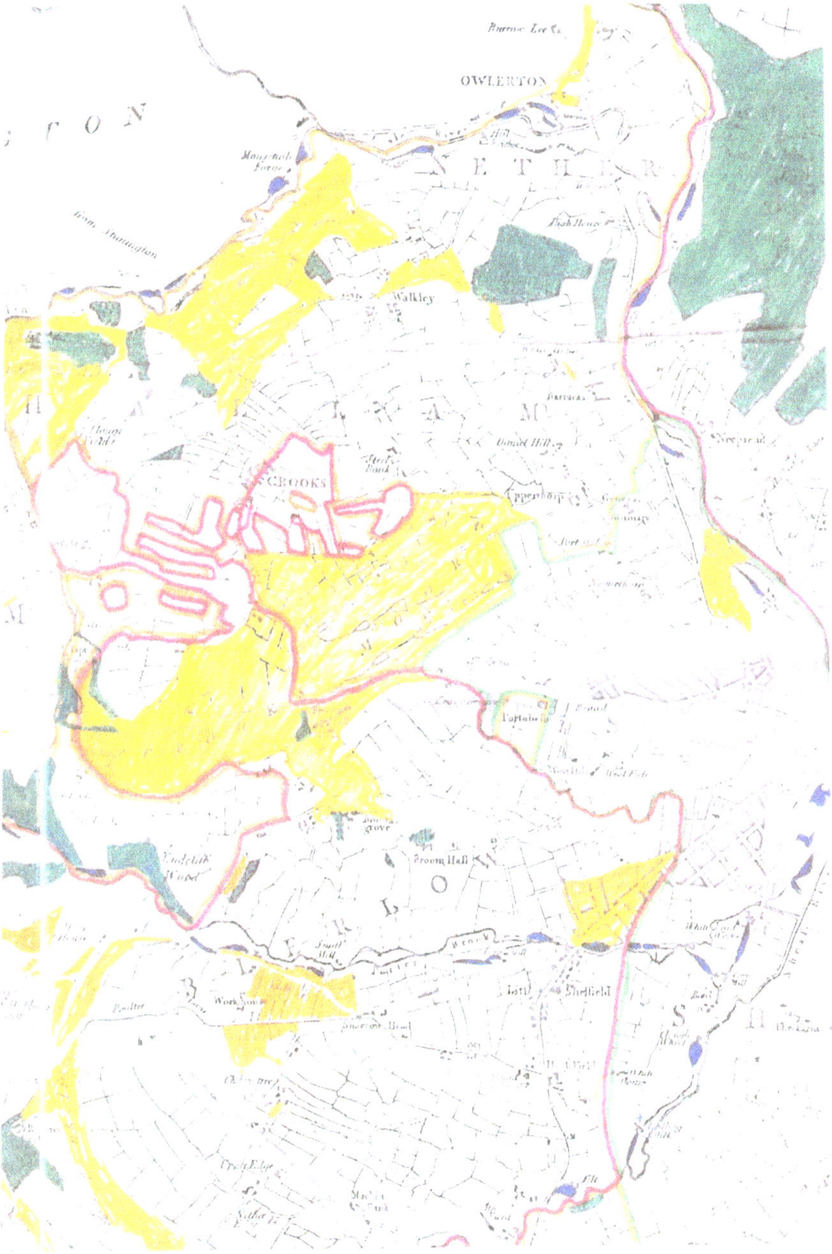

Fig. 18.3 Crookes in 1795. The hamlet of Crookes was farmed to an open-field system, with commons, pastures and woods beyond, even though the boundary (marked in *red*) between the townships of Ecclesall and Nether Hallam wove in and out of the strips and across Crookes Moor

Fig. 18.4 Stanedge Pole. This ancient boundary stone marks the western edge of Sheffield on the moors high above the city centre

Head to Little Sheffield. The majority of the allotments created under the 1788 enclosure award[2] are still recognisable as the gardens of houses along Psalter Lane, Williamson Road and Kingfield Road. Parliamentary enclosure of commons in an urban setting was often prompted not by thoughts of agricultural improvement but by making land on the edges of towns available for building. When Little Sheffield Moor was enclosed in 1788 it was immediately built upon, as William Fairbank's map of Sheffield in 1797[3] reveals (Fig. 18.4).

18.3 Moor and Common

It is noticeable that in and around Hallamshire the terms 'moor' and 'common' were interchangeable. Today, we think of moors as remote, unproductive land stretching over the Pennines, but in northern England 'moor' was also used regularly to denote a common pasture on the edge of a town, regardless of its size and

[2] Sheffield Archives, CA 363.

[3] See Footnote 1.

appearance. A well-known Sheffield suburb retains the name of Crookesmoor, which lay just beyond the central township. The administrative history of Crookes—an Old Norse name for 'a nook or corner of land'(Smith 1961)—is complex; farms and fields were divided in a perplexing way between the townships of Nether Hallam and Ecclesall Bierlow. When the sub-manor and township of Ecclesall Bierlow was enclosed in 1779–88, a few years before William Fairbank's map of the parish of Sheffield was drawn in 1795,[4] the commissioners and surveyors had great difficulty in drawing an exact boundary line between Ecclesall and Nether Hallam. This ran along the village street in Crookes, whose Old Norse name means 'a nook or corner of land', set high on a ridge, but then it twisted and turned round the headlands and strips of the 'townfields' in bewildering fashion. Crookes was farmed under a three-field system as a hamlet that straggled the township borders. Crookesmoor was also divided between the two townships: 170 acres belonged to Ecclesall and 165 acres to Nether Hallam. Clearly, communal farming in the townfields and commons of Crookes could not be organised at township level. The hamlet was the local unit that mattered.

18.4 Hamlets

The popular meaning of hamlet is a small group of farmsteads and cottages, but the term was also used in a more technical sense to describe a small, compact area with its own fields and common. In parts of northern England, we occasionally find that men brought their agreed rules and regulations about farming practices to the manor court for ratification. Although we are only dimly aware of hamlets in the records, their memory survived well into Victorian times. When the first edition of the six-inch Ordnance Survey map was published in the mid-1850s for Yorkshire—much earlier than in most parts of the kingdom—the boundaries of these hamlets were marked, even though their townfields and commons had long since gone. In the northern parts of Hallamshire, for instance, the 6-inch Ordnance Survey map shows the boundaries of the hamlets of Holdworth, Onesacre, Worrall and Ughill, all of which were recorded in Domesday Book. Such hamlets were ancient institutions.

At the southern edge of Sheffield parish, Heeley provides another example of how a hamlet could have been an effective organising unit within a complex administrative system. This 'high clearing' was squeezed in between the large deer park of the lords of Hallamshire and the river Sheaf and the Meersbrook which marked that lordship's boundary. Until 1904, Heeley was a detached part of the distant township of Nether Hallam, an extraordinary relationship that must date from the time when Ecclesall Bierlow was carved out of the area designated simply as Hallam. Heeley had two small townfields and a common. On the hillside

[4] See Footnote 1.

to the south was Heeley Green, which by 1795 had attracted a number of cottages under the name of Upper Heeley. The smaller greens - Mill Green, Newfield Green and Hanbank - had also attracted cottages by the late eighteenth century as the Hallamshire cutlery industry expanded and the population soared.

18.5 Moors, Greens and Commons

The western townships had extensive moors, which stretched over to the county boundary high on the Pennines, and they also had numerous greens. Documentary evidence for greens in this district is hard to come by. The earliest records of their names date from the late sixteenth century, but we are left with the suspicion that by that time at least some of them were centuries old. Perhaps they were noted in documents only after farmhouses or cottages had been erected upon them? We cannot tell whether their origins were associated with medieval assarting or with the recovery of the national population in Elizabethan times, or both. By the eighteenth century, when the first maps and field-books become available, most of these greens and small commons had irregular groups of buildings attached to them.

The manor of Ecclesall covered the whole bierlow or township. A list of the 'Commons and Waste Grounds' of the manor of Ecclesall in 1587 (Wigfull 1930) included Whirlow Green and Whiteley Wood Green, together with 'the Benty-haughe' and 'the Dale marshe', known later as Bents Green and Marsh Green. Oakes Green, or Broad Oak Green as it was also known, was not recorded, though it may have been subsumed under Bents Green, as it lay adjacent. The topographical nature of the names of these greens suggests that they pre-date the settlements that grew up around them.

The only name that incorporated 'common' in the 1587 list was Little Common, but other stretches of common land took their names from cliffs and hills. Graystones Cliff, Dobbing Hilll and Brockwell Hill were recorded, but curiously not several others that were included within the Ecclesall enclosure award of 1788: Brincliffe Edge Common, Dovehill Common, and the thin strips of Cherry Tree Common, Carter Knowle, Button Hill, and Millhouses Lane, which must surely have been used as commons in some way or other in earlier times. It is hard to see why some of these landscape features were referred to as greens, for example Bents Green, Oakes Green and Nether Green, while others of a similar size were known as commons, for instance Little Common. It was not a question of size, for their measurements overlap considerably. The most likely explanation is that, unlike the commons and moors, communal rights on the greens were restricted to the grazing of tethered livestock and the unrestricted wanderings of geese, ducks and poultry, as on Goose Turd Green in Attercliffe; no peat was available for fuel in these rough, grassy pastures, and perhaps nothing else was taken from them either.

Scattered manorial records give approximate dates to some of the encroachments onto the edges of greens and commons, for these squatters were tolerated by the officers of the manor on the grounds that their rents provided the lord with extra revenue. For example, in 1578, the Court Leet of the Manor of Sheffield noted that at Owlerton 'Robert Shawe hath builded a lytle house of the Lordes waste there conteyninge one Bay for the which he doth paye yearly the Rent of iiijd.'(Anon 1926) After the passing of the Act of Settlement (1665), however, such cottages became a concern for the overseers of the poor. Thus, in 1685, Christopher Lee, an Ecclesall labourer, and Robert Glossop of Ecclesall, mason, were charged with 'erecting a Cottage upon the Common in the night without lawful leave',[5] and in 1718 Thomas Colley was fined 'for erecting a Cottage at Crookesmoorside'.[6] Of course, a shrinking common meant reduced shares of the resources for those with common rights. This was not only rights of grazing but in the case of the manor of Sheffield in 1650, the right to get stone, turf, clods, earth and clay. In 1718, the Ecclesall manor court declared that none 'shall keep more Sheep upon the Commons in Summer time than they can keep on their Inlands in winter',[7] a common practice that was known to English lawyers as levancy and couchancy. Seven years later, a visitor saw a grinding wheel on common land by the River Sheaf and was told that 'this common or moor has been of late years much enclosed' (Historical Manuscripts Commission 1901) The parliamentary enclosure award of 1788 dealt with the moors that have already been mentioned and with the High Moors that covered 268 acres at the western edge of the bierlow, as far as the Derbyshire border.

As well as having a large share of Crookesmoor, and a small common at Owlerton, the freeholders and tenants of Nether Hallam township enjoyed common rights on the 25 1/2 acres of Ranmoor (whose name referred to its position on the edge or boundary with Ecclesall township, with the small Cockpit Green just beyond), and the steep slopes of Walkley Bank, Peyham Bank and Bell Hagg, which descend to the River Rivelin, totalling another 212 acres between them. Bell Hagg took its name from a coppice wood and this alerts us to the customary rights within some of the woods of the Manor of Sheffield. For instance, Harrison's 1637 survey noted that Scraith Bank, covering nearly 20 acres on the north-westerly tip of Brightside Bierlow, was 'a Common and spring wood of 21 years growth. The wood thereof belongeth to the Lord but the Feed to the Tenants'(Ronksley 1637).

The sparse population of the remaining township, Upper Hallam, was scattered in isolated farms or small hamlets, mostly arranged around greens—Birks Green, Brookhouse Green, and Nether Green—and Fulwood, the 'foul or dirty wood' that had long since been cleared by medieval assarting. The presence of the lords of Hallamshire was keenly felt here, for the entire township, together with much of the township of Stannington in the Chapelry of Bradfield beyond the northern border of the parish of Sheffield, formed the huge deer hunting ground known as Rivelin Chase, which encompassed 6,863 acres, of which 5,531 lay within Upper Hallam. Lodge Moor (which took its name from the hunting lodge) and Stanage

[5] West Yorkshire Archive Service, Quarter Sessions records, QSI/24/7.

[6] Sheffield Archives, WWM C/10/2.

[7] See Footnote 6.

Moor stretched across the bleak moorlands to the county boundary. On the edge of the moors, the medieval lords of Hallamshire had a deer keeper's lodge at Redmires, a cattle-rearing farm at Fulwood Booth, and a 429-acre pasture named after the Hawley family but now known as Hollow Meadows; nearby was Fulwood Grange, which the lords had donated to the canons of Beauchief Abbey. But as hunting was not a regular pastime, the freeholders and tenants of the neighbouring farms were granted grazing rights within the chase. Other parts were rented to framers from further afield. In the 1440s, for example, the accounts of Henry Wrasteler, the forester of Rivelin Chase, included 'pasturing-rents for divers plough-cattle on the moor from strangers there'. He also noted sales of wood, charcoal and tiling-stones and a quarry for grindstones. In what appears to have been a roundabout way of receiving rents, fifteen people were presented at the manor court for 'pannage of swine' and two men were fined for felling and carrying away 'two loads of green wood of the lord, out of Le Firth of Revelinge'. (Thomas AH 1914; Hall 1928) In 1723 the manor court fined people 'For putting cattle upon the commons called Riveling Wood, having no common right to do so' and 'For burning bracking on the above Common'.[8]

18.6 Disputes and Enclosures

In the centuries before maps were made to accompany parliamentary enclosure awards, boundary disputes on the moors were a common occurrence. When such a dispute between the inhabitants of Sheffield, Bradfield and Hathersage was resolved by arbitration in 1724, the depositions of old men called as witnesses provide a glimpse of the common rights that were claimed at the time. John Hall, aged 63, 'saith that about 45 or 46 years agoe when he was 18 He kept 300 Sheep for Wm Greaves of Hatherzidge parish for 2 Summers'. Other witnesses spoke of the grazing of cattle, and Joseph Halgreave, aged 63, remembered 'Burning Brackin upon Moscarr'. George Brownehill said that about 20 years ago he had looked after Hathersage cattle at Moscar for a couple of years and that 'they made him a Cabbin' to live in, and Lawrence Green, aged 72, recalled the digging of turves about 60 years previously.[9] When Upper and Nether Hallam and neighbouring parts of the Chapelry of Bradfield were enclosed in a long-drawn out process between 1791 and 1805, allotments were made to those farmers who had rights of herbage, but seventy others had their claims for compensation dismissed. An agent acting on their behalf claimed that 1,000 acres on the 'Black Moors' were 'Rocks where the Poor burn Fearne and raise £120 by the Ashes'. He also claimed that poor people kept flocks of geese and 'many Galloways for Grinders to carry goods'. The claimants, including thirty-five cutlers, eighteen grinders, seven

---

[8] Sheffield Archives, Ronksley collection, 4165.

[9] Sheffield Archives, ACM S60.

husbandmen, two shoemakers, a gentleman, a schoolmaster, a tailor and a wheelwright, were dismissed as being 'most of them Wheelfellows—probably not many Inhabitants'.[10] The grazing of geese and galloways were not customary rights. These practices had arisen with laxer manorial control after the removal of the deer in the late seventeenth century and because of the spread of cutlers' grinding wheels up the Rivelin Valley as trade and population increased.

This same enclosure provoked South Yorkshire's sole example of violent opposition because it included the township of Nether Hallam's share of Crookesmoor. The violence took place against a background of radical political activity in Sheffield, but the grievance at the loss of a common in a rural township may have had more to do with the demolition of the popular racecourse there. On 23rd July 1791, the Revd James Wilkinson (Vicar and Magistrate), Mr Joseph Ward (Master Cutler) and Mr Vincent Eyre (agent to the Duke of Norfolk and a Town Trustee) requested military aid when the enclosure commissioners were driven away by a mob that 'menaced them with the greatest personal danger' and which proceeded to burn farming property, to break the windows of several houses, and to menace the lives and property of the freeholders friendly to enclosure; they 'openly avowed their intention of laying open the enclosures in the neighbourhood already made … and burning the houses of all the freeholders who have countenanced the late enclosures', that is of the Ecclesall part of Crookesmoor. Four days later, Light Dragoons arrived from Nottingham. By 9 p.m. a crowd of 'many hundreds' freed the prisoners in the town gaol and marched on Broom Hall, the home of the Vicar, where they broke all his windows, destroyed much of his furniture and library, and set his haystacks on fire before they were dispersed by the dragoons. The mob then returned to the town and broke Vincent Eyre's windows. Next day, more soldiers arrived from York and the leaders of the mob were arrested. One of them, John Bennett, was convicted of arson and hanged at York. The authorities prevailed, but for a long time afterwards Vicar Wilkinson was taunted by children chanting 'They burnt his books/ And scared his rooks/ And set his stacks on fire.' (Stevenson 1989) Further enclosures proceeded with little opposition. When the enclosure award for Attercliffe-cum-Darnall was approved in 1819 the last of the commons of the ancient parish of Sheffield disappeared from view.

18.7 Conclusions

The enclosed commons of the rapidly growing urban part of Sheffield parish were soon covered with domestic and industrial buildings. In the west, the moorland fringes were converted into large, rectangular fields which can still be recognised as green pastures, though some have since been sub-divided and others have reverted to moorland. One of these fields took the name of Rape Piece because

[10] Sheffield Archives, ACM S466.

rape was the first crop to be grown there, followed by black oats, turnips and potatoes. (Sheffield Clarion Ramblers' Handbook 1950) A more significant and permanent change to the moorland landscape came between 1830 and 1854 with the construction by the Sheffield Waterworks Company of three reservoirs at Redmires and two in the Rivelin Valley to supply the needs of the rapidly expanding town. But the present appearance of the greater part of the moorland of Sheffield parish dates from the 1860s onwards, when improved gun technology and the technique of driving the grouse towards the shooters in specially constructed butts transformed the sport. (Hey 2007) Gamekeepers ensured that this former common land became the exclusive preserve of the grouse shooters until, after a long and bitter campaign ramblers received the 'right to roam' with the passing of the Countryside and Rights of Way Act in the year 2000.

References

Hey D (2010), A History of Sheffield (Lancaster: Carnegie, 3rd edition, 2010); D. Hey, The Fiery Blades of Hallamshire: Sheffield and Its Neighbourhood, 1660–1740 (Leicester University Press, 1991)

Ronksley JG (ed.) An Exact and Perfect Survey and View of the Manor of Sheffield With Other Lands. By John Harrison, 1637 (Hull Academic Press, facsimile edition, 2009), pp. 189, 192, 201–2, 335

Smith AH (ed.) (1961) The Place-Names of the West Riding of Yorkshire, part 1 (Cambridge University Press), p. 210

Ronksley (ed.) An Exact and Perfect Survey and View of the Manor of Sheffield With Other Lands. By John Harrison, 1637 (Hull Academic Press, facsimile edition, 2009), pp. 199–200

Ronksley (ed.) An Exact and Perfect Survey and View of the Manor of Sheffield With Other Lands. By John Harrison, 1637 (Hull Academic Press, facsimile edition, 2009), p. 2

Smith AH (ed.) (1961) The Place-Names of the West Riding of Yorkshire, part 1 (Cambridge University Press), p. 207

Smith AH (ed.) (1961) The Place-Names of the West Riding of Yorkshire, part 1 (Cambridge University Press), p. 196

Smith AH (ed.) (1961) The Place-Names of the West Riding of Yorkshire, part 1 (Cambridge University Press), p. 193

Wigfull JR (1930), 'Her Majesties Manor of Ecclesall', Transactions of the Hunter Archaeological Society, IV, part 1, pp. 28–45

(Anon.) (1926) 'The Court Leet of the Manor of Sheffield', Transactions of the Hunter Archaeological Society, III, part 2 pp. 143–54

Historical Manuscripts Commission, 29: Portland Manuscripts, VI (1901), pp. 143–6

Ronksley (ed.) An Exact and Perfect Survey and View of the Manor of Sheffield With Other Lands. By John Harrison, 1637 (Hull Academic Press, facsimile edition, 2009), p. 335

Thomas AH 'Some Hallamshire Rolls of the Fifteenth Century', Transactions of the Hunter Archaeological Society, I, part I (1914), pp. 31–53

Hall TW (1928) A Descriptive Catalogue of Sheffield Manorial Records, II (1928), p. 2

Stevenson J (1989) Artisans and Democrats: Sheffield and the French Revolution. Department of History, University of Sheffield, p 1787–97

Sheffield Clarion Ramblers' Handbook (1950–51), p. 142

Hey D (2007) The Grouse Moors of the Peak District. In: Barnwell PS and Palmer M (eds.) Post-medieval landscapes, Bollington, Windgather Press

Chapter 19
Traditional Uses, Destruction, Survival and Restoration of Common Land: A South Yorkshire Perspective

Melvyn Jones

Abbreviations

ACM	Arundel Castle Manuscripts
FC	Fairbank Collection
SA	Sheffield Archives

19.1 Traditional Commons and Their Use in South Yorkshire

Five 'landscape character areas' have been identified in South Yorkshire: the Dark Peak, the Southern Pennine Fringe, the Yorkshire Coalfield, the Southern Magnesian Limestone and the Humberhead Levels (Countryside Commission 1998), (Fig. 19.1). They vary geologically, have quite different landscape characters and in the past were characterised by different types of common land.

The western upland area, now characterised as the **Dark Peak** area, is composed of exposed, bleak plateaus with dramatic gritstone edges on the highest parts of the Millstone Grit country. Large stretches are covered with heather moorland and blanket bog, without a sign of habitation. Deep narrow valleys fringe the area. Settlement is largely dispersed. To the east of this moorland country is the narrow **Southern Pennine Fringe** character area which embraces both the lower eastern fringes of the Millstone Grit country and the higher western parts of the Coal Measure country and includes the western half of the modern city of Sheffield. It is crossed by the steep sided valleys of the Upper Don and its tributaries.

M. Jones (✉)
Sheffield Hallam University, c/o 4 Kirkstead Abbey Mews, Thorpe Hesley, Rotherham S61 2UZ, UK
e-mail: m.jones570@btinternet.com

I. D. Rotherham (ed.), *Cultural Severance and the Environment*, Environmental History 2, DOI: 10.1007/978-94-007-6159-9_19,

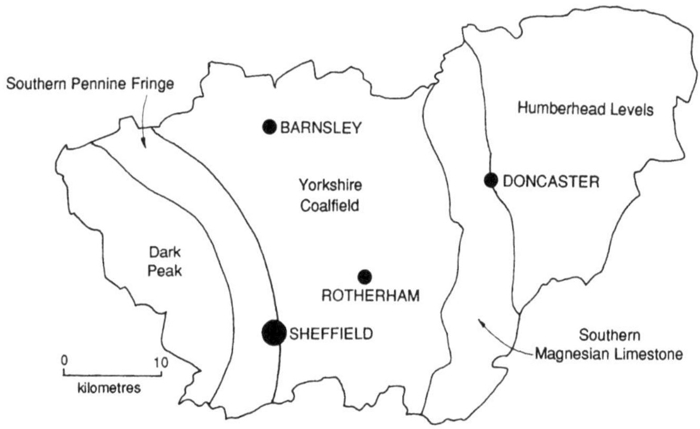

Fig. 19.1 South Yorkshire: landscape character areas

These two landscape character areas were once largely covered by Penistone parish, by Ecclesall Bierlow in the ancient parish of Sheffield, and by Bradfield township in the parish of Ecclesfield. The three areas contained thousands of acres of commons; as moorland, rough grassland or wood pasture. For example, in an undated charter of about 1297 Thomas de Furnival, the lord of the manor of Hallamshire, made a number of grants of common in the largely uninhabited uplands and river valleys in the western part of the manor. He gave to the men of Stannington, Morewood, Hallam and Fulwood the right of herbage and foliage (to gather green and dry wood) throughout his forest of Riveling (i.e. Rivelin Chase, his private hunting forest) all the way from Malin Bridge at the confluence of the rivers Rivelin and Loxley to Stanage Edge at a height of more than 1,421 feet (435 m) more than seven miles (11 km) to the south-west. To the inhabitants of Ughill, Nether Bradfield, Thornset and Hawkesworth he gave common rights in an area stretching across another part of the moorlands. In the same charter he granted common rights to the people of Wigtwizzle on their own moors (Hunter 1875, pp. 55–56). This vast area of common moorland in Sheffield manor was estimated to cover 18,689 acres (7,563 ha) in Harrison's survey of 1637 (Ronksley 1908, p. 336) and it was still used as common pasture until Parliamentary Enclosure in the early nineteenth century.

The **Yorkshire Coalfield** area (i.e. the exposed coalfield) is characterised by a rolling landscape of low escarpments and broad valleys and vales. In the medieval period this was a well settled and intensively farmed area, dotted with villages and hamlets. The pattern of common land that once existed in the townships and parishes here is very varied and full of interest. Two examples will exemplify this statement. Great Houghton township in the parish of Darfield provides an example of a very simple pattern of common land that once existed and part of which still survives (Jones 1999). The township is in the form of a strip of land extending for 5 km from north to south across the grain of the land from just over 40 m to about 100 m above sea level. The pattern of land use at the height of the operation of the

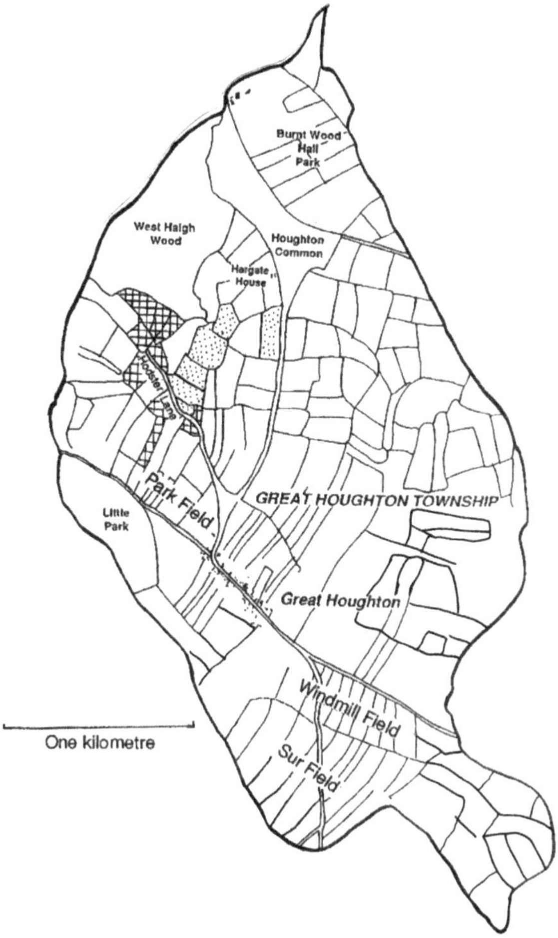

Fig. 19.2 Field patterns in Great Houghton township in the mid-nineteenth century. The former location of the wood *Odstorth* is shown by cross-hatching and the enclosures bearing the name *royd* are stippled. *Source* Tithe map and 1850 OS 6-inch map

open field system is clear to see even as late as the middle of the nineteenth century (Fig. 19.2). The lowest land in the south on the flood plain of the River Dearne was used as meadow land (place-names *holmr* and *ing*). Three open fields can certainly be recognised from surviving field names and field patterns (Park Field, Windmill Field and Sur Field) and there is evidence of possibly two others (Old Field and South Field). Beyond the open fields stretching to the northern boundary of the township once stretched a great common. Both the Tithe Map of the 1830s and the OS 6-inch map of 1850 show that most of this common land had been gradually enclosed, possibly since as early as the late medieval period, and is marked by a complex pattern of irregularly shaped fields including six enclosures that include the name Odster (=Odstorth), *storth* being the Old Norse name for a wood, and six enclosures carrying the *royd* name, the most common South Yorkshire name for a medieval enclosure from the waste. West Haigh Wood, an ancient wood first recorded in 1337, was no doubt once part of the common and

Fig. 19.3 Hoyland Nether township, pre-enclosure map of 1771. Besides the village of Hoyland Nether in the centre of the township there were hamlets at Upper Hoyland (**a**), Hoyland Lane End (**b**), Harley (**c**) and Elsecar (**d**). *Source* Sheffield Archives, Fairbank Collection FC/Wath 13L

then separated from it and managed as a coppice wood. The surviving fragment of common has characteristic concave boundaries and a funnel-shaped exit into Moor Lane. It is a Local Nature Reserve covering 62 acres (25 ha) and is a mix of bracken and acid grassland and open birch woodland. This pattern of meadowland on a river floodplain, open fields on higher well-drained land and common land on the highest land in a township or parish was repeated in a succession of townships in the Dearne valley between Barnsley and Mexborough (see for example, Harvey 1974; Hey and Rodwell 2006).

In contrast to the common land in Great Houghton township which was in one large block beyond the open fields, that in Hoyland Nether township on the eve of Parliamentary Enclosure at the end of the eighteenth century (Fig. 19.3) formed a complex network running across almost the whole of the township. In 1771, 23 years before Parliamentary Enclosure entirely obliterated the common land, it covered 360 acres (146 ha). Of this, 230 acres (93 ha) was in the shape of Hoyland Common that formed a very long, narrow, rectangular common where the township adjoined Tankersley parish. From Hoyland Common, green lanes ran across the township linking it to Hoyland village, other minor settlements and a series of smaller commons and greens. For example, a funnel shaped green led from the common to the hamlet of Hoyland Lane End and then continued via a common called The Lowe to Upper Hoyland and Hoyland village, the latter being arranged around a large green. There were three exits from the eastern and southern edges of this green. On the east one led via Ulley Bank and Cobcar Green on the township boundary and another further south by a green lane, to Stubbing Green and Elsecar Green on the western boundary of the township. The third exit led via Broadcar southwards and south-westwards back to Hoyland Common. It is clear that this complex system of commons and greens, not only formed a valuable agricultural resource for the commoners of Hoyland but also the major road system around the township.

A substantial number of the medieval commons in the coalfield country and in the Pennine Fringe zone were wooded commons (Jones 1998). Some disappeared quite early but some remained until enclosure in the eighteenth and early nineteenth centuries. In an inquisition *post mortem* of 1332 on the death of Thomas de Furnival it was recorded that his properties in the manor of Sheffield included pastures in among other places 'Greno, Billy Wood, Ryvelyngden and Baldwinhousteads' (Curtis 1918). Ryvelyngden was the valley of the Rivelin occupied by his private hunting forest which continued in part to be wood pasture until the Parliamentary Enclosures but by 1600 the other three, Greno Wood, Beeley Wood and Bowden Housteads Wood, were all enclosed coppice woods. But some wood pastures continued in existence for more than two centuries longer. There are graphic descriptions of a number of wooded commons in the late seventeenth and early eighteenth centuries. In 1650, Loxley Chase was referred as 'one Great wood called Loxley the herbage common and consisteth of great Oake timber'. About 10 years earlier, another wooded common, Walkley Bank, was said to have 'a great store of rough Oake trees and some Bircke (birch) woods'.

The **Southern Magnesian Limestone** area is in the form of a narrow ridge a few miles across from west to east. It is elevated and quite distinct from the coalfield character area to the west but to the east it merges imperceptibly with the lowlands forming the Humberhead Levels. In South Yorkshire this ridge is broken only at Sprotbrough and at Maltby. The main settlement form is the nucleated village, separated from each other now by extensive tracts of rolling agricultural land. Parishes are much smaller here than to the west and to the east and commons consequently were much smaller. Enclosure from the late medieval period onwards into the eighteenth century was followed by a tidying up process under the Parliamentary Enclosure acts and very little common has survived. Registered common land only survives in Maltby parish (see below).

The **Humberhead Levels** is mainly a flat landscape. The clay lowland to the west of the Don in the parish of Fishlake is no more than 6 m above sea level and to the east stretching across the county boundary into Nottinghamshire and Lincolnshire the land rises no more than 2 m above sea level, 'level almost as the surface of the sea itself' as Hunter (1828, p. 150) put it. The oldest settlements are located on what in early times would have been along riverbanks as at Sykehouse and Fishlake, and small islands among the extensive fenland formed by the protruding bed-rock and deposits of sand and gravel as at Lindholme, Arksey and Hatfield.

The area covering the townships of Fishlake (including Sykehouse), Thorne and Hatfield (including the settlements at Hatfield Woodhouse, Dunscroft, Stainforth and Tudworth), comprised in the medieval and early modern periods the manor of Hatfield. This was the property of the de Warenne family of Conisbrough castle from shortly after the Norman Conquest until 1347 when it was forfeited to the Crown and in whose ownership it remained until the early seventeenth century. 70,000 acres (more than 28,000 ha) of the manor was a private forest or chase, Hatfield Chase, which became a royal forest in which deer were protected by forest laws until it was disafforested in 1629.

The fen-edge settlements in the Humberhead Levels in the medieval and early modern periods were surrounded by thousands of acres of fen made up of carr (scrub woodland) and peat bog utilised as commons. The uses that the commoners made of the fens were legion. First there was grazing although there is evidence that commoners in the manor of Hatfield were not allowed to pasture their sheep and goats in Hatfield Chase because they competed with the deer for short grasses (Byford 1997, p. 92). Turf from the peat moors was obviously a perquisite for all commoners, not only as a fuel but also as a building material. Reeds and sedges were cut to thatch cottages, outhouses and barns; rushes and other marsh plants were cut for hay; osiers in the willow carrs were made into wicker baskets and eel traps and bigger scantlings of wood would be used to make barges, turf barrows and sledges; and the wildlife was a food source: wildfowl and waders, birds' eggs and fish. There were also the remains of a prehistoric wildwood in the form of tree trunks and tree stumps buried in the peat deposits on Thorne and Hatfield Moors. No doubt the bog oaks and bog pines unearthed during turf cutting were rescued by the commoners. And, of course, there were the deer. An inventory of 1607, said there were 1,000 deer in the chase and the diarist Abraham de le Pryme writing at the end of the seventeenth century said that deer in the Levels were 'as plentiful as sheep upon the hill' and venison was as 'abundant in a poor man's kitchen as mutton' (Hunter 1828, p. 157). No doubt many commoners were tempted over the centuries to stock their larders with venison.

Unprofitable though the thousands of acres of common fenland may have been to the Crown, to the inhabitants of the Levels they provided an invaluable natural resource. But a combination of the financial difficulties of King Charles I and the very low level of income coming from his lands in the Humberhead Levels led to his agreement in 1626 to the drainage scheme proposed by Cornelius Vermuyden. This was the beginning of the end for the fenland commoners in the Humberhead Levels.

19.2 Destruction of Common Land in South Yorkshire

As late as the early nineteenth century when the Parliamentary Enclosure awards were made, the Dark Peak area still contained vast areas of unenclosed common moorland. For example, the 1826 Parliamentary Enclosure award in Bradfield township dealt with 13,626 acres (5,514 ha) of common and waste. The figures for Langsett township and Thurlstone township to the north were 3,014 acres (1,220 ha) and 5,001 acres (2,024 ha) respectively. On enclosure the privatised moorlands were largely given over to sheep grazing and grouse shooting.

On the coalfield and in the Magnesian Limestone belt the Parliamentary Enclosures constituted the final act in a process that had been taking place since late medieval times in both the open fields and on the commons. In Hoyland Nether, the pattern of common land in 1771 on the eve of the Parliamentary Enclosure has already been described in some detail but what the map also reflects in its field patterns and field names is the relentless piecemeal enclosure that had been taking place for centuries. If we take field names alone, this illustrates the process very clearly. There are dozens of *royds* and *closes* and *leys* and a sprinkling of *greaves* and *stubbings* and more recent names such as Moorside, Moor Field and Common Piece. At Hoyland, too, Parliamentary Enclosure changed the village itself. In 1771, the village was clustered around an irregularly-shaped green, but with enclosure in 1794 and the imposition of a street up the middle of the green and the allocation of parts of the green to different owners, the green village became a street village (Fig. 19.4).

In the eastern lowlands the story of Cornelius Vermuyden's great drainage scheme of 1626 has been told many times (e.g. Hunter 1828; Stonehouse 1839; Tomlinson 1882; Dunnin 1997). In short in his scheme, one third of the drained land was to continue to belong to the Crown, one third to Vermuyden and one third to the tenants in those manors where reclaimed land was situated. The initial scheme was completed in 18 months and involved changing the course of the River Don so that it fed into Turnbridge Dike and then into the Aire, and draining the straightened River Torne and the straightened River Idle into the Trent at Althorpe (Fig. 19.5). It has been calculated that about 12,400 acres (*c*. 5,000 ha) of fen were reclaimed (Metcalfe 1960). One of the consequences of Vermuyden's initial drainage work was increased flooding in the Fishlake and Sykehouse areas and this led to the construction of the Dutch River to Goole in 1635. In 1633, the commoners of Hatfield Chase (which had been disafforested in 1629) were granted a further 4,035 acres (1,633 ha) to compensate them for loss of common rights. In 1636, Vermuyden sold his holdings in the Levels. Acts of sabotage continued for a considerable period causing drained land to be flooded.

However, Vermuyden's initial drainage scheme, later remedial work, the practice of warping (controlled flooding of land to leave a layer of silt), Parliamentary Enclosure acts, and the use of pumps over the last two centuries to facilitate further drainage, all conspired to produce one of the most fertile arable areas in England. The turf moors at Thorne and Hatfield were not exempt from commercial

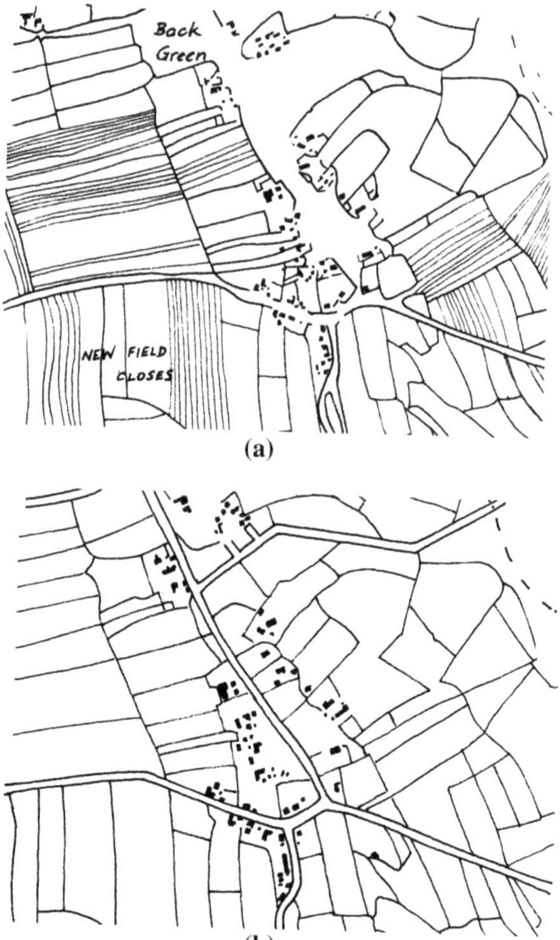

Fig. 19.4 **a** Hoyland Nether village before parliamentary enclosure in 1771, and **b** immediately after enclosure in 1794. *Source* Sheffield Archives, Fairbank Collection FC/Wath 13L

development. Turf cutting there had been reduced drastically through the more widespread use of coal for most of the nineteenth century, but from the 1880s until the 1920s the turf was commercially exploited on a large scale for peat for litter for animals. This activity then declined markedly until the 1960s when it was resuscitated as the market for peat for horticultural purposes expanded. Commercial peat cutting finally ceased on Thorne Moors in 2001 but continues on Hatfield Moors.

19.3 Survival and Restoration of Common Land

Despite the piecemeal enclosure of common land throughout the medieval and early modern periods and the wholesale destruction of commons during the period of Parliamentary Enclosure, and in what later developed into a highly urbanised

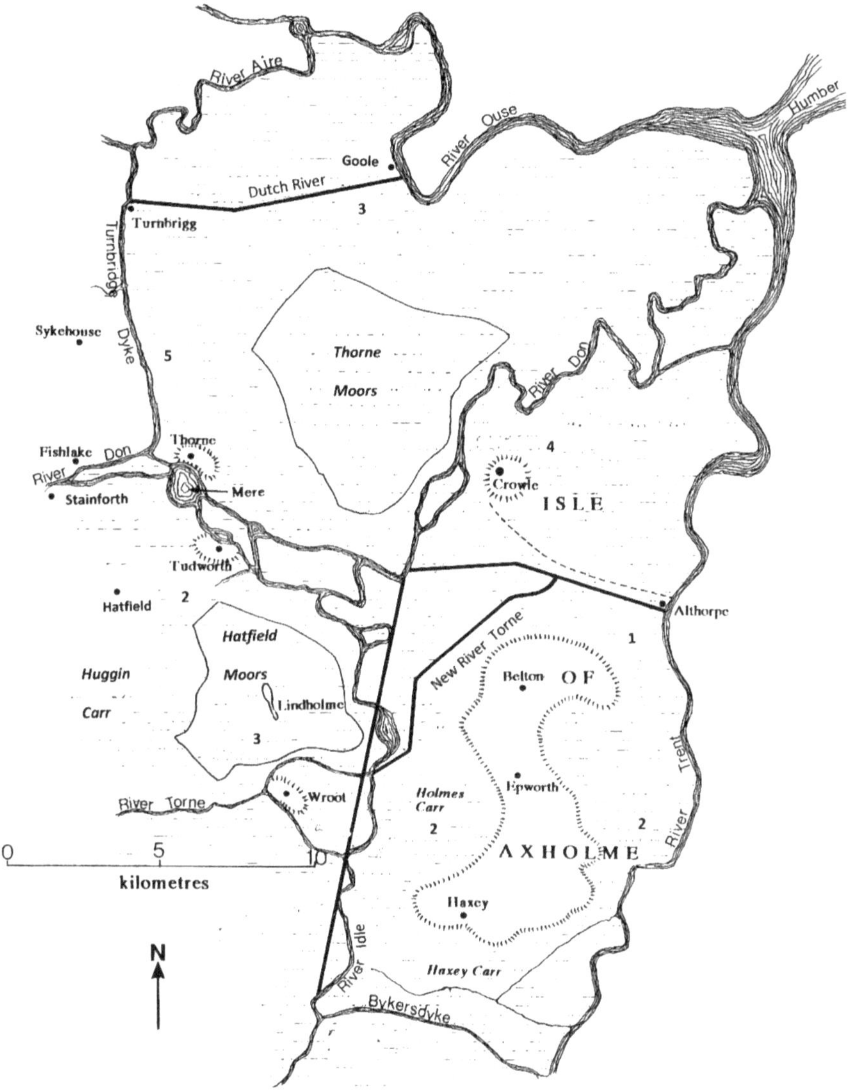

Fig. 19.5 The Humberhead Levels of South Yorkshire and north Lincolnshire showing the main features of Cornelius Vermuyden's 1626 drainage scheme. Spot heights are in metres. *Source* after Stonehouse (1839) and Dunston (1909)

and industrialised region, there is a surprising number of surviving fragments of common land, registered and un-registered, in South Yorkshire, most managed either as public open spaces and/or nature reserves. Many are Local Nature Reserves (LNRs), one is a National Nature Reserve (NNR) and six are Sites of Special Scientific Interest (SSSIs).

Fig. 19.6 Part of Loxley Common as shown on the 1894 6-inch OS map showing the regular layout of parliamentary enclosure fields to the west of Long Lane and the still wooded area surrounding the outcrop of Loxley Edge to the east of Long Lane

Only a minority of the areas of surviving common land are formally registered areas of common land or village greens: 29 fragments of common land and 29 village greens, in most cases tiny fragments of village greens. One of the areas of registered common land in Doncaster is Thorne Moors, where at its registration in 1970, named tenants still had the rights of turbary, piscary and auceptary (fowling). And in 1968, on the registration of Royston village green in Barnsley a named tenant still enjoyed the right to graze 18 cattle, 18 sheep, 18 pigs and 18 geese and enjoyed the rights of estovers, turbary, piscary and pannage.

Some of the largest surviving areas of former moorland common lie within the Peak District National Park, such as Burbage Moor and Howden Moors, the former an area of green space owned by Sheffield City Council and popular with walkers and climbers, the latter being part of the National Trust's High Peak estate. Other fragments of moorland common within Sheffield city boundaries include Ughill Moors and Whitwell Moor. Whitwell Moor lies on the interfluve that rises to 359 m (1,179 feet) between the valleys of the Little Don and Ewden Beck. It belongs to the Broomhead estate and access management work has been undertaken by the estate in conjunction with Sheffield City Council and the Steel Valley Project.

Gorse-bracken-birch heathland has also survived at Townend Common at Stocksbridge, at Hunshelf Bank and at Loxley Common (all unregistered), all within Sheffield City boundary; and at Houghton Common (a registered common), referred to earlier, in Barnsley. Perhaps the best documented and most interesting of these from the historical, archaeological and ecological points of view is Loxley Common and the coterminous Wadsley Common (Fig. 19.6). Wadsley Common is a registered common, Loxley Common is not. The area in question covers the valley side to the north of the River Loxley with Loxley Edge, a sandstone escarpment outcropping in the eastern half of the site and with a boulder-strewn slope below it. Loxley Common is variously referred to in historical documents as Loxley Common, Loxley Chase and Loxley Firth. Loxley Common was mentioned in a late thirteenth century document in which Thomas de Furnival, the lord of Hallamshire, granted common rights to the inhabitants of the area (Hunter 1875, p. 56). In an inquisition *post mortem* in 1332 on the death of Thomas de Furnival it was recorded that the pannage of the woods (*pannagium boscorum*), i.e. the pasturing of pigs on acorns, at Rivelyngden and Lokesley was worth forty shillings (Curtis 1918). Just over 300 years later in 1637 it was recorded in John Harrison's survey of the manor of Sheffield as 'A Common Called Loxley wood and firth' and covered 1,517 acres (614 ha) (Ronksley 1908, p. 336). In 1650 as noted earlier, the common was referred to as 'one Great wood called Loxley the herbage common and consisteth of great Oake timber'. Two old oak pollards still survive on the site.

An enclosure bill for 'Loxley Chace' went through Parliament in 1784 and the final award was made in 1789. Most of the area was laid out as farmland, with the straight-sided rectangular fields bounded by dry-stone walls. Two absolutely straight enclosure roads, Myers Lane and Loxley Road, bound the northern and southern edges of the former Loxley Chase and another straight enclosure road, Long Lane, divides the former Loxley Chase from Loxley and Wadsley Commons. While the western half of the area to the west of Long Lane was used after enclosure as farmland, the eastern half of the area was the scene of quarrying until the 1920s for the Loxley Edge Sandstone, a much sought-after building stone and of the mining, from drift pits, of coal and ganister (a fine-grained sandstone used to make refractory bricks for lining furnaces) which ceased about 1940 (Bailey 2007). These industrial activities inevitably scarred and reduced the quality of the lowland heath character of the site.

Most of Loxley Common had been in the ownership of the Payne family since 1800. In 1913, three surviving family sisters presented 75½ acres (30.5 ha) of the common to the City of Sheffield provided it retained its semi-rural character and was used as a public open space. In the 1970s it was designated a Site of Scientific Interest and it became a Local Nature Reserve in 1999, because of its status as lowland heath, an increasingly rare and threatened habitat.

What is also interesting is that a number of coal measure woodlands have gone almost full circle in their function, not commons in their original sense but important public open spaces. For example, Bowden Housteads Wood, as noted above, was listed in an inquisition *post mortem* in 1332 as pasture, i.e. wood pasture. By the late sixteenth century it was a spring wood (i.e., a. coppice-with-standards) which it remained for more than 400 more years before being purchased by Sheffield City Council from the Duke of Norfolk for use as a public open space. Other woodlands, once wood pastures, have also become public open spaces. Three Rotherham woods, Wath Wood, now a heavily planted section of a former wood pasture in the parish of Wath-upon-Dearne, Birch Wood, formerly part of Rawmarsh Common and Creighton Woods, early nineteenth century plantings on the former Swinton Common, are also now public open spaces. In the coalfield country, other former commons, or at least fragments of them, have re-appeared as places frequented by the public as public parks. Locke Park includes part of the former Barnsley Common, Firth Park in Sheffield includes part of Hinde Common and Ecclesfield Park occupies part of the former Ecclesfield Common.

Surviving commons on the Magnesian Limestone belt are scarce commodities. There are three registered rare survivors all in Maltby parish: Wood Lee Common, Far Common and Low Common. Far Common and Low Common are particularly interesting. Far Common is a Local Nature Reserve and Low Common and Pieces Bank combined are a Site of Special Scientific Interest. All of these commons belong to Lord Scarborough's Sandbeck Estate and 75 acres (30 ha) of the three commons are registered common land. The SSSI which covers 15 acres (6 ha) is mainly neutral grassland with a diversity of flora not found elsewhere in South Yorkshire. The three commons support, among other uncommon species, common spotted-orchid, marsh orchid, bee orchid and greater butterfly orchid.

Fragments of the once large peat bogs at Hatfield and Thorne are all that remain of the once many thousands of acres of common land in the Humberhead Levels. Long exploited for domestic and commercial peat extraction they still contain areas of outstanding ecological and archaeological importance. Thorne and Hatfield Moors in South Yorkshire, Crowle Moor in Lincolnshire and Goole Moor in East Yorkshire together comprise the Peatlands National Nature Reserve, an area of raised bog wilderness covering 7,134 acres (2,887 ha). The National Nature Reserve was designated in 2001 and expanded in 2005. As already noted, commercial peat production finally ceased on Thorne Moors in 2001 and within the boundaries of the SSSI on Hatfield Moors in 2004. The mainly cut-over peatlands are being restored by damming drains and so re-wetting the abandoned peat workings to re-create the original boggy conditions. Five thousand species of plant and animal have been recorded on the reserve. On Packard's Heath on Hatfield Moors a flock of Hebridean sheep has been introduced to control the spreading willow and birch scrub.

19.4 Conclusions

The varied geology and associated dramatic variations in landscape character; from high bleak moorlands in the west to extensive lowlands just a few metres above sea level in the east gave rise in South Yorkshire by the medieval period to marked differences in population density and settlement patterns. Associated patterns of farming organisation were equally differentiated with tenants having access to commons of vastly different sizes and character. Over the centuries many of these commons were drastically reduced in size, most of them being finally enclosed during the great period of Parliamentary Enclosure between 1750 and 1830. But just sufficient fragments have survived, some formally registered as commons, others not, and almost all managed to preserve their character, to provide the twenty-first century visitor with a flavour of what were once common landscape elements and an integral part of the agricultural system. Thankfully John Clare's lines in his poem 'Enclosure' have *not quite* come to pass in South Yorkshire:

Ye meadow-blooms, ye pasture-flowers, farewell!
  Ye banished trees, ye make me deeply sigh –
  Enclosure came, and all your glories fell:

References

Bailey AE (2007) On Common Ground. Privately published

Byford D (1997) Open field farming in Fishlake and Hatfield: the evidence of the Court Books, 1582–1808. In: Elliott B (ed) Aspects of Doncaster: discovering local history. Wharncliffe Publishing Limited, Barnsley, pp 87–110

Countryside Commission (1998) Countryside Character, Volume 3: Yorkshire & the Humber. Countryside Commission, Cheltenham

Curtis E (1918) Sheffield in the fourteenth century: two Furnival Inquisitions. Trans Hunter Archaeol Soc 1:31–53

Dunnin M (1997) The drainage history of the Humberhead Levels. In: van de Noort R, Ellis S (eds) Wetland heritage of the humberhead Levels. Humberhead Wetlands Project, University of Hull, Hull, pp 19–30

Dunston G (1909) The Rivers of Axholme with a history of navigable rivers and canals of the district. A. Brown & Sons, Hull

Harvey JC (1974) Common field and enclosure in the Lower Dearne valley: a case study. Yorkshire Archaeol J 46:110–127

Hey D, Rodwell J (2006) Wombwell: the landscape history of a South Yorkshire coalfield township. Landscapes 7(2):24–47

Hunter J (1828) South Yorkshire, vol 1. Reprinted by E. P. Publishing, Wakefield

Hunter J (1875) Hallamshire: the history and topography of the Parish of Sheffield. In: Gatty A (ed) Definitive (first published in 1819), 3rd edn. Virtue & Co, London

Jones M (1998) The rise, decline and extinction of spring wood management in south-west Yorkshire. In: Watkins C (ed) European Woods and Forests: Studies in Cultural History. CAB International, Wallingford, pp 55–72

Jones M (1999) Landscape history and initial archive survey and assessment. In: Grimethorpe Woodlands Project, Hallam Environmental Consultants, Sheffield

Metcalfe B (1960) Geographic aspects of the reclamation and development of Hatfield Chase. Unpublished M.A. Thesis, University of Leeds

Ronksley JG (1908) An exact and perfect survey of the manor of Sheffield and other lands by John Harrison 1637. Robert White & Co, Worksop

Stonehouse WB (1839) History and Topography of the Isle of Axholme. Longman, Rees, Orme & Co, London

Tomlinson J (1882) The level of Hatfield Chase and parts adjacent. Wyman & Sons, London

Chapter 20
Abandoned Landscapes of Former German Settlement in the Czech Republic and in Slovenia

Petr Mares, Robin Rasin and Primoz Pipan

Abbreviations

CTU	Comparable territorial unit from LUCC Czechia/Slovenia project
AL	Arable land (land use category)
BuA	Built-up areas (land use category)
FA	Forest areas (land use category)
LUCC	Land use/cover changes
OA	Other areas (land use category)
PC	Permanent cultures (land use category)
PGL	Permanent grasslands (land use category)
WA	Water areas (land use category)

20.1 Introduction

Despite the fact that the landscape of Czech borderlands and the landscape of Slovenian remote Kočevje region are more than 500 km apart, several similarities can be identified, incurred by common historical stories. These similarities make both landscapes exceptional and different from the rest of the respective nation states

P. Mares (✉)
Institute for Environmental Studies, Charles University in Prague, Benatska 2, 120 00 Prague 2, Prague, Czech Republic
e-mail: petr.mares@scientica.cz

R. Rasin
Department of Social Geography and Regional Development, Charles University in Prague, Albertov 6, 120 00 Prague 2, Prague, Czech Republic

P. Pipan
Anton Melik Geographical Institute, Scientific Research Centre of the Slovenian Academy of Sciences and Arts, Gosposka ulica 13 SI-1000 Ljubljana, Slovenia

I. D. Rotherham (ed.), *Cultural Severance and the Environment*, Environmental History 2, DOI: 10.1007/978-94-007-6159-9_20,

and for that they deserve our attention and scientific focus. The uniqueness of both areas can be seen in settlement development and in forming of the landscape after World War II. War and post-war development has resulted in migration and transfer of indigenous German settlers. It means that both landscapes have lost their long-term stewards, whose job was only partly (or not at all) taken over by re-settlers. Nevertheless, major areas in both the Czech borderlands and in Slovenian Kočevje region have remained abandoned and have been developing with only minor landscape management practices. Therefore, the current shape of landscapes in these areas provides a fantastic base for societal, humanist, environmental, or agricultural research of land-abandonment and its consequences. The aim of the paper is to present different scientific views at these abandoned landscapes and to combine various outcomes of diverse studies undergone in the presented landscapes.

After short historical introduction to the question of Germans in Czechia and Slovenia, the paper deals with three scientific approaches to these, in Central Europe unique, abandoned landscapes.

Firstly, land use development and its characteristic features are analyzed. Similarities in land use development of both regions—Czech borderland and Slovenian Kočevje region—are under investigation. In the focus of the scientific investigation is also the importance of changes that occurred in the studied areas from the perspective of the whole landscape(s) of the both countries.

Secondly, landscapes are perceived as entities that carry landscape memory and landscape heritage. The importance of the landscapes, of their structures and landscape elements is discussed. Attention is mostly drawn to the relict landscape elements that remained in the landscapes as a memento of former human activity in presently abandoned and unmanaged landscapes.

Thirdly, environmental approach and landscape-ecology view is applied on the selected regions. This part of the research is aimed at plant species diversity in secondary forest and the vegetation underneath the trees. Development of the secondary forest and the succession process are discussed. The selected region where this part of the research was carried out represents a landscape which enables us to monitor processes developing after human activities in the area came to an end.

It is obvious that such different approaches to landscape science work with distinct methodology and are focused on different questions. Therefore, after the introduction of both landscapes, exemplary studies and exemplary outcomes of the three landscape science approaches are presented. In the conclusion, results are combined and the ties among them as well as the societal importance of the landscapes are discussed.

20.2 Historical Approach

Landscapes of both countries—the Czech Republic and Slovenia—are on the whole very heterogeneous. This historically led, together with the countries' complicated position on the border between the "West" and the "East", to the

Fig. 20.1 Czechia (CZE) and Slovenia (SLO) in Europe

Table 20.1 Relevant facts about Czechia and Slovenia

	Area (km^2)	Area of majority of Germans (km^2) before WW II	Today's number of inhabitants	Share of Germans before WW II (%)	Share of Germans today (%)
Czechia	78,866	~ 22,000	10,251,079	31.2	0.01
Slovenia	20,273	~ 800	2,010,377	8 (year 1910)[a]	0.04[a]

[a] According to ŠIRCELJ (2003)

inconsistent colonization and establishing of several cultural regions within the countries.

We can see basic comparison of both countries bellow (Fig. 20.1 and Table 20.1).

The landscape of Czechia can be generally viewed as fertile lowlands of main rivers and basins and highlands and lower hills within the country, all this bounded by the ring of Variscian and Carpatian mountains, which are relatively high and which almost exactly follow the border of the Czech Republic. The lowest point is 115 m a.s.l. at the Labe River, the highest peak is Sněžka with 1,602 m a.s.l. in the Krkonoše Mountains.

When the Slav tribes came to Central Europe from the area of the Visla River in the second half of the sixth century, they occupied and started to use primarily the most convenient areas of lowlands. Forested, impervious areas of border mountains lay for a long time aside from the main colonization efforts, serving mainly as a natural protection against invasions of enemies. In the course of the thirteenth century, the king as well as different landlords became aware of the economical potential of these—by that time uncultivated—areas. Mining, wood industry and business connected with the control over trade routes were good reasons for their interest in these areas. It was also important to have border areas settled and thereby under control. To do the colonization work, colonizers from abroad were invited, partly because of the overpopulation in source areas, partly because of their skills in special professions. The German colonizers came from different parts of Germany and Austria and settled in new villages around north, west and south parts of the Czech lands. Until the end of the eighteenth century, a belt of exclusively German villages arose within Czech borderland, somewhere German "islands" could be found even in the space of dominantly ethnically Czech areas. In contact zones, Germans got mixed with Czechs.

The coexistence of Czechs and Germans went on without significant difficulties until the end of World War I, which was also the end of the Austro-Hungarian Monarchy. After establishing the Republic of Czechoslovakia in 1918, Czech Germans found themselves completely cut off the core area of Germans. This was hard to accept for many of them and some separating tendencies had to be managed even by the Czechoslovak army in some places (Andrýsek et al. 2003). The borderline was fastened, but social tensions between Czechs and Germans intensified (Fig. 20.2).

The worldwide economic downturn in the 1930s, made the coexistence of the two nations in one state even worse. The Czech government, which was at that time in the middle of realizing a great project of building the defence line all around the potentially dangerous borders, used local people as employees at the

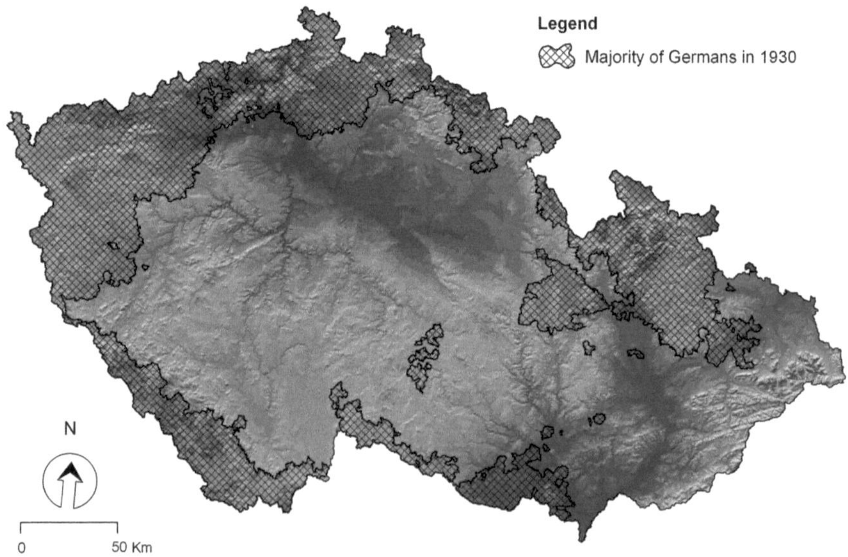

Fig. 20.2 Areas of majority of Germans (more than 50 % in cadastres) in Czechia in 1930 *Source* LUCC Czechia Project

building sites of bunkers, to prevent growing unemployment and poverty in peripheral regions. However, their decision was not to employ other nationalities but Czech in this strategic issue. Because of that, Germans felt to be regarded as an unreliable ethnic minority in a foreign country and their potential loyalty to Czechoslovakia disappeared forever.

The economic crisis also brought about the "Nazi" movement in Germany, whose ideas were broadly accepted by most Germans in Czechoslovakia. Hitler took advantage of the pro-Nazi feeling of Germans in the Czech borderland area and in September 1938 managed to force (with the approval of Great Britain, France and Italy) the government of Czechoslovakia to give-up all the border regions to Germany.

That was the first step leading later to the general brake down of the landscape continuity here. Czech people, as well as Jews and also German anti-Nazis were expelled to the rest of Czechoslovakia, in total around 200,000 people (Kural et al. 2002). Their property, houses and fields were confiscated and taken by already residing or newly arriving Germans, but still, the impact on the landscape and its management was not very significant. Another and exactly the opposite situation occurred after the end of WW II. A great hatred of Czechs toward their occupants, which formed during the war and grew after the return of people from Nazi concentration camps, culminated into transfer of almost all Germans out of the historical space of the Czech crown. For Czechs, this short period in their history is not still easy to speak about. More than the prevention against later nationality conflicts (how the transfer was officially approved by post-war international agreement), it was done with thoughts of revenge. First actions were uncontrolled violent expulsions without distinguishing between those guilty and those who were not, it was a sad paradox that also people actively fighting against the Nazis during the war were also expelled, just because of their German nationality.

In a result of the War, over 3 millions of Czech Germans (Glotz 2006) left or had to left their homes, mostly from mountainous boarder areas. The Czechoslovak government tried to resettle the empty space and created many stimulating factors to invite people from the rest of the country to come here and continue the life of these regions. However, resettlement was never done properly and another issue is that the newly coming settlers were not the average sample of population—mainly individuals from the very bottom of the society structure took advantage of the possibility of obtaining a house (mainly fully equipped) and fields for almost zero price (it is logical, because more able people had their own property and could not easily leave everything behind and move to a completely different region). That was how the source areas of resettlement were especially regions of eastern Slovakia and other less developed areas, which later resulted in many social and economical problems in the resettled border regions. The resettlement was problematic also from the point view of managing the land—the harsh conditions of mountainous landscape required special skills in agriculture (these places were originally settled by Germans who had come from the Alps) so in the first years the new arrivants (if they continued the farm work at all) had serious problems to do the agriculture.

It was a fickleness of fate that some parts of newly (although not fully) resettled areas were force-depopulated once more: somewhere because of the establishing of the Iron Curtain, elsewhere because of delimiting of "military training areas".

The so-called "Iron Curtain", how Winston Churchill labeled[1] the strongly protected belt of the frontier between ideologically different worlds of capitalism and socialism, ran in Czechia along its western and southern frontier (frontier with West Germany and Austria). Its creation was directly connected with the overtaking of the power in Czechoslovakia by the communist party in February 1948 (Jílek and Jílková 2006). Its purpose was to hermetically close the space of countries of the "Eastern Block", lead by the Soviet Union, and to separate them from the influence especially of neighbouring western countries. From our "landscape" point of view, it is important that parts of the Iron Curtain formed also a belt of "border prohibited zone", in some places more than 15 km wide and used exclusively for the purposes of the protection of the border, and a wider "border zone", serving as a buffer area. Based on a secret government order from 1951, all the villages within the "prohibited zone" were settled-out and destroyed (just a few years after their previous "resettlement"), and all "politically unreliable" persons from the "border zone" had to move out (Jílek and Jílková 2006).

Within the Czech Republic, nine strictly prohibited "military training areas" were established or re-established after WW II. Many of them were located in former German areas; for example the most important—the Doupov area (330 km^2, area of former twenty-two villages) and Boletice (219 km^2, area of former eight villages). From these "military training areas" all people were moved out and villages served for the training purposes, the houses often also directly as targets for tanks and cannons. Now the military areas represent unique study plots for researching principles of succession, thanks to the long-term influence of heavy military technique keeping some parts of the landscape in a completely "bare" condition (Fig. 20.3).

Today, most of the border zone is freely accessible again, even some parts of the "military training areas", so it is possible to observe the signs and remnants of the complicated historical development, which are still present in the landscape.

The Slovenian landscape, on the contrary to Czechia, lies in the narrow transitional region between the Alps, the Pannonian basin, the Dinaric Alps, and the Adriatic Sea. In spite of its small size (20,273 km^2) it is famous for its great natural diversity, variability and transitional characteristics. The lowest point is the coastline (0 metre a.s.l.) on the Adriatic coast, the highest peak is Mount Triglav (2,863 m a.s.l.) in Julian Alps. As 44 % of Slovenia is covered by limestone and dolomite karsts, its average inclination is 13° and the average altitude is 557 m a.s.l., it is not a favourable area for agriculture (Perko 2001).

[1] It was during his speech for Westminster Collage in fifth of May 1946: "From Stettin in the Baltic to Trieste in the Adriatic, an iron curtain has descended across the continent…".

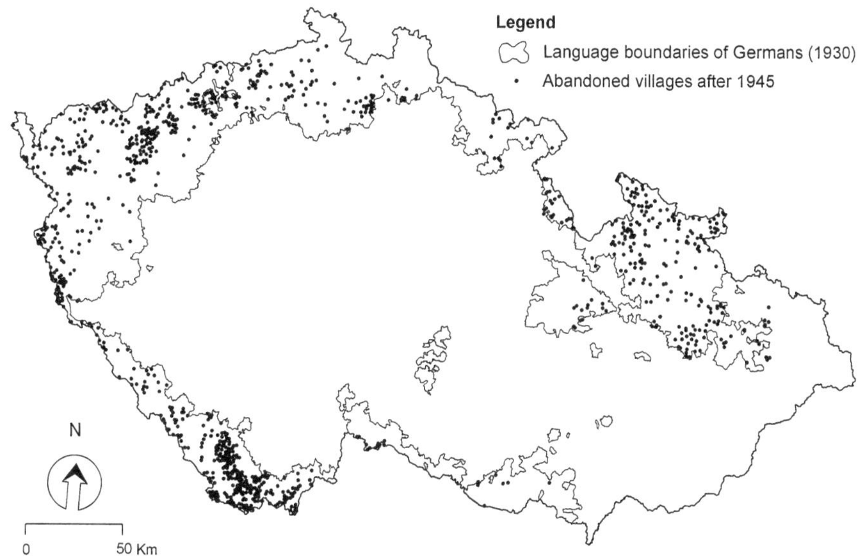

Fig. 20.3 Abandoned villages after 1945 within the area of former German settlement in Czechia (Kučera 2007)

In Slovenia, the most consistent and intensive German settlement (c. 880 km^2) could be historically found in the southern part of the country, in the Kočevje region (see Fig. 20.4).

Germans from around the town of Kočevje (Kočevars) were the descendants of German colonists from Tyrol and Charintia, which settled there during the times of the Ortenburgs lords in the fourteenth century in order to cultivate the region. Their colonisation lasted until the end of the fourteenth century; however, afterwards also Slovenes were settled there until the twentieth century. Kočevars were clearing the forests in order to gain new land for agricultural production which would enable the survival of an increasing amount of population. The main economic activities were cattle breeding and forestry. Agriculture was less developed due to the less favourable natural conditions (hilly and dry karst area). In the second half of the sixteenth century, there were already 9,000 inhabitants living in 137 Kočevars hamlets around the Kočevje region.

Because of harsh social and economical conditions there, the Austrian Emperor Friderik III (1492) granted the Kočevars the right for free trade with wooden ware, cattle and canvas. Peddlers made a positive contribution for economic development of the Kočevje region. On the other hand this was also the reason for increasing emigration and diminishing agriculture. Because of the agrarian crisis with its peak at the end of the nineteenth century, more and more Kočevars were employed as seasonal workers outside of the area which is nowadays Slovenia. Moreover, between 1857 and 1941, more than half of all the 23,000 Kočevars emigrated to USA. During the era of Habsburg monarchy, Kočevars enjoyed a high degree of linguistic, cultural and political autonomy in predominantly

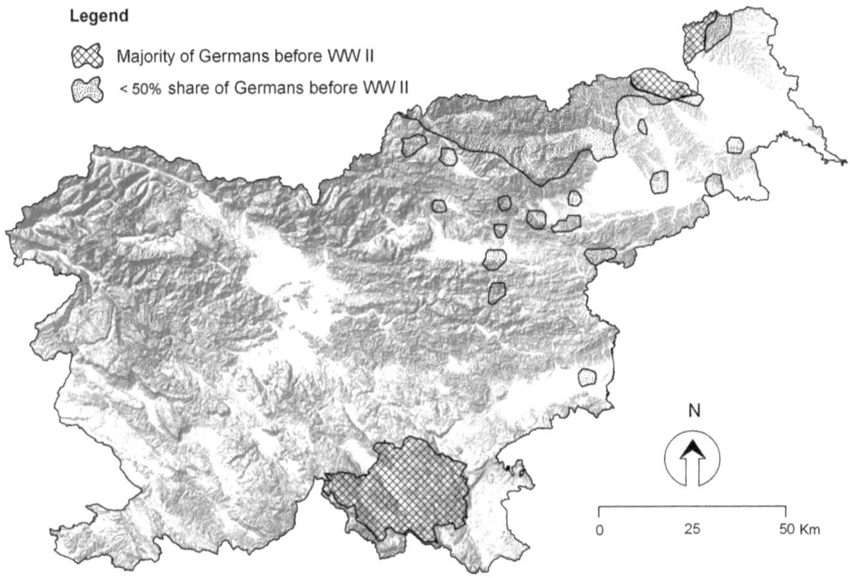

Fig. 20.4 Areas of German presence within the territory of today's Slovenia before WW II (according to Ferenc 2005). Circled is the area of Germans in Kočevje region (the area of interest in this article)

Slovene population inhabited the wider territory of Carniola. After the dissolution of the monarchy in 1918, the Kočevje region became part of the new Kingdom of Serbs, Croats and Slovenians, where Germans were not any more the favoured nation. Great changes such as the abolition of German schools, the use of Slovene language in public and Slovenisation of local geographical as also personal names caused a great deal of dissatisfaction among them (Urbanc 1998).

This was the fruitful ground for Nazi ideas in the 1930s. At the beginning of WW II in Yugoslavia in 1941, the Germans in the Kočevje region organised themselves politically and militarily. Their expectations for the incorporation in the German occupation zone or even annexation to the German Reich like in the case of eastern part of Slovenia (The lower Styria region), were not realized. Instead, the situation there was similar to that of Germans in South Tyrol administered by Italy. The leadership of Kočevars made a contract with Germany and Italy in 1942, which enabled them to relocate to the new lands in the Reich under the condition of signing the option announcement. According to the contract, 12,147 (95 %) of Kočevars signed the option announcement (Ferenc 2005). Kočevars from 176 hamlets were moved to eastern Slovenia in the area near rivers Sava and Sotla from where approximately 35,000 Slovenes were expelled to Germany and Serbia by the Nazi authorities.

During WW II the majority of the abandoned Kočevar hamlets were destroyed by the Italian army in military operations. After the war, Kočevars stayed without their former homes and homeland. Most of them either fled to Austria, others were expelled there by Yugoslav authorities; some died in Yugoslav prison camps. Most of Kočevars and their descendents today live in Austria where they publish a magazine called Neue Gottscheer Zeitung.

After the war, the destroyed and burned villages of Kočevars were not renewed. In fact the new political regime deliberately destroyed most of the German cultural heritage in the Kočevje region. Out of 123 churches, just 28 remain today (Urbanc 1998).

After 1945 the Kočevje region was mystified and people were discouraged from going there due to various reasons. In the second half of 1945 many karstic chasms in Kočevje region were used as unmarked graves for numerous war prisoners which had been killed by the communist authorities (Ferenc 2005). Due to its remoteness, the region was used for punishment camps. In 1951, there were 20 camps on 5,000 ha where the convicts/prisoners sentenced to forced labour were either working as forest workers or providing help by haymaking and building of the forest roads (Ferenc 2005).

As in the case of Czech borderland, also some areas south of Kočevje were restricted for public due to newly built secret underground installations of military importance during the era of the Socialist Republic of Slovenia. They were planned to provide a safe place for the leaders of the regime in case of a possible armed conflict. The security was so strict, that not even the Yugoslav federal security authorities had access to those areas. Due to this, these areas were a key factor in preparation for the Slovenian 10 day war of independence of Yugoslavia in 1991. Some smaller areas of former Kočevars hamlets south of Kočevje as Gotenica (Göttenitz) and Škrilj (Skrill) have even today restricted access as they are used as a training ground for the Slovene army and police force.

20.3 Land Use Change

As a main basis for the analysis of the land use changes, Land Use/Land Cover Changes (LUCC) database, created by our teams both for Czechia and for Slovenia, was used. It utilizes and puts together long-term historical data about the structure of land use in all cadastral units of both countries. The very precise cadastral data can be dated back to the first half of the nineteenth century; our database completed from this 170 year long period four time horizons with fully comparable data: 1845-1948-1990-2000 for Czechia Bičík at al LUCC Czechia Project and 1825-1900-1961-2002 for Slovenia Gabrovec and Petek, LUCC Slovenia Project.

Areas of Czechia and Slovenia were divided into so-called "Basic territorial units" (CTUs), which are in principle cadastral units, only slightly modified to be stable in their size within four observed time horizons.[2] Every CTU carries

[2] For more detailed methodological description see Bičík et al. (2001) for Czechia and Gabrovec and Petek (2007) for Slovenia.

Table 20.2 Comparison of LUCC databases created for Czechia and for Slovenia

	Total area of country (km²)	Number of CTUs	Average size of CTU (km²)	Time horizons	Number of Land Use classes
Czechia	78,866	8,903	8.6	1845-1948-1990-2000	7
Slovenia	20,273	2,635	7.7	1825-1900-1961-2002	7

information about the absolute area of seven Land Use classes: arable land (AL), permanent cultures (PC), permanent grasslands (PGL), forest areas (FA), water areas (WA), built-up areas (BuA) and other areas (OA). For the comparison of Czech and Slovenian databases see Table 20.2.

To enrich information about detail landscape structure, we also made a direct map comparison of the former (year 1845, historical maps) and present (year 2000-2010, field mapping work) state of land use in selected study areas for Czechia. To date, we have analysed twenty-one model cadastres all around Czechia, several of them also in the border regions of former German settlement. With the use of GIS tools, we carried out a complete analysis focused on the changes of the individual plots between 1845 and today.[3]

As we proved in our previous works focused on land use change (for more see Mareš and Štych 2005), we can observe strong specializing tendencies between individual regions in Czechia in the period of last 170 years. The regions, different from each other from the point of view of natural predispositions as well as of their position in the socio-economic complex, have been each orienting to some special function most suiting to their actual potential. This regional differentiation can be illustrated as a long-term opening of the scissors of differences between regions (see Fig. 20.5, which shows the development of percentage share of forest areas in separate regions defined according to altitude).

We also showed that similar polarization in land use development can be seen between regions defined according to "social-economic exposedness", "classes of nature protection" or "typology of countryside". It means that within Czechia, there have been recently establishing some wider areas with unified sort of land use, in contrary to former micro-heterogeneity which was closed inside every cadastre and showed itself as a typical mixture of all the land use forms in these small functional units.[4] However, the key factor directing the location and exact

[3] For methodology see Mareš and Štych (2003).

[4] This regional specializing tendency corresponds with results of Krausmann (2006), who found that in times of traditional agriculture most flows of energy and materials were closed in the framework of local or micro-regional cycle, while today the local cycle is opening and the energy flow to/from local unit is several times higher.

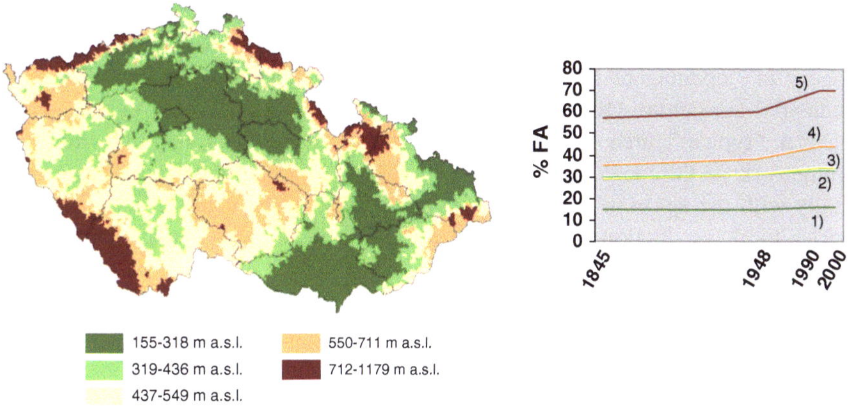

Fig. 20.5 Development of the percentage share of forest areas (FA) in zones of increasing altitudes in Czechia between 1845-1948-1990-2000. *Source* LUCC Czechia Project

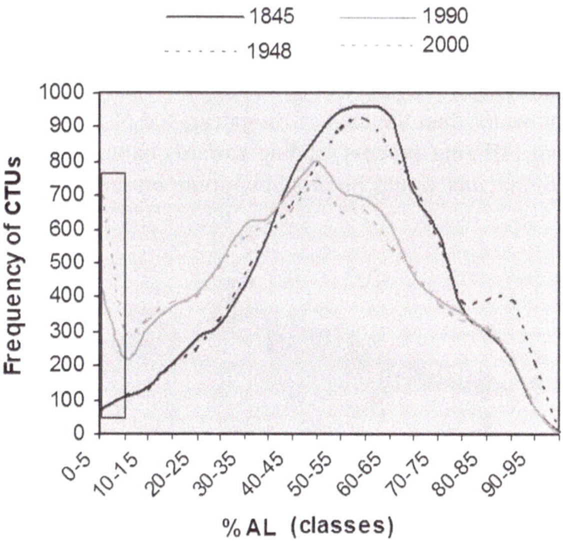

Fig. 20.6 Frequencies of CTUs in classes of percentage share of arable land (AL) in CTU for years 1845-1948-1990-2000 (for Czechia). The rectangle on the left side of the graph shows the change which is illustrated in the map in Fig. 20.9. *Source* LUCC Czechia Project

shapes of specializing regions is not easy to find, land use development is always caused by a large complex of driving forces.

It is without any discussion that one of the key factors influencing the change of land use from the perspective of whole Czechia was the great social change which took place in Czech borderland after WW II. For illustration of the importance of this change look at Fig. 20.6. There is a histogram classifying all the 8,903 CTUs into classes according to a percentage share of arable land for years 1845-1948-1990-2000. It is evident that the main peak in the middle of the graph has been decreasing in time (there have been less and less CTUs in Czechia with "average" or "normal" share of arable land), and on the other hand, a significant increase has been taking place at the left-side part of the graph (more and more CTUs with extremely low share of arable land). This change goes very well together with the theory of "opening of the local cycle" during the transition from traditional to industrial agriculture (Krausmann 2006)—in 1845 every cadastral unit needed to have a "typical" area of arable land necessary for self-production within the framework of the local cycle; together with the integration of society and especially with the agrarian and traffic revolution some areas could give up intensive agricultural production and concentrate on other—more suitable—forms of activities.

But an important question is: "Where exactly has the change been taking place?" Since we have all the CTUs with all their datasets in the format of GIS shapefile, we can demonstrate every selected part of the change directly on the map. In Fig. 20.7, see CTUs that entered the interval of 0–10 % of arable land (the interval marked by rectangle in Fig. 20.6) between 1948 and 2000.

From Fig. 20.6 we can see that between 1948 and 2000, arable land was disappearing almost exclusively in the areas of former German settlements along the Czech borderline. Comparing this map with the frontier of German linguistic area in Fig. 20.2, it is obvious that the expulsion of Germans was not the only driving force influencing the abandonment of agricultural land, but also other factors as altitude or periphery of concrete places played an important role (the German area is wider than the area of disappearance of arable land in this period). Perhaps the specializing process leading towards extensive use of peripheral areas in higher mountains would be probably going on anyway, the outgoing of three million of

Fig. 20.7 CTUs that entered the interval 0–10 % of arable land between 1948 and 2000 (the change that is delimited by the rectangle in Fig. 20.6). *Source* LUCC Czechia Project

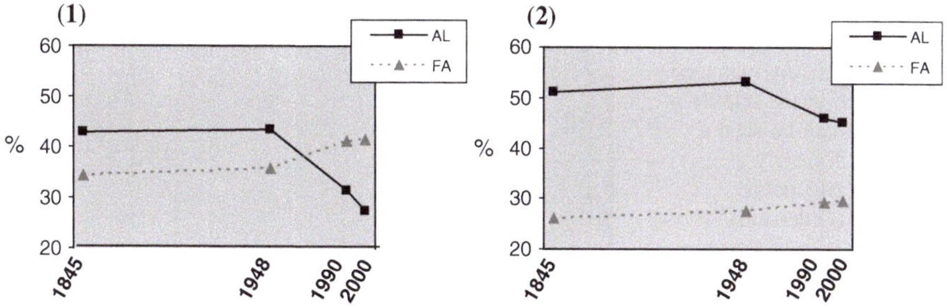

Fig. 20.8 Development of the percentage share of Arable Land (AL) and Forest Areas (FA) between 1845-1948-1990-2000 inside the former German linguistic area (1) and in the rest of Czechia (2). *Source* LUCC Czechia Project

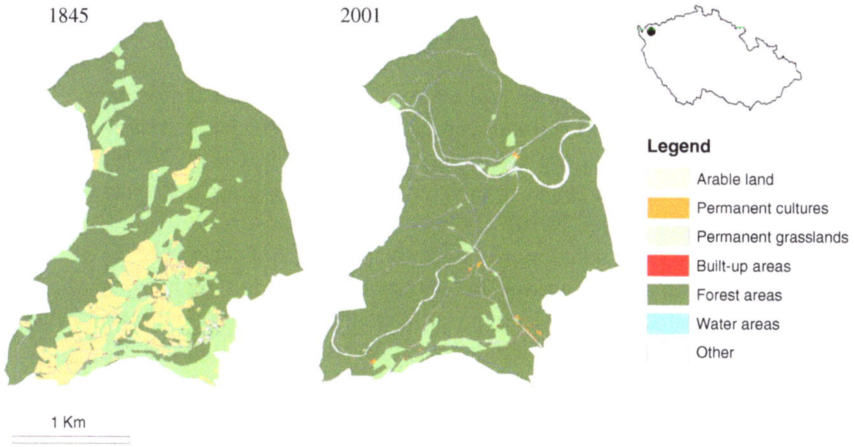

Fig. 20.9 Land use change in the model cadastre Oldřichov (Olderstoh; north-west part of Czechia) between 1845 and 2001. Today the cadastre is almost entirely forested, just several huts used as second-homes are distributed around the forest. *Source* LUCC Czechia Project

Germans (together with the further military interests in these areas) put just a definitive push to this tendency, which accelerated the process and led to much more significant result. Because the final state of the landscape after Germans left was too harsh, it was partly dimmed by the re-settling process, but not already in the whole area—the least convenient places stayed already aside the attention and they remained so until today.

In Fig. 20.8 we can see, that a general trend of the decreasing of the share of arable land and increasing of the share of forest areas[5] is common during studied period for former German regions as well as for the rest of Czechia, but in former German areas the *intensiveness* of this process is higher. In these regions, arable

[5] The sign of a phenomena named by some authors a "forest transition" (see Mather 2002).

Fig. 20.10 Spruce (*Picea abies*) forest at the place of former intensively managed arable land; in the middle of the picture can be seen a stone wall formerly separating two fields. Oldřichov (Olderstoh), north-west part of Czechia. (photograph: Petr Mareš 2009)

land and forest areas even changed their positions in the dominance of the share of land use class after 1948.

As a final illustration of the real state of the abandoned landscape of Czech border areas, we show the map comparison from model cadastre Oldřichov (Olderstoh; north-west part of Czechia) in 1845 and 2001 (Fig. 20.9); and also a picture of today's spruce forest in the place of former intensively managed arable land in this cadastre (Fig. 20.10).

In Slovenia, the German minority never played any important role. Even though Slovenian German settlement was, in contrary to the one in Czechia, very small (in Kočevje just around 800 km^2), we can still define significant connection between the end of German settlement and the turn of land use development. The graph in Fig. 20.11 shows the different tendency of arable land; the last line (year 2002) has

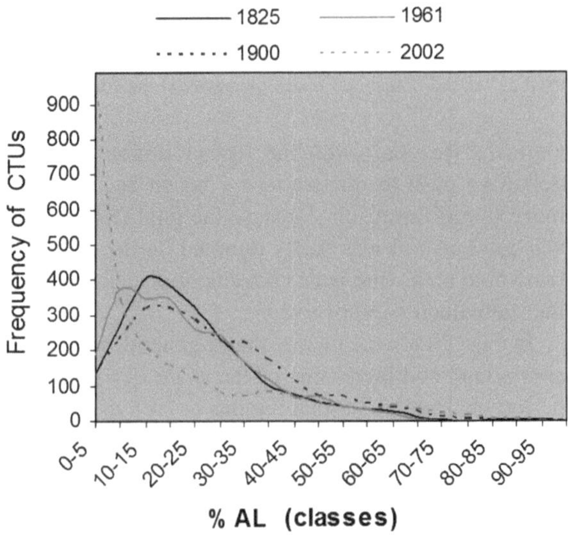

Fig. 20.11 Frequencies of CTUs in classes of percentage share of arable land (AL) in CTU for years 1825-1900-1961-2002 (Slovenia). The rectangle on the left side of the graph shows the change which is illustrated in the map in Fig. 20.12. *Source* LUCC Slovenia Project

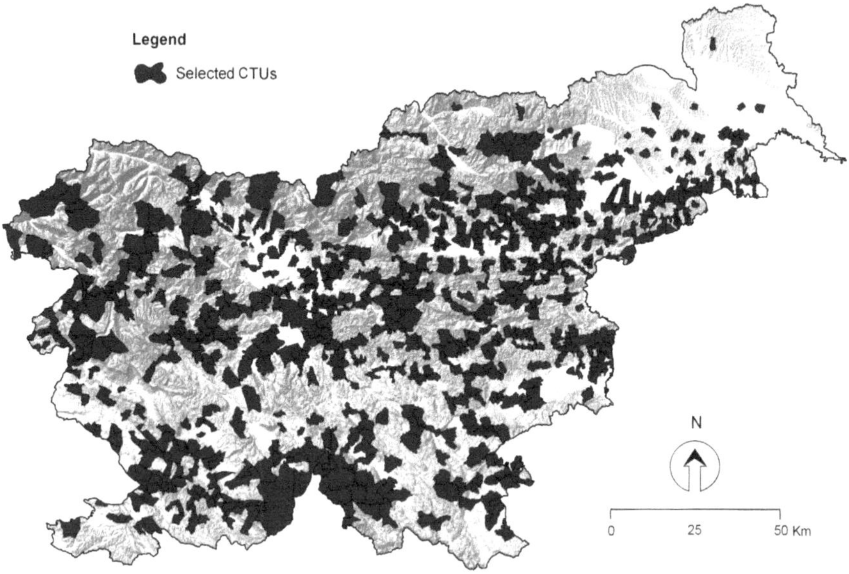

Fig. 20.12 CTUs that entered the interval 0–10 % of arable land between 1925 and 2002 (the change that is delimited by the rectangle in Fig. 20.11). *Source* LUCC Slovenia Project

the shape typical for greatly specialized features (extreme asymmetric distribution—in the meaning of Hampl (1998). When we display again the change visible at the very left side of the graph (CTUs that between 1900 and 2002 entered the interval of 0–10 % arable land) on the map (Fig. 20.12), we find out that among other places with diminishing arable land, the Kočevje area is quite clearly defined as a whole.

Figure 20.13 shows that also the comparison of the development of percentage share of arable land and forest areas, made for the Kočevje region and the rest of Slovenia, bear strong similarities to the Czech findings: the development in the region slightly correlates with the development in all the other areas, but is more

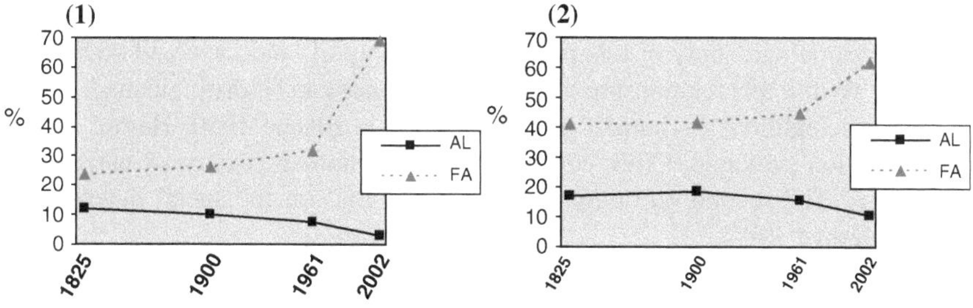

Fig. 20.13 Development of the percentage share of arable land (AL) and forest areas (FA) between 1845-1948-1990-2000 inside the former German language area of Kočevje (1) and in the rest of Slovenia (2). *Source* LUCC Slovenia Project

Fig. 20.14 Look of a formerly prohibited military area of Kočevska Reka (photograph: Primož Pipan, 2007)

extreme. Arable land developed from 11.9 to 3 %, forest from 23.8 to significant 68.8 %.

The Slovenian Kočevje region has now attributes of a really abandoned landscape. Particularly during the last period examined, (1961–2002), the change towards land abandonment has been accelerating and led to an almost definite result (Fig. 20.14).

20.4 Landscape Structure, Landscape Memory and Landscape Heritage

However significant and demonstrative the results of quantitative methods are, we feel that landscape cannot be evaluated only by charts, indexes and coefficients. That puts landscape in a position of an object that is a part of us, of our space, it reduces it in quantitatively perceived "thing". Such a view at landscape is incomplete. A large number of scientists are interested in a mutual relation between society development and landscape development (e.g. de la Blache 1950; Cosgrove 1985; in a way Hampl 1998; Cílek 2005). Therefore, an apparent need for description and study of this mutual relationship of landscape and society has evolved. Within this relationship a two way influence can be distinguished—one is natural preconditions → human society (e.g. de la Blache 1950; Hampl 1998—determination principle; Cílek 2005), the other is human society → natural preconditions. The second one might depend (not only) on the social development (e.g. Cosgrove 1985).

An accelerating pace of human society development has its negative consequences on rapidly changing landscapes around the world. On the other hand, landscape and society have been in *a never ending process of change* since ever (Antonson 2009). For that, landscape can be viewed as "*a palimpsest, a document that has been written on and erased over and over again…*" Crawford (1953, p. 51). Moreover, stable elements can be identified in the changing landscapes. Such elements are present in landscapes over several time periods and they usually are of a societal importance. These landscape elements are studied within concepts of landscape memory and landscape heritage.

Landscape memory investigates single objects/places within the landscape and seeks for stories and events related to those objects. Furthermore, landscape memory studies a relative importance of selected landscape objects for an individual or a society. In contrast, landscape heritage studies landscape as a whole and it views landscape as a heritage passed over from previous generations. Therefore a landscape is a witness and carries on traditions of landscape management etc. At the same time, some landscapes are distinguished as being "more traditional" than others. This classification is under the critique of scientists such as Germundsson (2005).

Abandoned landscapes of the Czech borderland and of the Kočevje regions provide a lot of suitable places where landscape memory can be studied and landscape features that evolved as by-products of human activity in the landscape can be closely investigated. To demonstrate this, an example from south–west Czechia is presented. The series of aerial photographs show identical place in two time periods. The selected place is an abandoned village of Pleš (Plöß). The village was founded in the sixteenth century and over 700 Czech Germans lived there in more than 100 houses by 1930. After the war, the inhabitants were forced to abandon their homes and no new settlers have come since… In the years to come, the village literally and technically disappeared. However, in the structure of the landscape footprints of past human activity can be identified. The top aerial photos have their origin in 1947—single houses as well as field plots in the village proximity can be seen. Bottom aerial photos show the state of the landscape in the same area today. Most eye-catching is the absolute disappearance of the village. However, managed landscape of 1947 as well as current "wild" landscape has some linear landscape features and elements in common. Such linear features are the old plot boundaries and paths (indicated in the photos) that persisted in the landscape and they are part of the landscape memory (Fig. 20.15).

Similar examples could be found in many places where villages were abandoned. Structure of current landscape is an example of an alternative view of the landscape memory phenomenon. Usually, landscape elements (trees, paths etc.) are seen as an anchor points for individual or collective memory. In other words, it is our memory, historical or mythological events that are related to landscapes. In contrast to that, landscape structure in places of abandoned settlement can be viewed as landscape memory, but it is the landscape that is the subject that *remembers*. The fact that landscape structure can be perceived as landscape memory is insufficiently distinguished. At the same time it is this premise on which landscape archaeology is based. The term "landscape memory" is, however, not used in that discipline.

Questions therefore arise if such abandoned landscapes were labelled as landscape heritage, as that concept is closely related to "traditional landscapes" and

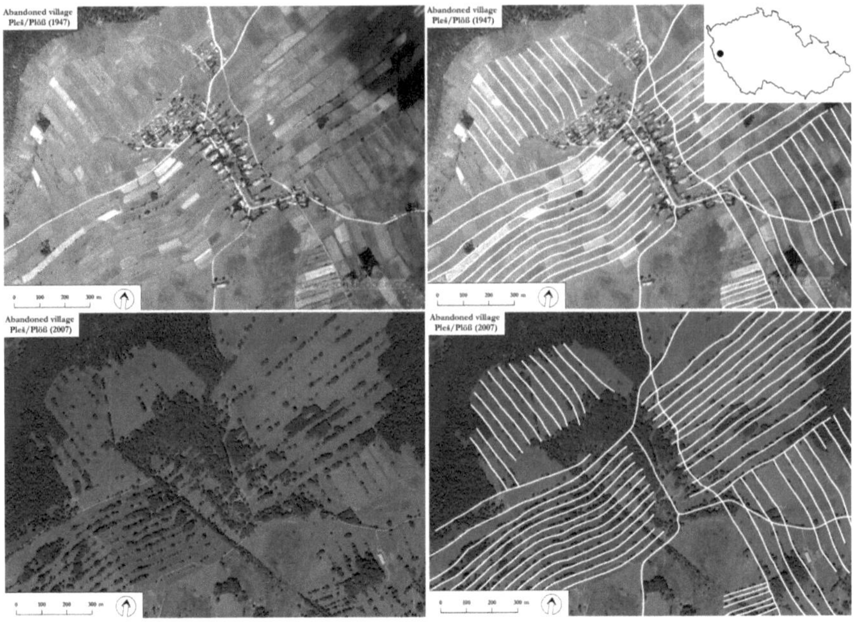

Fig. 20.15 Relict linear landscape features—plot boundaries and paths, an example of the abandoned Pleš (Plöß) village. *Source* LUCC Czechia Project

landscapes that have been constantly managed in non-varying way. On the one hand, landscape of abandoned villages represent landscape heritage—a heritage of human activities in landscape, on the other hand landscape heritage has generally a different connotation.

Abandoned settlement and its landscape structure provide us with a new insight into landscape memory and landscape heritage. It also shows us how deeply the human activity and landscape management is written in landscape. Such landscapes are indeed a memento and reminder of war and post-war consequences in Europe. One way or the other, these sometimes overlooked landscapes deserve our higher attention because they are of a great societal importance and they offer an enormous number of interpretation possibilities.

20.5 Environmental and Geo-Ecological Landscape Research

The previous parts were dedicated to the analysis of land use changes, landscape structure and landscape memory were also partly in focus of our investigation. In this last part, plant species diversity and environmental value of abandoned landscape will be discussed. In many places of former settlement a natural succession has taken place and secondary forest has evolved. Secondary forests are generally viewed as less diverse than ancient forest and it is also believed that secondary forest host fewer forest herb species (e.g. Dzwonko and Loster 1989; Matlack 1994). Hence, very little is known about the vascular plant diversity in a secondary forest that was formed by the processes of natural succession. It is the abandoned landscape in which natural succession took place and has been going on for more than 60 years. Therefore, it is a suitable place for analysis of plant species diversity in secondary forests.

Research results (Vojta 2007) from abandoned landscape of western Czechia reveals a relationship between former land use and current trees and scrubs diversity. Interestingly enough, a higher share of *Acer platanoides* and *Populus tremula* can be found in places, where used to be originally grassland (meadows, pastures). Among the dominating trees, few exemples of *Pyrus communis* or *Prunus avium* can be found. It proves that there used to be fruit trees in the grasslands. Forest on the edges of former villages is dominated by *Acer campestre*, *Corylus avellana* and fruit trees such as *Prunus avium*, *Pyrus communis*. In the centres of former villages, *Fraxinus excelsior* and *Sambucus nigra* can be found (Vojta 2007).

Also the distribution and occurrence of a plant species in underbrush reflect former land use of the landscape. Underneath the scrubs vegetation typically meadow species grow, such as *Trifolium alpestre*, *Festuca rupicola*, *Potentilla tabernaemontani*, *Cynosurus cristatus* etc. Together with the named species, there are rare and endangered plants which have disappeared from the surrounding landscape due to agricultural intensification. Such plants are *Ophioglossum vulgatum*, *Gentianopsis ciliata*, *Thalspi caerulescens*. Moreover, in this plant society there are first forest species—*Mercurialis perennis*, *Daphne mezereum* etc. Scrub vegetation forms in plant species rich biotopes that are often overlooked and not researched. Furthermore, this plant diversity is telling us a story of both former land use and former settlement. The described plant associations are enriched with a species that were planted in old gardens (e.g. *Narcissus poëticus*) (for more details see Vojta et al. 2010).

20.6 Conclusion

Landscapes of both Czech borderland and Kočevje region have experienced during their recent history major changes: from man managed, agricultural landscape to landscape that seems abandoned and shabby. However, such attributes are not fair to the landscapes that are actually very valuable. Rare and endangered plant species have found a refuge in scrubby biotopes. Moreover, landscape structure and tree species distribution tell a lot about succession process in places, where used to be settled and managed landscape. The abandoned landscapes have to be viewed as natural laboratory, in which unique and hard-to-be modeled processes biodiversity changes can be studied and analyzed.

Furthermore, abandoned landscapes carry landscape memory and their structure refers to old settlement and human activities. Therefore, the landscapes are of a high interpretative and socio-cultural value, which does not necessarily have to be disclosed at the very first sight.

Both Czech as well as Slovenian landscapes of former German settlements are very specific areas within their countries. Since these areas are of inconvenient natural conditions, they have been naturally developing towards extensive use and peripheral meaning. However, the crucial break in their historical continuity, the end the presence of the indigenous settlers, meant extremely fast change of land use resulting into abandonees of the whole regions.

Acknowledgments This work was supported by the Grant Agency of the Czech Republic, project no. 205/09/0995 "Regional differentiation and possible risks of land use as a reflection of functional changes of landscape in Czechia 1990-2010".

References

Andrýsek P, Bahenský P, Rábon M (2003) Utajené pevnosti—Československé opevnění z let 1936–1938 na Jižní Moravě. Fortifikace, sv, Brno, p 18

Antonson H (2009) Landscapes with history: addressing shortcomings in Swedish EIAs. Land Use Policy 26:704–714

de la Blache PV (1950) Principles of human geography. In: de Martonne E (ed) H. Holt & Co, New York, p 511

Bičík I, Jeleček L, Štěpánek V (2001) Land use changes and their societal driving forces in Czechia in 19th and 20th centuries. Land Use Policy 18(/):65–73

Cílek V (2005) Krajiny vnitřní a vnější. Dokořán, Praha

Cosgrove D (1985) Social formation and symbolic landscape. Barnes and Noble, Totawa

Crawford OGS (1953) Archeology in the field. Dent & Sons, London

Dzwonko Z, Loster S (1989) Distribution of vascular plant species in small woodlands on the Western Carpathian foothills. Oikos 56:77–86

Ferenc M (2005) Kočevska - pusta in prazna. Nemško jezikovno območje na Kočevskem po odselitvi Nemcev. Modrijan, Ljubljana

Gabrovec M, Petek F (2007) Land use changes in different slovenian landscapes. In: Gabrovec M (ed) Man in the landscape across frontiers. 2008, IGU/LUCC Central Europe conference 2007, Založba ZRC, ZRC SAZU, Ljubljana

Germundsson T (2005) Regional cultural heritage versus national heritage in Scania's disputed national landscape. Int J Heritage Stud 11:21–37

Glotz P (2006) Vyhnání—České země jako poučný případ. Paseka, Praha—Litomyšl

Hampl M (1998) Reality, society and geographical organization—search for integrating order. Faculty of Science, Charles University in Prague, Praha—Prague

Jílek T, Jílková A (2006) Železná opona—Československá státní hranice od Jáchymovska po Bratislavu. Nakladatelství Miloš Uhlíř—Baset, Praha

Krausmann F (2006) Forest transition in Osterreich: eine sozialokologishe annaherung. Mitteilungen der Österreichischen Geographischen Gesellschaft 148:75–91

Kučera Z (2007) Zanikání sídel v pohraničí Čech po roce 1945 – základní analýza. Historická geografie 34, Praha, s. 317–334

Kural V, Radvanovský Z, Bartoš J, Biman S, Hoffman J, Kocourek L, Kocourková K, Macek J, Řeháček K, Toms J, Šamberger Z (2002) Sudetypod hákovým křížem. Albis International, Ústí nad Labem

Mareš P, Štych P (2003) Evaluation of historical changes of land-use in Czechia exemplified on selected study areas. In: Jeleček L, Chromý P, Janů H, Miškovský J, Uhlířová L (eds) (2003) Dealing with diversity. 2nd international conference of the european society for environmental history in Prague. Charles University in Prague, Praha, pp 269–276

Mareš P, Štych P (2005) Historical changes of Czech landscape 1845–2000 and their natural and social driving forces studied on different spatial levels. In: Milanova E, Himiyama Y, Bičík I (eds) (2005) Understanding land-use and land-cover changes in global and regional context. Science Publisher Inc., Enfield, pp 106–134

Mather AS (2002) The reversal of land-use trends: the beginning of the reforestation of Europe. In: Bičík I et al (eds) (2002) Land use/land cover changes in the period of globalization. proceedings of the IGU-LUCC international conference. KSGRR PřF UK, Praha, pp 23–30

Matlack G (1994) Plant demography, land-use history, and the commercial use of forests. Conserv Biol 8:298–299

Perko D (2001) Terrain. In: Fridl J, Kladnik D, Orožen Adamič M, Perko D, Zupančič J (eds) (2001) National atlas of Slovenia. Institute of geography, Anton melik geographical institute at scientific research centre of the Slovenian academy of sciences and arts, Ljubljana

Šircelj M (2003) Verska, jezikovna in narodna sestava prebivalstva Slovenije. Popisi 1921–2002. Posebne publikacije. 5 prebivalstvo. Statistični urad Republike Slovenije, Ljubljana

Urbanc M (1998) Kočevarji. In: Hrvatin M (1998) Mala gora, Kočevski rog in Poljanska gora. In: Perko D, Orožen Adamič M (eds) (1998) Slovenija—pokrajine in ljudje. Založba Mladinska knjiga, Ljubljana

Vojta J (2007) Relative importance of historical and natural factors influencing vegetation of secondary forests in abandoned villages. Preslia 79:223–244

Vojta J, Kopecký M, Drhovská L (2010) Opuštěná krajina Doupovských hor. Živa 2:70–72

Datasets

Bičík I et al. LUCC Czechia Project: Charles University in Prague, Faculty of Science, Department of Human Geography and Regional Development; www.lucc.cz, www.lucc.ic.cz

Gabrovec M, Petek F LUCC Slovenia Project: Anton Melik Geographical institute at Scientific Research Centre of the Slovenian Academy of Sciences and Arts

Chapter 21
Land Management and Biodiversity Through Time in Upper Ribblesdale, North Yorkshire, UK: Understanding the Impact of Traditional Management

Helen Shaw and Ian Whyte

21.1 Introduction

The role of anthropogenic land use in the maintenance of culturally-derived ecosystems has been central to the development of thinking in the ecosystems approach (CBD 2000; Defra 2007, 2010). It is now widely recognised that in Europe, where there is a long cultural history of land use, the highly valued semi-natural habitats of the upland commons rely on traditional management techniques for their maintenance and survival. Similarly the gradual greening of the Common Agricultural Policy as a post-productivist environmental payment provides added incentive to combine policy for social and ecological systems and to highlight the value of traditional management. However, whilst elements of traditional land management are being implemented in new land management regimes some ecologists point out that traditional management can be misunderstood or used as rhetoric whilst the contribution to ecology is limited or, worse, continues to cause biodiversity loss (e.g. Ratcliffe 2002). The Millennium Ecosystem Assessment finding that "*Over the past 50 years, humans have changed ecosystems more rapidly and extensively than in any comparable period of time in human history*" (MEA 2005) has the potential to exacerbate this problem by focussing on recent decades of damaging anthropogenic practice and upholding traditional management as a tool in a social-ecological approach to ecosystems management. Without a fuller understanding of the variability and sustainability of traditional

H. Shaw (✉)
Lancaster Environment Centre, Lancaster University, Lancaster LA1 4YQ, UK
e-mail: dr.helen.e.shaw@gmail.com

I. Whyte
Department of Geography, Lancaster Environment Centre, Lancaster University, Lancaster LA1 4YQ, UK
e-mail: i.whyte@lancaster.ac.uk

I. D. Rotherham (ed.), *Cultural Severance and the Environment*, Environmental History 2, DOI: 10.1007/978-94-007-6159-9_21,

management practices there is a danger of developing a one-dimensional view (c.f. Renberg et al. 2009).

In England, it must be remembered that the Romantic movement in the nineteenth century and the national park movement in the mid twentieth were already responding to a feeling that the landscape was changing, and not for the better (Winchester 2006; Wordsworth 1835). Socio-economic and political influences on the English uplands have a long history from widespread monastic control in medieval times, to post-Reformation lay estates and improvement which have changed upland economies, farm size and social structures and, in doing so, have influenced human activity in ways that are still not well understood (Whyte and Winchester 2004). There is still debate regarding how climatic shifts during historic times affected land use and settlement, possibly triggering periods of land abandonment in upland areas (e.g. Parry 1978; Tipping 2002). Although the large-scale and long-term effects of changing human influence on upland habitats and vegetation are recognised, the influences of these more subtle socio-economic, political and environmental drivers on the diversity of pastoral plant communities over recent centuries are still not well appreciated. Beyond *c.* 50 years ago we often do not know in any detail the effects on biodiversity of past changes in land management.

The ecosystems approach does, however, call for a scientific framework to "encompass the essential structure, processes, functions and interactions among organisms and their environment" (Secretariat of the Convention on Biological Diversity 2004, p. 1), and advises the use of traditional management *where relevant*. Detailed understanding of changes in ecosystems and associated cultural communities over time is therefore an essential tool in the implementation of the ecosystems approach (Shaw and Whyte 2008). Such data can be gained from palaeo-historical analyses and these allow us to understand the trajectory of ecosystems and the human cultures that are intimately linked to them in relation to sustainable and resilient land use.

Given the appeal of upland cultural landscapes it is now, arguably, within their pastoral systems that integrated palaeo-ecological/historical research has still to make a fuller contribution. The scope of palaeo-ecology and pollen analysis to undertake this challenge is dependent upon the interpretation of subtle shifts in non-arboreal biodiversity for which recent developments in the understanding of the pollen vegetation relationship (e.g. Hjelle 1997, 1998; Gaillard et al. 1992; Hellman et al. 2009) are an essential part of this. However, as will be demonstrated here, much can be gained from less detailed pollen data and a simple illustration of the changes in key taxa over time can also be enlightening.

Where historical land use and vegetation change have been investigated interesting results have emerged. For example around Lake Vänern in Sweden eighteenth-century land use is found to be more important than later and current land use in determining the spatial diversity of plants in grassland. Here land use currently lacks the heterogeneity and small-scale dynamics provided by historical land use (Gustavsson et al. 2007). Such site-specific hysteresis effects are important in understanding current ecological systems and future trajectories. So, although a return to 'traditional' land management in order to provide both culturally valued landscapes and biodiversity benefit is beguiling, closer examination reveals that our understanding of traditional management may lack a detailed grasp of important long-term processes and be based upon a limited appreciation of the role and relevance of baseline conditions. Ironically, it is in the semi-natural areas of the uplands, where there is most opportunity to respond to calls to use an ecosystems approach, that the 1950s baseline may be least appropriate: In southern Scotland shifts in grazing management are demonstrated via pollen analysis in the Cheviot Hills at the transition to extensive sheep farming leading to a loss of diversity in the last *c.* 200 years (Tipping 2000; Davies and Dixon 2007). The pre-1950s traditional farming methods were already decreasing biodiversity in Scotland: What then was the pattern of change in the English uplands?

This chapter discusses preliminary results from a research project that has used history and palaeoecology to reveal evidence of changes in ecology under shifting traditional land management regimes from upper Ribblesdale, North Yorkshire in the uplands of northern England. Evidence combining palaeo-ecological and historical approaches demonstrates the nature of the drivers of ecological and landscape change and the current lack of detailed information on consequent shifts in ecosystem biodiversity. We argue that it is important in the light of international and European policy to understand better how different histories have shaped landscapes and how traditional land management practices have influenced the sustainability and resilience of ecosystems.

21.2 Historic Drivers of Landscape Development and Vegetation Change in the Uplands of Northern England

In broad terms the uplands of Northern England have undergone a familiar pattern of Holocene vegetation development which palaeoecology describes well (e.g. Atherden 1999; Pennington 1970). This was followed by Neolithic, Bronze and Iron Age woodland clearance for increasingly widespread agricultural exploitation (King 1986). Driven by population increase and demand from elsewhere within the Roman Empire Romano-British communities implemented further deforestation. However, when examined at the local scale, site specific variation occurs even within regions (Skinner and Brown 2001). The subsequent collapse of Roman rule in northern Britain saw a reduction in the extent of arable land and a partial regeneration of woodland (e.g. Dark 2005; Higham 1986). Further woodland clearance occurred during early medieval times associated, in some parts of the north, with Scandinavian settlement but evidence for vegetation change in the landscape becomes patchy and site specific (Dumayne-Peaty 1999). In high medieval times, a major expansion of pastoral farming on monastic and lay estates linked to the wool trade is evident via an increase in grass and pastoral pollen indicators (Kershaw 1973). As Atherden (2006) notes, however, there has been little detailed palaeoecological research on medieval and subsequent periods. Similarly, in landscape research, Iron Age and Romano British archaeology has taken precedence over studies of more recent periods (Whyte 2006).

Thus, although the broad patterns of socio-economic change in this region have long been appreciated the more finely nuanced variations which occurred over shorter timescales have not. A plethora of possible changes caused by the impact of varying management practices at estate and farm level may have affected biodiversity in the past and today. For example, following the Dissolution of the Monasteries in the 1530s there were important changes such as sectoral shifts between cattle and sheep rearing from the sixteenth century to the early twentieth century which may have produced subtle shifts in pastoral diversity. How did these management shifts in turn affect vegetation patterns and how can such vegetation changes be identified in the pollen record? Given the generalised approach of shifting to a mixed sheep and cattle grazing regime to enhance biodiversity targets in pastoral land it is at this more detailed level that a combination of palaeo-ecological and historical evidence can be most fruitful.

21.3 Ribblesdale: Historical and Environmental Change

Upper Ribblesdale in North Yorkshire lies between the mountains of Pen-y-Ghent and Ingleborough. The valley floor is broken up by glacial drumlins providing areas of arable and improved pasture on their slopes and wet meadows in the marshy areas between them. Above this on the limestone plateaus are extensive limestone pavements separated by areas of free-draining loessic soils. The area contains a number of small hill farms with enclosed and common pastures predominantly grazed by sheep. There are a number of deserted settlements on the higher limestone plateaus that may date to the Iron Age, Romano-British or medieval periods. In other cases farms were abandoned much later e.g. the hamlet of Thorns which was progressively deserted in the later nineteenth century (Johnson 2009). As is typical in the English uplands farmstead abandonment has been accompanied by farm amalgamation rather than land abandonment so that grazing management has been continuous. The area contains a number of SSSIs with rich limestone floras and, importantly, a major part of the eastern slopes of Ingleborough is now a National Nature Reserve and forms the Ingleborough complex Natura 2000 Special Area of Conservation. Examples of the implementation of conservation management are areas of traditional hay meadow management at Colt Park and shifts to extensive cattle grazing on Sulber Pasture whilst the enclosure at South House Moor is being re-wilded with the aim to demonstrate alternative land use strategies (Evans n.d). Given the implementation of both traditional grazing and re-wilding concepts within this area, understanding the site history and potential will be of benefit.

21.4 Historic Landscape and Environmental Change in Ribblesdale

The limestone plateaus around Ingleborough were extensively settled in Bronze Age, Iron Age and Romano-British times before a contraction occurred in early historic times (Higham 1986). A renewed expansion of settlement occurred in early medieval times, epitomised by the excavated tenth-century farmstead at Ribblehead (Johnson 2009). This was linked with large grants of land to monastic orders, particularly Furness, Fountains and Jervaulx Abbeys. Under the management of these religious houses a number of upland vaccaries (cattle ranches) and sheep farms (bercaries) were developed. A major crisis occurred in the early fourteenth century due to crop failure, famine, livestock disease and Scottish raids (Briggs 2005), followed in 1348-9 by the Black Death, leading to less intensive management with the abandonment of cultivated land and the reversion of land to scrub. The leasing out of monastic lands and then their seizure by the Crown at the Dissolution in the early sixteenth century created a landscape of small farms, limited areas of arable, improved pasture and meadow and extensive upland commons which survived, with modifications, into modern times (Johnson 2009). This broad-scale history, however, hides many smaller-scale changes which impacted on pasture management and the landscape. These included the effects of the build-up of population in the sixteenth and early seventeenth centuries, the development of the system of meadows with isolated field barns in the late sixteenth and early seventeenth centuries (Gambles and Pierre 2010), the progressive intake of land by enclosure from common pastures, initially by the creation of stinted pastures, later through parliamentary enclosure, the development of cattle farming in the seventeenth and eighteenth centuries with the rise of the droving trade, the creation of sporting estates in the nineteenth century and the massive expansion of sheep numbers between the 1950s and the foot and mouth outbreak of 2001.

21.5 Pollen Analysis

In upper Ribblesdale, previous pollen analyses of sites on Ingleborough (Gosden 1965; Swales 1987) have provided vegetation history with reference to pollen zones but are only poorly dated by modern standards of palaeoecology and lack sufficient detailed analysis of near-surface peat to provide information on vegetation change over the historic period. Some previous research in the wider landscape has highlighted the potential for this period. Piggott and Piggott (1963) describe, at nearby Malham, a slow stepped decline in woodland cover during the Neolithic and Bronze Age, that later increased in the Iron Age, Romano British, Norse and monastic periods (Piggott and Piggott 1963, p. 333) together with the possible recent proliferation of Ash (*Fraxinus excelsior*) woodland due to short periods of cessation of grazing. Similarly Mackay and Tallis (1994) identify woodland regeneration following periods of falling population driven by epidemics such as the Black Death in their more recent, radiocarbon dated pollen analysis in the Forest of Bowland, Lancashire, whilst on Extwistle Moor, Lancashire fluctuations in tree abundance in the landscape up to the medieval period are evident, followed by the extension of summit heaths during the Little Ice Age (Bartley and Chambers 1992).

Pollen evidence is presented here from an analysis of a core from Wife Park near Selside SD 77382 73648 (Fig. 21.1). The pollen diagram spans a period from at the early fourteenth century to the present day. A remarkable increase in grass pollen and a reduction in the structural diversity of the plant community are evident as a gradual process of ecological change from the medieval to the present. From 1300–1700 approximately, the pollen diagram shows mixed vegetation with a balance between arboreal, shrub, dwarf shrubs, herbs and grasses. The arboreal element is not great and is likely to be extra-local, due to the nature of arboreal pollen transport mechanisms, created perhaps by small areas of wood or scattered trees, including smaller trees and shrubs. There is a mixed ground flora with herbaceous species, heath, sedge and grassland elements.

A remarkable decline of the arboreal element including small trees and shrubs is evident in the loss of tree pollen at 22 cm interpolated from radiocarbon dates to

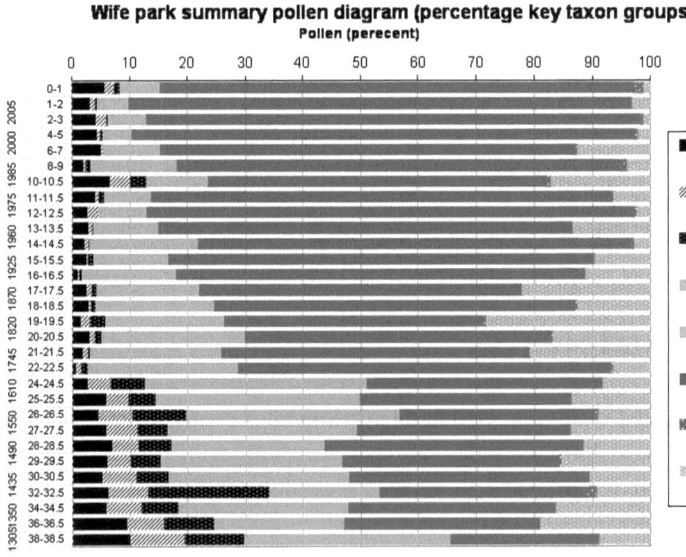

Fig. 21.1 Pollen percentage diagram for key taxa from Wife Park, Ribblesdale. Pollen data are presented as a percentage of the total land pollen counted (excluding pollen from conifers assumed to be a planted population). The peat core is from a small basin between two drumlins and represents the local plant community around the basin (Jacobson and Bradshaw 1981). The top 40 cm has been used in this study with peat dated via four radiocarbon dates at 20, 24, 32, and 40 cm. The chronology is approximate and is determined from a visual best fit through the calibrated radiocarbon dates

the early eighteenth century and this element of the biodiversity does not recover. A gradual increase in herbaceous pollen percentages precedes this woodland decline and herbaceous taxa remain as an important but gradually declining part of the pollen spectra for the next 300 years. The heathland pollen percentage also undergoes two rapid stepped declines in the early seventeenth and late nineteenth centuries; the diversity of heath pollen taxa as well as the overall percentages are in decline until in the recent peat only *Calluna vulgaris* pollen is found. These declines in diversity of major taxa reduce the community to the modern grass dominated one.

Of particular interest is the loss of the arboreal component with a marked decline in trees and woody shrubs evident between the pollen samples at 22 and 24 cm. The timing of this fits with documentary evidence from the historical record showing a lack of woodland management due to weak manorial control leading to a decline in trees in the late seventeenth and early eighteenth centuries. The majority of the shrub pollen component is pollen from *Corylus avellana* and *Salix* species. Dispersal of pollen from *Corylus* may be extra-local: however, Dark (2005) notes that *Salix* pollen only disperses locally. Given the richer heath diversity in this time period the presence of small Salix shrubs within the local landscape probably indicates lower density grazing. In the present landscape arboreal species are limited to a few areas and gill woodlands, for example on the limestone pavement at Colt Park (SD 774 776) on the lower slopes of Ingleborough approximately 1.5 km to the south west of Wife Park.

Calluna vulgaris pollen may have dispersed over the extra-local area and may have originated from plants on the slopes of Pen y Ghent. However, the presence of grains of *Empetrum nigrum* and *Erica tetralix* pollen below 16 cm may demonstrate a very local heathland presence (e.g. Shaw 2006) on the adjacent drumlins. The loss of heath may relate to the decline of grouse shooting in the area since the early twentieth century but in the locality of Wife Park is more likely to relate to gradually increasing stocking densities of sheep.

Sedge pollen percentages maintain around 10 % of total land pollen from *c.* 1300 to *c.* 1750, then increase in dominance before an overall reduction with more rapidly fluctuating percentages in recent decades. The loss of sedges may be due to drainage and pasture improvement. The drainage of pastures has a long history, but increased markedly since the 1950s (Holden et al. 2007). It is possible that the sedge pollen signal is responding to the differing water levels in the soil, but could also be responding to nearby areas of bog that were cut for peat fuel.

21.6 Discussion

The pollen data show a gradual long-term decline in diversity from the fourteenth to the twentieth centuries which intensified in the last two centuries. The decline is marked by a stepped reduction in elements of structural diversity in the vegetation, with first the arboreal component, then the heaths, then herbaceous species decreasing.

The pastoral impact on the post-medieval landscape is well represented in the pollen diagram. The system of cattle-grazed stinted pastures in the eighteenth and nineteenth centuries is depicted by increases in herbaceous species thus supporting the decisions of the NNR management and the Limestone grazing project. However, despite the maintenance of herbaceous flora there was a sustained loss of diversity over these recent centuries. The loss of diversity extends back well beyond the intensification of livestock production since the 1950s linked to the influence of the CAP on modern hill farming. The loss of the arboreal component represents a reduction in diversity linked to poor management and a shift away from the reliance on multiple local resources.

These are important factors in understanding, reinstating or supporting traditional land use as these changes have occurred within a period of pre-1950s traditional land use. Certainly, although there has been an overall dominance of pastoral open landscape rather than major fluctuations in tree cover, the pollen and historical evidence show that the landscape has been subject to change and that pre-1950s diversity was already declining.

Management has, throughout the period studied, been focussed upon the development of the landscape for agricultural gain and human use. Sheep husbandry, at varying levels of intensity, first developed on a large scale by the monasteries, fostered the development of short-cropped, thin-soiled, limestone pasturage and hay meadows which were botanically rich and may have included a dwindling wood-pasture component, perhaps indicating the long-term trajectory of this system to one of reduced structural biodiversity. The enclosure of pastures for cattle grazing was probably followed with only slow biodiversity declines as illustrated in the Wife Park core. But the next phase of improvement, from the late eighteenth century onwards, involving the widespread drainage of the landscape, led to another phase of diversity reduction with the decline in sedges and a dramatic loss of herbaceous diversity. The recent decades of sheep grazing have reduced overall diversity further, but as part of a process of longer-term vegetation change.

Additionally, underlying the pastoral variations there are a range of changes associated with the reduction in a rural population. Peat cutting has been widespread in the past as a source of fuel, quarrying in the landscape is evident at the small scale, with local surface quarries used for extracting stone for walling and large quarries still active around Horton in Ribblesdale. Limestone pavements were often stripped of their clints for Victorian rockeries. Limestone was also used widely to sweeten the more acid pastures and limekilns are evident throughout the landscape. A wider range of natural resources was used, including harvesting *Juncus* from rush pastures and the controlled use of timber is evidenced in manorial documents. Some of these activities—perhaps the harvesting of rush pasture—may have been beneficial for nature conservation targets, but many were not, although contributing to the economic development and social cohesion of the area. The challenge will be to identify and reinstate traditional management techniques that improve diversity and maintain landscape structure, but avoiding some of the previous activities that provided economic stability at the expense of the environment.

21.7 Conclusions

The gradual 'improvement' of agriculture includes a series of shifts away from a localised subsistence production system with multiple products from the ecosystem, to one reliant on grazing and more recently dominated by sheep. These shifts correlate with a gradual loss of diversity within the pre-1950s period of traditional management. Traditional land management is perhaps adaptable rather than sustainable. Hill farming has remained resilient in the post-medieval English uplands, but has become focussed on the pastoral element whilst other important aspects of local management have been lost. Whilst land abandonment has been dominant in other European uplands production has continued to dominate in the English uplands; however, the remnants of traditional management are much reduced and sheep rearing has become less intensive. Thus the system has shifted to one perhaps culturally resilient but less desirable for biodiversity.

Whilst traditional land management can have benefits for biodiversity we must be careful to reinstate a range of elements of productive ecosystem services. The model for Ingleborough is taking this approach assisted by the status of a large tract of the area as a National Nature Reserve with some arboreal planting; however, using the history of this environment as an example the results illustrate the need to examine in detail the landscape history each area and to reintroduce a range of management activities and productive ecosystems services to ensure robust and sustainable future management. The results from Ribblesdale highlight that it is those parts of the system that lose economic value and purpose that decline. There is indeed a danger, as Ratcliffe (2002) highlights for the Lake District, that traditional management is misunderstood and that limited elements are reinstated based upon current economic models of hill farming, thus bringing limited biodiversity benefits.

Acknowledgments This research is funded by a research grant from the Leverhulme Trust. We thank the relevant farmers and landowners, Natural England and the Yorkshire Dales National Park Authority for access to land and advice in locating peat deposits.

References

Atherden M (2006) Re-thinking Craven's limestone landscape: vegetation history. North Craven historical research group workshop Oct 2006—re-thinking Craven's limestone landscape, pp 21–23. http://northcravenhistoricalresearch.co.uk/Outreach/YDCONFERENCEOCT06.pdf

Atherden M (1999) The vegetation history of Yorkshire: a bog-trotters guide to God's own country. Naturalist 124:137–156

Bartley DD, Chambers C (1992) A pollen diagram, radiocarbon ages and evidence of agriculture on Extwistle Moor, Lancashire. New Phytol 121:311–320

Briggs C (2005) Taxation, warfare and the early fourteenth-century crisisin the Nprth, Cumberland lay subsidies 1332–1348. Econ Hist Rev 58(4):639–672

CBD (2000) Ecosystems approach. Convention on Biological Diversity COP 5 Decision V/6. In: The conference of the parties 5 decision V/6 Nairobi, 15–26 May 2000

Dark P (2005) Mid- to late-Holocene vegetational and land-use change in the Hadrian's Wall region: a radiocarbon-dated pollen sequence from Crag Lough, Northumberland, England. J Archaeol Sci 32(4):601–618

Davies AL, Dixon P (2007) Reading the pastoral landscape: palynological and historical evidence for the impacts of long-term grazing on Wether Hill, Cheviot foothills, Northumberland. Landscape Hist 29:35–45

Defra (2007) Securing a healthy natural environment: an action plan for embedding an ecosystems approach. http://archive.defra.gov.uk/environment/policy/natural-environ/documents/eco-actionplan.pdf

Defra (2010) Delivering a healthy natural environment: an update to "Securing a healthy natural environment: an action plan for embedding an ecosystems approach" http://archive.defra.gov.uk/environment/policy/natural-environ/documents/healthy-nat-environ.PDF

Dumayne-Peaty L (1999) Continuity or discontinuity? Vegetation change in the Hadrianic-Antonine frontier zone of Northern Britain at the end of the Roman occupation. J Biogeogr 26(3):643–665

Evans P (N.D.) Ingleborough national nature reserve (NNR) Ingleborough complex SAC (Natura 2000): a brief review of the site. http://www.natura.org/DOC/uk_limestone_summary.pdf. Accessed 15 March 2012

Gaillard M-J, Birks HJB, Emanuelsson U, Berglund BE (1992) Modern pollen/land-use relationships as an aid in the reconstruction of past land-uses and cultural landscapes: an example from south Sweden. Veg Hist Archaeobot 1(1):3–17

Gambles D, St. Pierre T (2010) Hay time in the Yorkshire Dales. Yorkshire Dales Millennium Trust, Clapham

Gosden MS (1965) An investigation into the origin and nature of some organic deposits of the Ingleborough region. Unpublished Ph D thesis, University of Leeds, Leeds

Gustavsson E, Lennartsson T, Emanuelsson M (2007) Land use more than 200 years ago explains current grassland plant diversity in a Swedish agricultural landscape. Biol Conserv 138:47–59

Hellman S, Bunting M-J, Gaillard M-J (2009) Relevant source area of pollen in patchy cultural landscapes and signals of anthropogenic landscape disturbance in the pollen record: a simulation approach. Rev Palaeobot Palynol 153(3–4):245–258

Higham N (1986) The northern counties to AD1000. Longman, London, pp 243–250

Hjelle KL (1997) Relationships between pollen and plants in human-influenced vegetation types using presence-absence data in western Norway. Rev Palaeobot Palynol 99(1):1–16

Hjelle KL (1998) Herb pollen representation in surface moss samples from mown meadows and pastures in western Norway. Veg Hist Archaeobot 7:79–96

Holden J, Shotbolt L, Bonn A, Burt TP, Chapman PJ, Dougill AJ, Fraser EDG, Hubacek K, Irvine B, Kirkby MJ, Reed MS, Prell C, Stagl S, Stringer LC,Turner A, Worrall F (2007) Environmental change in moorland landscapes. Earth-Science Reviews 82(1):75–100

Jacobson GL, Bradshaw RH (1981) The selection of sites for palaeoecological studies. Quatern Res 16:80–96

Johnson D (2009) Ingleborough. A landscape history. Carnegie, Lancaster

Kershaw I (1973) Bolton Priory. The economy of a northern monastery. Oxford University Press, Oxford

King A (1986) Romano-British farms and farmers in Craven, North Yorkshire, vol 158. In: Manby TG, Turnbull P (eds) Archaeology in the Pennines. British Archaeological Reports, Oxford, pp 181–193

Mackay AW, Tallis JH (1994) The recent vegetational history of the forest of Bowland, Lancashire, UK. New Phytol 128(3):571–584

MEA (2005) Millennium ecosystem assessment, 2005. Ecosystems and human well-being: synthesis. Island Press, Washington. http://www.millenniumassessment.org/documents/document.356.aspx.pdf

Parry ML (1978) Climatic change, agriculture and settlement. Dawson, Folkestone

Pennington W (1970) Vegetation history in the north-west of England: a regional synthesis. In: Walker D, West RG (eds) Studies in the vegetational history of the British Isles: essays in honour of Harry Godwin. Cambridge University Press, Cambridge

Piggott CD, Piggott ME (1963) Late-glacial and post-glacial deposits at Malham, Yorkshire. New Phytol 62:317–334

Ratcliffe D (2002) Lakeland. Collins new naturalist library No. 92. Collins, London

Renberg I, Bigler C, Bindler R, Norberg M, Rydberg J and Segerström U (2009) Environmental history: A piece in the puzzle for establishing plans for environmental management. J Environ Manag 90(8):2794–2800

Secretariat of the Convention on Biological Diversity (2004) The ecosystem approach, (CBD Guidelines) Montreal: secretariat of the convention on biological diversity, 50 p

Shaw HE (2006) A palaeoecological investigation of long-term stand-scale ecological dynamics in semi-open native pine woods: contributing to conservation management in east Glen Affric. Unpublished Ph D thesis, School of Biological and Environmental Sciences, University of Stirling, Stirling. http://hdl.handle.net/1893/201

Shaw H, Whyte I (2008) Shifting ecosystem services through time in the North West uplands and implications for planning adaptation in the future. Aspects Appl Biol (Shaping a vision for the uplands) 85:99–106

Skinner C, Brown AG (2001) Mid Holocene vegetation diversity in eastern Cumbria. J Biogeogr 26(1):45–54

Swales S (1987) The vegetational and archaeological history of the Ingleborough Massif, North Yorkshire. Unpublished Ph D thesis, University of Leeds, Leeds

Tipping R (2000) Palaeoecological approaches to historic problems: a comparison of sheep-grazing intensities in the Cheviot Hills in the Medieval and later periods, vol 293. In: Atkinson J, Banks I, MacGregor G (eds) Townships to farmsteads. Rural settlement studies in Scotland, England and Wales. British Archaeological Reports, Oxford, pp 30–43

Tipping R (2002) Climatic variability and marginal settlement in upland Britain: a re-evaluation. Landscapes 3(2):10–28

Tyler AN (1994) Environmental influences on gamma ray spectrometry. Unpublished Ph D thesis, University of Glasgow, Glasgow

Tyler AN, Sanderso DCW, Scott EM, Allyson JD (1996) Accounting for spatial variability and fields of view in environmental gamma ray spectrometry. J Environ Radioact 33(3):213–235

Whyte I (2006) Upland Britain: cultural processes and landscape change through time. Int J Biodivers Sci Manag 2(3):138–141

Whyte I, Winchester A (2004) Society, landscape and environment in upland Britain. Society for Landscape Studies Supplementary Series 2, Birmingham

Wordsworth W (1835) Guide to the Lakes. In: Bicknell P (ed) Select editions. Hudson and Nicholson, Longman and Co, Kendal, London

Winchester A (2006) Common Land in Upland Britain: tragic unsustainability or utopian community resource? In: Bosbach F, Engels JI, Watson F (eds) Umwelt und Geschichte in Deutschland und Grossbritannien: Environment and History in Britain and Germany. Prince Albert Studies, vol 24. K.G. Sauer, Munich. http://commons.ncl.ac.uk/?q=system/files/CoburgWinchester.pdf

Chapter 22
Policing the Commons in the Vale of York, *c*.1550–*c*.1850

Brodie Waddell

22.1 Introduction

The Vale of York was a wealthy, densely populated, low-lying agricultural region throughout the sixteenth, seventeenth and eighteenth centuries. It was not known for extensive and highly-regulated common fields like those of the East Midlands, nor for huge common moors such as were found in the fells of north-western England. Moreover, its highly commercialised agriculture and growing prosperity made it the antithesis of those isolated regions where one might expect that 'medieval' regulations and 'feudal' institutions would remain relevant in the early modern period. As such, it is perhaps surprising to find that not only were 'commons' a vital part of the region's economy, but also that they were carefully managed and policed by local communities through the mechanism of the manor court.

The primary focus of this study is a group of around sixty settlements, most of which are situated along the River Ouse. It draws on over one hundred sets of manorial regulations as well as a variety of estate papers, county records, enclosure awards, and ordinance surveys to together provide evidence of past land use. Each section addresses one of the major collectively-managed features on the early modern landscape. Specifically, this chapter examines common fields, common meadows, other common-land (such as woodlands, marshes, and 'wastes'), and shared 'infrastructure' (such as roadways, drains and fences).

B. Waddell (✉)
Department of History, Classics and Archaeology Birkbeck, University of London,
28 Russell Square, London WC1B 5DQ, UK
e-mail: b.waddell@bbk.ac.uk

I. D. Rotherham (ed.), *Cultural Severance and the Environment*, Environmental History 2, DOI: 10.1007/978-94-007-6159-9_22,

22.2 Common Fields

Communally managed open fields were an important part of the early modern landscape in the Vale of York, where most townships used a multi-field system to cultivate a significant amount of their arable crops and to pasture their animals after the harvest. However, compared to the extremely extensive open-field agriculture found in some other areas—including much of the Midlands[1]—this region had a relatively modest quantity of unenclosed tillage. Indeed, for some settlements, there is little evidence of post-medieval common fields at all.

Of the 52 townships along the River Ouse, at least 37 definitely had early modern 'town fields' and most of the others were small hamlets whose fields may have been subsumed within those of their larger neighbours (Table 22.1). Yet, it is clear that a considerable amount of enclosure was underway or had already been completed by this period. In some cases, townships had lost all of their arable commons well before the era of parliamentary enclosure began in the mid-eighteenth century. There were also some townships where only a patchwork of small common fields survived amongst large numbers of closes; however, there were a number of other settlements that retained very substantial open-field systems into the late eighteenth century or beyond. So, it seems that some arable common land survived in most townships in the Vale until at least the eighteenth century, even if it accounted for only a minority of the total cropland in some of these places by this time.

The residents of the many villages with common fields devoted a considerable amount of energy to protecting and sustaining this shared resource, and this policing was nearly always conducted through the manor courts. They had two primary objectives: managing access and preventing damage.[2]

Access to the common fields was probably the more important of these two concerns. Of course, only tenants of the manor were to have use of the fields, while sub-tenants and non-residents were normally excluded. Hence, the long list of bylaws passed by the jurors of Escrick in October 1753, included one forbidding any inhabitant from pasturing livestock on the town fields unless they had 'good right' to do so.[3] Locals often attempted to block outsiders from even indirect access by, for example, fining tenants 'for taking into the field other peoples sheep to winter' at Hemingbrough in 1715.[4] Timing was obviously an important restriction on access as well. It was only after the harvest had been completed, during the season described by a Wistow custumal as 'the time of eatage or averish' that animals were permitted in the open fields.[5] Likewise, gleaning was

[1] Thirsk (1967), pp. 5–7; Rackham (2000), pp. 3–5, 164–179.

[2] For the evolving nature and functions of post-medieval manor court in England more generally, see Waddell (2012).

[3] Borthwick Institute for Archives (hereafter BIA), PR/ESC 20.

[4] BIA, Ware 7, unsorted (presentments, 13 May 1715).

[5] BIA, CC/Ab/12/Caw-Wis/4/1.

Table 22.1 Common fields along the Yorkshire ouse

From *c*.1550 to *c*.1750	
Settlements with no evidence of early modern common fields	13 (25 %)
Settlements with early modern common fields	39 (75 %)
After *c*.1750	
Settlements with no evidence of common fields after *c*.1750	23 (43 %)
Settlements with common fields after *c*.1750	30 (57 %)
... minimal fields	... 14 (47 %)
... moderate or substantial fields	... 10 (33 %)
... extent of fields unknown	... 6 (20 %)

Source In addition to manorial records and enclosure awards (too numerous to be listed here), evidence has been drawn from English (1985), Kain et al. (2004), Tillott (1961), pp. 498–506, Allison (1976), Beresford (1948–1951). All figures should be regarded as approximate given the inexact nature of the surviving evidence

only permitted after the corn stacks had been taken in.[6] Manorial jurors thus regularly presented individuals whose horses, cattle, swine or sheep were found unlawfully grazing in these areas. Peter Russome, amerced 12d at Drax in 1627 'for his horses trespassinge in Newland field', was just one of many who were punished for such offences throughout the period.[7]

Protecting the fields from the damage that could be caused by negligence or misuse was the other main priority. This aspect of local management was essential if the value of the land was to be maintained. The manorial records of Bishopthorpe, stretching across the seventeenth and eighteenth centuries, provide examples of some of the most common regulations of this type. Jurors here demanded that every man with holdings in the common fields to 'diligently look and weed his corn land in due and convenient tymes and season', and that no one risk harming the crop by attempting to 'drive any louse beasts through Bishopthorpe Southfield'.[8] They also punished individuals for specific acts of neglect and abuse, such as fining Margaret Teasdale 6d for leaving 'her backfront gate into the field open', which would have allowed livestock to wander in, and fining William Robison 6s 8d 'for ploughing the Comon baulk away in the far South field'.[9] However, in Bishopthorpe and elsewhere, the most frequent point of contention was the maintenance of the drainage required to keep the common fields from becoming waterlogged and of the fences necessary to protect them from unlawful grazing. Hence, there were endless presentments for insufficient dikes at the land ends in the North Field or along the Far South Field, insufficient fences against the

6 BIA, Rev IV, unsorted (pains and presentments, 4 May 1625).

7 Hull History Centre (hereafter HHC), U DDEV/32/6. Unfortunately, they rarely state whether the 'trespass' was due to pasturing out of season or pasturing 'without right'.

8 BIA, Rev IV (pains and presentments, 4 May 1625).

9 Ibid (presentments, 21 Oct 1702, 14 Nov 1756).

Hard Corn Field, or between the Barley Field, the Fallow Field and the Common, and so forth.[10]

So, although the land itself was privately owned and the crops belonged to particular individuals, communities policed their town fields throughout the year, imposing penalties for a wide range of offences in order to sustain this vital resource.

22.3 Meadows and Ings

Common meadows, like town fields, produced both private profits (an annual crop of hay) and collective benefits (a rich post-harvest pasture). Nearly all of the townships that bordered rivers in the Vale of York—at least those north of the fens and marshes of the Humberhead Levels—possessed a strip of riverside meadows called 'Ings' that were generally grazed in common.[11] The seasonal flooding experienced by these lands provided them with fertile alluvial soil which made them exceptionally valuable. In some townships, only the old enclosures were worth more per acre.[12] Common meadows were sometimes enclosed over the course of the seventeenth century, as seems to have happened to the Common Ings at Barlow. In most cases, however, they continued to be collectively managed until the coming of parliamentary enclosure and occasionally long after. For example, the South Ings at Acaster Malbis continued to be communally pastured even in the early twentieth century.[13]

Controlling access was always a key priority. The long-running court rolls of the manor of Acaster Malbis show this process at work. According to an Elizabethan bylaw, for example, no horses or cattle were permitted in the South Ings after Whitsunday, and any livestock found there would cost their owners a fine of 12d per animal.[14] Over three centuries later, in 1907, the manors 'Byelaws and Rules for the Regulation of the South Ings' continued to address these issues. They ordered that no stock was to be let onto the Ings unless possessed by the owner at least 14 days, and also set the stint for grazing on the 'fog' (the

[10] BIA, Ware 7 (Hemingbrough, presentments, 24 Apr 1676, 16 Oct 1690, 25 Oct 1705); BIA, Rev IV (Bishopthorpe, presentments, 18 Nov 1715, 24 Oct 1735, 14 Nov 1756), and passim. Ditches and dikes were doubly important because they could also serve as fence-like barriers.

[11] Of the 32 townships on the Ouse above the Rivers Aire and Derwent, only three (Newton-on-Ouse, Selby, and Newland) did not have Ings. For sources, see Table 22.1 above, and Smith (1961–1963).

[12] Harris (1959), pp. 11–12, Harris (1961), p. 116.

[13] York Minster Archives (hereafter YMA), Hailstone BB13 (Barlow, presentments, 7 Oct 1636); BIA, SELBY/WILLS (Richard Heminlay, inventory, 12 Sep. 1695); BIA, Wenlock 7/7 (Acaster Malbis, by-laws, 1907).

[14] North Yorkshire County Record Office (hereafter NYCRO), ZDV(F), unsorted (pains, 3 Apr 1573).

nourishing grass that springs up after the hay harvest) according to each tenants acreage.[15] Other communities laid down restrictions on access by non-residents, such as the prohibition passed by Linton-on-Ouse in 1745 against letting grass in the Ings (or other commons) to any 'outer town person' for a month after the fog.[16] Furthermore, individual attempts to quietly appropriate and enclose parts of the Ings were vigorously resisted at Wistow in the late sixteenth century, because even a minor encroachment would have denied the manorial tenants access to part of their common right.[17] Unsurprisingly, however, the most frequently presented offence concerning access to the meadows was for 'trespassing' livestock, grazing 'without right' or at unlawful times.[18]

Maintaining the value of common Ings required additional regulations. Damaging the hay by making a pathway 'over the Carr Ings for either foot or hors in time of middew' was forbidden, and negligent acts such as 'leaving the Inges yard open' were also liable to penalties.[19] Individuals who failed to repair the fences that shielded meadows from wandering livestock often received fines from the manor court, just as they did in the case of common fields, but greatest concern seems to have been associated with preventing these lands from being overwhelmed with water. At Brotherton in the 1660s, for instance, the jurors ordered that every man should keep a sufficient fence against the Ings, but also should fill any breaches in the banks that protected the meadowland from river floods.[20] There were also numerous orders to scour ditches, 'grip' lands and maintain 'gotes' on the Ings in order to keep them reasonably well-drained.[21] Indeed, even in the midst of enclosure manorial jurors continued to impose bylaws to this effect. So, two years after the passage of an Act of Parliament in 1806 to enclose the open meadows at Kelfield, the jury commanded that the 'main drain in the Ings [was] to be completed of the width directed by the Commissioners Award'.[22] All of these measures were designed to prevent destructive incidents like that which unfolded at Barlow in 1636, when Peter Halliday was fined 10 s for 'drowning the common Inges'.[23]

---

[15] BIA, Wenlock 7/7.

[16] West Yorkshire Archive Service at Bradford (hereafter WYASB), DB10/C2.

[17] BIA, CC/Ab/12/Caw-Wis/4/1.

[18] See, for example, presentments for calves, sheep, oxen, and horses in the Ings at Everingham: HHC, U DDEV/10/31 (presentments, 23 Oct 1633, 8 Apr 1634; 27 Oct 1634). Or the presentments for overgrazing in the Ings at Acomb in 1624: Richardson (1969/1978), I, pp. 140–141.

[19] BIA, PR/ESC 20 (Escrick, pains, 25 Oct 1753); YMA, Hailstone BB13 (Barlow, presentments, 6 Oct 1636).

[20] NYCRO, ZEC, unsorted (Brotherton, pains, 16 Apr 1667, presentments, 25 Oct 1675).

[21] HHC, U DDBH/3/2 (Deighton, pains, Oct 1585); BIA, PR/ESC 20 (Escrick, pains, 25 Oct 1753); NYCRO, ZDV(F), unsorted (Acaster Malbis, pains, 2 Oct 1606).

[22] HHC, U DDPR/8/3.

[23] YMA, Hailstone BB13.

22.4 Common Woodlands, Marshes and Pastures

The common woods and 'wastes' enjoyed by early modern townships in lowland Yorkshire varied greatly the size and function. Common woodlands seem to have been relatively scarce. Of the 52 townships along the Ouse, only Cawood and Wistow had substantial commonable woods listed in their parliamentary enclosure awards. These two settlements shared five Haggs, consisting of nearly 1,000a in 1641, and additional Outwoods.[24] Here, the tenants defended their rights of access from repeated attempts at lordly enclosure. The custom, as described in both the 1560s and the 1700s, was that the lord of the manor could temporarily fence two of the haggs for 7 years after making a 'wood fall' but then had to re-open them to common for the tenants' cattle and pannage for their swine.[25]

Other commons—including marshes, carrs and unspecified 'pastures'—were much more prevalent than woodlands. In fact, they appear to have been nearly universal, with almost every Vale of York township possessing such lands until the late eighteenth century. They accounted for upwards of 40 % of the surface area in Howdenshire, where the adjoined lands of Bishopsoil and Wallingfen amounted to nearly 9,000 acres in the late eighteenth century and were shared between 30 different settlements.[26] Likewise, at Reedness and Swinefleet on the southern banks of the Ouse, around 3,200 acres of marshy common pasture covered more than half of acreage of these two townships.[27] However, in most settlements along the Ouse north of its junction with the Aire, such commons accounted for between a tenth and a third of the surface area at the time of their enclosure.

These commons may have been on ill-drained marginal land, but they contributed to the rural economy by providing pasturage, fuel, building material and much else, so villagers predictably invested time and expense in policing them carefully. Pasture was only to be enjoyed by residents and even they could not over-stock the grounds with unreasonable numbers of animals. Hence, inhabitants of Thorpe in Balne were ordered to not 'gist any strangers Cattell upon Thorpe Marsh', and those of Selby were prohibited from putting more 'goods' upon the common than they could winter.[28] Similarly, manorial jurors developed a wide array of rules to govern the collection and extraction of valuable resources from these 'wastes'. At Deighton in 1584, they commanded that 'none but tenants of the lorde of this manor shall digg any turfe on the Common', imposing a heavy fine (6s 8d) for disobedience.[29] A few miles away, at Acomb, the jury not only repeatedly restricted the collection of turves on the moor to two loads per

---

24 BIA, CC/Ab/8/5.

25 HHC, DDBH/19/33-34; BIA, CC/Ab/12/Caw-Wis/4/1; BIA, MD 112, p. 38.

26 McDonagh (2007), pp. 5–6, 23.

27 English (1985), p. 115.

28 Doncaster Archives (hereafter DA), DD/DC/A2/2 (Thorpe in Balne, bylaws, 1669); HHC, U DDLO/2/8/3 (Selby, presentments, 26 Apr 1716).

29 HHC, U DDBH/3/2.

messuage and one per cottage, they also equally limited the cutting of whins.[30] This prevented the moor from being depleted by the ruthless stripping of resources for sale outside the township. One can see the danger inherent in the possibility of unconstrained commercial exploitation of commons in the pains laid at Selby in the late seventeenth century: the growing shipping trade led to a ban on taking earth for ballast from East Common, and the demands of the pottery industry meant that potters had to pay an annual fee of 2s 6d for the privilege of digging clay on the common and even then only within the bounds marked out by the freehold jury.[31]

Communities also sought to protect their commons from damaging negligence by imposing rules similar to those made concerning fields and meadows. Bylaws against putting 'scabbed' or diseased horses in the shared pastures were very widespread, and at Brotherton only stallions of suitable height were allowed to mix with the rest of the town herd on the marsh.[32] These efforts to preserve the health and quality of grazing livestock were augmented by close monitoring of the state of the fencing around the commons, both to keep lawful animals in and to keep unlawful ones out.[33] The state of the land itself was also an issue that received attention. Spoiling the common marsh by carrying loads of wood and coals across it to the river, or by leaving obstructions on it, was thus punished repeatedly by the Brotherton jurors in the late seventeenth and early eighteenth century.[34]

22.5 Common Infrastructure

The inhabitants of these lowland townships not only policed the many parts of the landscape conventionally described as 'commons', they also asserted control over the shared infrastructure essential to local economic life. This included a wide range of key features such as roads, paths, bridges, drainage works, embankments, hedges, fences, gates, stiles, and village pinfolds. Some elements—for example, most roadways—were legally part of common land of the manor in the same way as 'wastes' and marshes, whilst others—including many fences and ditches between closes—were supposed to be entirely private property.[35] Yet, as will be

---

30 Richardson (ed.), *Acomb*, I, pp. 24, 29–32, 37, 46, 59–60, 88, 91, 105, 109–110 (presentments and pains, 1567–1602).

31 HHC, U DDLO/21/171-172 (7 Oct 1669, 25 Apr 1682).

32 For examples of the former, see Richardson (1969/1978), I, p. 9 (pains, 6 May 1555); NYCRO, ZFR in MIC 3603 (Burton Salmon, pains, 1659); DA, DD/DC/A2/2 (Thorpe in Balne, bylaws, 1669); BIA, Ware 7 (Hemingbrough, presentments, 13 May 1715); York City Archives (hereafter YCA), Acc. 135, N.P. 1/1 (Nether Poppleton, presentments, 1750); BIA, PR/ESC 20 (Escrick, pains, 1753). For the latter, see NYCRO, ZEC in MIC 1554 (Brotherton, pains, 1667).

33 BIA, Rev VIII (Wistow, pains, 1 Oct 1662); DA, DD/DC/A2/2 (Thorpe in Balne, bylaws, 1669).

34 NYCRO, ZEC in MIC 1554 (Brotherton 1667–1729).

35 Rackham (2000), p. 279.

seen, the 'private' status of the latter did not deter communities from regulating them when they felt that the public good was at stake. So, although local infrastructure accounted for little acreage and much of it was completely held in private hands, its central role in the success of the Vale's economy ensured that it was usually subject to vigilant collective oversight.

Major roadways fell under the purview of county magistrates, but smaller roads and tracks—the capillaries of regional economic circulation—were managed by the authority of the manor courts, and this continued long after turnpikes began appearing. Quantitatively, pains and presentments related to transport routes made up around 8 % of the total over the whole period, appearing in nearly all substantial sets of regulations (Table 22.2).[36] The manorial jurors demanded that individuals maintain the 'highways' adjoining their lands and the streets fronting their doors, as well as calling on all inhabitants to attend the 'common days work' prescribed by statute for repairing shared roads.[37] Similar rules applied to minor bridges and foot-stiles.[38] On occasion, common resources might be made available for these tasks, as at Wistow where the lord customarily assigned sufficient 'Bowes… for mending of the high Wayes'.[39] Ultimately, however, keeping travel and transport links open necessitated regularly punishing offenders for obstructing them with rubbish, carrion and overhanging trees, or for encroaching on them with sandpits, ploughland and other infringements.[40]

It seems that managing and monitoring boundaries was an aspect of local regulation that was even more important than maintaining roadways. Fences, hedges and walls demarcated different owners, occupiers and land uses as well as protecting crops from wandering animals. To that end, manorial juries in the Vale of York passed dozens of bylaws and presented hundreds of individuals for related offences, around one in seven of the total recorded in the sample (Table 22.2). At the Easter court at Acomb in 1555, for example, 25 men and women were fined for not repairing their fences, and at the Easter court in 1707 another 24 were fined for like

36 Note that this figure includes those related to bridges, stiles and river-landings which made up around 1 % of the total.

37 For examples of the former, see HHC, U DDEV/32/6 (Drax, presentments, 1627); BIA, Ware 7 (Hemingbrough, presentments, 1691, 1699, 1700); HHC, U DDLO/2/8/3 (Selby, Oct 1709); NYCRO, ZEC in MIC 1554 (Brotherton 1729). For examples of the latter, see HHC, U DDBH/3/2 (Deighton 1587); YMA, Hailstone BB13 (Barlow, 1636); NYCRO, ZFR in MIC 3603 (Burton Salmon, pains, 1659); DA, DD/DC/A1/5/1-27 (Owsten, presentments, 1709, 1722); HHC, U DDBH/14/2 (Ryther, pains, n.d.).

38 BIA, Rev IV (Bishopthorpe, pains, 1617, presentments, 1765, 1774, 1795, 1805); BIA, Rev VIII (Wistow, pains, 1 Oct 1662); HHC, U DDLO/2/8/3 (Selby, presentments, 6 May 1702).

39 BIA, CC/Ab/12/Caw-Wis/4/1.

40 For examples of the former, see The National Archives (hereafter TNA), SC 2/211/61(Howden, 1 Oct 1616); NYCRO, ZFR in MIC 3603 (Burton Salmon, pains, 1659); HHC, U DDEV/32/6 (Drax, presentments, 1660); DA, DD/DC/A2/2 (Thorpe in Balne, presentments, 1784). For the latter, see HHC, U DDEV/10/31(Everingham, 8 Apr 1634); HHC, U DDLO/1/15 (Barlow, Oct 1638); BIA, Ware 7 (Hemingbrough, presentments, 18 Apr 1700).

Table 22.2 Pains and presentments, *c*.1550–1850

	Unlawful grazing	Unlawful appropriation	Drainage and banks	Roads, paths, etc.	Fences and hedges	Other	Total offences
No.	463	117	649	197	331	657	2,414
%	19.2	4.8	26.9	8.2	13.7	27.2	100

Source 23 manors in the Vale of York. For more detailed figures and full references, see Waddell (2011), Appendix 1

misdemeanours.[41] Hedge-breaking, which was punished at Howden, Bishopthorpe, and many other places, similarly threatened the integrity of local boundaries, though it was also considered to be equivalent to theft as it was usually an attempt to illicitly collect fuel to warm a cold hearth.[42] But focusing solely on firm barriers of this sort would ignore the fact that unfenced boundaries were policed too. Manorial, parochial and civil perambulations exemplified this facet of communal regulation. The most famous was the annual riding of the bounds of York, a procession that circumnavigated around all of the extensive 'average grounds' outside the city walls in which the citizens enjoyed common pasturage rights until 1826.[43] Less grand were those that took place in small settlement like Barlow, where eighteenth-century perambulations helped to fix in the tenants' minds the unmarked division between their common and that of the township of Camblesforth.[44]

Finally, a ubiquitous and complex grid of drains and flood defences in lowland Yorkshire made possible the relatively high level of agricultural productivity enjoyed by this region.[45] To describe it as merely an important part of the rural landscape would be an understatement, for it was perhaps the most critical component in the whole system of infrastructure. Some indication of its centrality can be gleaned from the remarkable number of words, including several particular to local dialects, used to describe the various parts of this system—not just dikes, ditches and embankments, but also 'gotes', 'runners', 'descenders', 'grips' and 'dreiners'. The Commissions of Sewers for the West and East Ridings policed some aspects of the drainage system in several places in southern Yorkshire the seventeenth century, but this issue was primarily managed by individual localities.[46] Indeed, it received a very large proportion of the regulatory energy of the manor courts, accounting for

41 Richardson (1969/1978), pp. 8–9, 190.

42 TNA, SC 2/211/61 (Howden, 9 Apr 1616); BIA, Rev IV (Bishopthorpe, presentments, 21 Oct 1702).

43 Tillott (1961), p. 501.

44 East Riding of Yorkshire Archives and Local Studies Service (ERYALSS), DDCL/124. The Manorial Documents Register at TNA indicates that records of manorial perambulations also survive for other Vale settlements such as Skelton, Wressle and Tadcaster. For more on the significance of this practice, see Hindle (2008).

45 Describing the area around Selby, it was noted that 'Strict attention is necessary in keeping the ditches clean, and letting the water off the fields, which are greatly hurt by rain water stagnating upon them': Brown (1799), p. 42.

46 DA, DX/BAX/65159/1; Sheppard (1966), pp. 13–27.

over a quarter of the pains and presentments recorded in the sample (Table 22.2). These overwhelming consisted of banal but necessary orders to scour clogged dikes or remove obstructions from drains, and it was not unusual for ten or fifteen such presentments to be made at a single meeting of the court.[47] By end of the period, sometimes this had become the only offence handled by the manor, as at Everingham between 1790 and 1857.[48] Related problems also arose, though in smaller numbers. At Wistow, for example, one finds tenants with holdings in the North Field ordered to 'gripp' their lands there in 1662, and three people—including a gentleman—fined 39 s per rood for insufficient banks against the Ouse in 1688.[49] It was also the custom here to supply manorial timber for making and repairing 'clows' (outfall sluices with floodgates) and other 'common' infrastructure.[50]

As a whole, then, the web of highways, footpaths, river landings, fences, hedges, drains and banks that stretched across the Vale of York cannot be ignored when attempting to understand the history of communal landscape management in the region. Residents appear to have regarded it as just as much of a 'common' resource as town fields, open meadows and manorial 'wastes'. As a result, it was subject to constant surveillance and communal policing by the manor courts, sometimes long after common land had been enclosed by Acts of Parliament.

22.6 Conclusions

Taking a step back from the minutiae of individual townships and the manor courts, several features of early modern 'commons' in the Vale become apparent.

First, one must note the changing prevalence and size of land subject to common rights. In *c*.1550, almost every settlement seems to have had some form of common land. Of the 52 townships along the Yorkshire Ouse, only two do not have any evidence of common fields, meadows or wastes in the early modern period.[51] By late eighteenth century, however, the proportion had declined. Only 41 experienced parliamentary enclosure, suggesting that a little under one fifth had been become entirely enclosed at some point during the sixteenth, seventeenth or early eighteenth centuries (Table 22.1).[52] Moreover, many others suffered

---

[47] For examples of large groups of presentments of this sort, see HHC, U DDBH/3/2 (Deighton, Oct 1585); HHC, U DDEV/32/6 (Drax, presentments, 29 Apr 1628, 24 Apr 1661); BIA, Ware 7 (Hemingbrough, 24 Apr 1676); WYASB, DB10/C2 (Linton-on-Ouse, pains and presentments, 12 Mar 1745).

[48] HHC, U DDEV/10/31; HHC, U DDEV/22/6.

[49] BIA, Rev VIII.

[50] BIA, CC/Ab/12/Caw-Wis/4/1.

[51] These are Beningbrough and Airmyn, both of which were dominated by a single landholder from at least the sixteenth century. Closer investigation may reveal that they too had commons.

[52] This total of 41 includes one (Acaster Malbis) that did not experience parliamentary enclosure but that did retain common land long after *c*.1750.

piecemeal enclosure, leaving them with only a modest amount of common land. Still, this process should not be exaggerated. A preliminarily analysis of these enclosure awards indicates that perhaps a third (33 %) of the total acreage in the 52 townships was commonable at this time, a figure that rises to at least two-fifths (41 %) if one includes only those that were not already fully enclosed.[53] This was, therefore, a landscape characterised by a substantial and gradually increasing quantity of 'closes' and 'intakes', but also one that featured many town fields—especially in the drier areas—and plentiful numbers of marshy common pastures, which could cover thousands of acres in places like Howdenshire.

Second, the day-to-day management of these commons deserves attention. It is clear that inhabitants carefully monitored the use and abuse of shared resources, meeting regularly to set down rules to protect this vital part of the regional economy. This policing operated primarily through the authority of the manor courts and, until mid-eighteenth century, higher levels of government only rarely intervened. Thus, not only was Walter King right to argue that early Stuart courts leet were 'still needful and useful', these institutions often remained essential to local agricultural life until the era of parliamentary enclosure.[54] Unsurprisingly, governing the commons took up a considerable portion of court time. Almost a fifth of pains and presentments in the sample dealt with unlawful grazing, ranging from horses straying into a town's cornfield before the harvests to individuals over-stocking the common marsh. Another 5 % related to the illicit appropriation of manorial resources, a category that included encroaching on wastes, collecting excessive amounts of turves or underwood, and digging up sand on common pastures. The hundreds of men and women fined for such offences experienced first-hand the vigilance with which tenants sought to protect their common rights.

Third, managing communal resources involved more than just overseeing commonable land. It also entailed keeping a close watch over the shared infrastructure of the rural economy—encompassing transportation networks, local boundaries, and drainage systems. Moreover, when measured quantitatively, this infrastructure accounted for nearly half (48.8 %) of the business of the manor court. The lowland nature of the Vale of York meant that drainage featured most prominently, but the other aspects were hardly neglected. It would be easy to dismiss the resulting innumerable minor acts of enforcement as mundane, and indeed they normally were, yet it they fulfilled an indispensable function, without which agrarian life would very quickly begin to disintegrate. They contributed to the relative wealth and stability the scores of settlements in this part of England by linking them to both internal and external markets, dividing them into countless different specialised fields and farms, and—perhaps most importantly—keeping them from simply washing away in one of the region's recurring great floods.

I will leave it to others to determine what we should learn from this history, how past attempts to preserve shared landscapes and communal resources might

[53] For sources, see Table 22.1.

[54] King (1990).

inform our current struggles to sustain ‘common goods’. To close, I merely note that we may have underestimated the resilience of local governing institutions in early modern England, or at least in this particular part of it. These centuries were certainly a period of change, but the manor court remained a key element in rural life long after its medieval heyday.

22.7 Primary Sources

Borthwick Institute for Archives (BIA)

CC/Ab	Records of the Church Commissioners of York, Archbishopric Estate records
MD 112	Book of Remarks of William Storr of Scalm Park
PR/ESC	Parish records of Escrick
Rev	Estates and Revenues of the Archbishop of York
Ware 7	Manorial Records of Ware and Company, Solicitors of York
Wenlock	Deeds and papers of the Wenlock family

Doncaster Archives (DA)

DD/DC/A Manorial records of Davies-Cooke of Owston

East Riding of Yorkshire Archives and Local Studies Service (ERYALSS)

DDCL Papers of Clark and Co., Solicitors, Snaith

Hull History Centre (HHC)

U DDBH	Papers of the Baines Family of Bell Hall, Naburn
U DDEV	Papers of the Constable Maxwell Family of Everingham, Caerlaverock and Terregles
U DDLO	Papers of the Estates of the Earls of Londesborough
U DDPR	Papers of the Preston Family of Moreby

North Yorkshire County Record Office (NYCRO)

ZDV(F)	Fairfax of Gilling (Newburgh papers)
ZEC	Various manorial records
ZFR	Redmire, Wormersley, Little Smeaton and Burton Salmon records

The National Archives (TNA)

SC 2 Special Collections: Manor Court Rolls

West Yorkshire Archive Service at Bradford (WYASB)

DB10 Deed Box 10, Miscellaneous collections relating to the Bradford district

York City Archives (YCA)
Acc. 135 Gray, Dodsworth and Cobb Collection

York Minster Archives (YMA)
Hailstone Papers of Edward Hailstone

Acknowledgment I am very grateful to the late Rena Fenteman and her trustees as well as the Borthwick Institute for Archives at University of York for their generous funding and assistance during my research. This chapter is a revised version of part of a longer study published as Waddell (2011). I thank the Borthwick Institute for permission to republish this here.

References

Allison KJ (1976) A history of the county of York, east riding, vol 3. Oxford University Press (Victoria County Histories), London

Beresford MW (1948–1951) Glebe terriers and open field, Yorkshire. Yorkshire Archaeol J 37:323–368

Brown R (1799) General view of the agriculture of the West Riding. Edinburgh (in press)

English B (1985) Yorkshire enclosure awards. Department of Education, University of Hull, Hull

Harris A (1959) The open fields of east Yorkshire. East Yorkshire Local History Series No. 9

Harris A (1961) The rural landscape of the east riding of Yorkshire, 1700–1850: a study in historical geography. Oxford University Press, Oxford

Hindle S (2008) Beating the bounds of the parish: order, memory and identity in the English local community, *c*. 1500–1700. In: Halvorson MJ, Spierling KE (eds) Defining community in early modern Europe. Ashgate, Aldershot, pp 205–227

Kain RJP, Chapman J, Oliver RR (2004) The enclosure maps of England and Wales, 1595–1918. Cambridge University Press, Cambridge. <http://hds.essex.ac.uk/em/>

King WJ (1990) Early stuart courts leet: still needful and useful. Soc Hist (Histoire Sociale) 23(46):271–299

McDonagh B (2007) Preparatory report on volume X: Howden and Howdenshire. Unpublished paper for Victoria County history

Rackham O (2000) The history of the countryside: the classic history of Britain's landscape. Flora and Fauna, Phoenix

Richardson H (1969/1978) Court rolls of the manor of Acomb. Yorkshire archaeological society record series 131, 137, Leeds

Sheppard JA (1966) The draining of the marshlands of South Holderness and the Vale of York. East Yorkshire Local History Series, No. 20, York

Smith AH (1961–1963) Place names of the west riding of Yorkshire. Cambridge University Press, Cambridge

Thirsk J (1967) The farming regions of England. In: Thirsk J (ed) The agrarian history of England and Wales. Cambridge University Press, London

Tillott PM (1961) A history of the county of York: the city of York. Oxford University Press (Victoria County Histories), London

Waddell B (2012) Governing England through the manor courts, *c*.1550–1850. Hist J (in press)

Waddell B (2011) Landscape and rural society in the vale of York, *c*. 1500–1800. Borthwick Papers, No. 120, York

Chapter 23
The Parliamentary Enclosure of Upland Commons in North–West England: Economic, Social and Cultural Impacts

Ian Whyte

23.1 Introduction

Parliamentary enclosure was one of the most important socio-economic changes to affect English communities in the later eighteenth and early nineteenth centuries, affecting up to 8.4 million acres (3.4 m ha) in England and Wales (Turner 1980). In midland, southern and eastern England much of this was open field arable and lowland commons. Enclosure in these areas had considerable social impacts. The nature and scale of these have been widely debated since the nineteenth century with opinions oscillating between Marxist and classical economic stances. Marxist interpretations have portrayed parliamentary enclosure as an instrument for oppressing the rural proletariat, with loss of common rights forcing smallholders to sell out because of the heavy costs that enclosure demanded, and turning cottagers and smallholders into biddable full-time wage labourers for the larger farmers, a form of social engineering. The alternative view is that the process was reasonably fair and did not severely disadvantage most of those in the lower strata of rural society with the main phase of decline of small farmers occurring well after parliamentary enclosure had occurred (Chambers and Mingay 1966; Mingay 1997; Neeson 1993; Snell 1985). Over much of northern England, however, parliamentary enclosure involved mainly upland common pasture. In the early nineteenth century, Cumberland and Westmorland had the highest proportion of their areas in unenclosed common of any English counties (Williams 1970). Open fields had mostly been removed before the later eighteenth century by piecemeal enclosure (Winchester 1987). In North West England (here taken as comprising the pre-1974 counties of Cumberland and Westmorland together with north

I. Whyte (✉)
The Lancaster Environment Centre, Department of Geography, University Campus, University of Lancaster, Bailrigg LA1 4XB, U K
e-mail: i.whyte@lancaster.ac.uk

I. D. Rotherham (ed.), *Cultural Severance and the Environment*, Environmental History 2, DOI: 10.1007/978-94-007-6159-9_23,

Lancashire), over 500,000 acres (202,342 ha) were enclosed under parliamentary act, much of it in upland or upland marginal areas (Whyte 2003). There was an initial burst of enclosure in the 1770s involving mainly smaller lowland commons which were capable of conversion to arable. A second major phase of enclosure occurred in the early nineteenth century during the period of high wartime food prices and included more marginal pasture at higher altitudes. The third surge of activity took place in the mid nineteenth century, especially following the General Enclosure Act of 1845, again involving land of relatively poor quality (Whyte 2003). It is important to appreciate, however, that not all common pasture in north-west England was enclosed by parliamentary act. There had been a good deal of piecemeal enclosure associated with population increases in the sixteenth and seventeenth centuries, either for cultivation (Appleby 1978) or as stinted cow pastures shared between small groups of farmers (Winchester 2000b). Equally much land completely escaped enclosure. In Westmorland for example, *c.*100,000 acres (40,468 ha) of land were enclosed under parliamentary act, but 129,000 acres (52,204 ha) remain as open common pasture today (Humphries 2008). The poor quality of some of the land involved probably helps to explain the failure to enclose at least some other commons.

A distinctive element which influenced the nature and impact of parliamentary enclosure in north-west England was the structure of rural society, especially the prevalence of customary tenures which conferred rights of occupation almost equivalent to freehold (Searle 1984). These had evolved in late medieval times out of Border tenures in which security of occupation was granted in return for military service against the Scots (Hoyle 1987). Contemporary estimates suggested that up to two thirds of the improved land might be occupied by customary tenants, of which there were some 10,000 in Cumberland alone. Cumbrian landownership was characterised by relatively few resident nobles and major gentry but lots of small owner-occupiers. The lives of these small farmers, idealised by Wordsworth, were compared to slavery by another commentator (Whyte 2003). They were, however, slaves who had some say in the regulation of matters affecting their lives, including the management of the commons through systems of levancy and couchancy or stinting (Winchester 2000a). Customary tenants, no matter how small, had a voice regarding whether parliamentary enclosure went ahead or not, unlike the leasehold tenant farmers of southern England. Sub-tenants, who leased land from the customary tenants, cottagers and labourers, might have had some *de facto* access to local commons but no say in their fate. The larger customary tenants actually stood to gain from parliamentary enclosure and in general they supported it (Searle 1993). One reason for this was overgrazing which appears to have been a serious problem on many commons by the later eighteenth century due to the larger customary tenants trying to profit from the cattle droving trade by overstocking. In some cases, they had even denied smaller customary tenants' access to the commons at all by using dogs to drive off their livestock (Searle 1984; Whyte 2003). By the late eighteenth century, local manorial courts were increasingly unable to regulate the management of the commons and enclosure was an obvious solution to the problem. The interests of the larger customary tenants regarding parliamentary enclosure often coincided with those of the manorial lords who could use their share of an enclosure award—a sixteenth, fourteenth even as much as a twelth—to create new farms which could be leased at economic rents (Whyte 2007)(Figs. 23.1, 23.2, 23.3, 23.4, 23.5).

Smaller customary tenants were sometimes panicked or coerced into agreeing to enclosure. Correspondence between Christopher Dobson, steward of the Edenhall estates, and his master Sir Philip Musgrave indicated that in 1770 the customary tenants of Bleatarn, south of Appleby, were anxious to enclose their commons. The reason stated by Dobson was that the Earl of Thanet, owner of the Appleby Castle estates, had bought the neighbouring manor of Great Ormside. It was thought, (correctly as it turned out), that he had done this specifically in order to enclose the common and, in doing so, gain a substantial block of land for leasing. The Bleatarn tenants believed that the enclosure of Great Ormside would cause people who had lost their common rights to use the Bleatarn common illegally and overgraze it. In other instances, the smaller customary tenants were against enclosure and successfully opposed it by means of a counter petition to parliament (Searle 1993). In 1767 for example, ninety-one customary tenants in Ravenstonedale petitioned against Sir James Lowther ('Jimmy Grasp All'—the nickname speaks volumes) when he pushed for enclosure (Whyte 2009). Customary tenants with anti-enclosure views could demonstrate their opposition more subtly at an earlier stage in the proceedings before a petition to bring in an enclosure bill had even been presented to parliament. Attempts at enclosure where an advertisement was placed in a local newspaper announcing that a petition was being prepared but where one never subsequently materialised may have been killed by this kind of opposition. Only occasionally, as at Cartmel, is there evidence of the larger landowners deliberately trying to mislead or cheat the smaller men (Whyte 2003). Some communities, like Broughton in Furness and Crosby Ravensworth are known to have considered the possibility of enclosure more than once at different periods (Whyte 2003). It is likely that most communities in the

Fig. 23.1 A typical parliamentary enclosure landscape of regular allotments and straight access roads, Orton

Fig. 23.2 Enclosure road built *c*.1820, Shap enclosure

Fig. 23.3 Ditto

Fig. 23.4 Parliamentary enclosure wall incorporating granite glacial erratic, Shap

Fig. 23.5 Well-built parliamentary enclosure wall with three lines of throughstones and well-laid capstones, Trawden

North West with sizeable commons met at least once during the later eighteenth and early nineteenth centuries to discuss the possibility of enclosure (Whyte 2003). In many cases, failure to act may have been due to a realisation of the likely costs (Whyte 2006).

Commons provided a range of resources the use of which, before enclosure, was regulated by manorial courts in the interests of all the commoners. The manorial courts were dominated by the larger customary tenants (Searle 1986). As well as grazing, the commons provided turbary (the right to cut peat), and access to a range of raw materials including wood, heather, bracken, stone, clay, sand and gravel. The recreational use of commons is less well attested. The diary of Thomas Rumney of Mellfell in Watermillock township on the north side of Ullswater, covering 1805 and 1806, indicates that he used the extensive commons adjoining his small estate for fox hunting, shooting wildfowl and walking for pleasure (Rumney 1936). In addition local horse racing meetings and many fairs were held on common land. Rumney's diary also shows that the commons provided spaces for interactions between people of different social status through they could sometimes be spaces of conflict as well as co-operation.

Once an enclosure act was passed by Parliament the first task of the newly-appointed enclosure commissioners was to determine the boundaries of the common to be enclosed, especially where they marked a division between different commons. This was not always easy as the boundaries between commons were not normally defined by continuous man-made barriers but by natural features and occasional boundary stones and cairns (Winchester 2000a). In 1806, when the lord of the manor of Dacre parish and his larger customary tenants wished to enclose their common the commissioners had to agree the line of its boundary with the neighbouring Watermillock common (Searle 1993, 1995). They invoked a provision from the 1800s General Enclosure Act, which permitted such boundaries to be redrawn as straight lines by the exchange of parcels of land in order to reduce the cost of ring fencing. When the commissioners and representatives of the two groups of commoners met on the ground they discovered that not only did the two parties of customary tenants differ regarding where they thought that the boundary ran, but that the Watermillock tenants could not even agree this among themselves (Rumney 1936). Enclosure commissioners often had to settle long-standing boundary disputes, either by agreement or through the Quarter Sessions, before they could begin their work. Intercommoning between communities was sometimes a long-established practice with animals ranging over huge areas and this could allow the tenants of one community to gain a legal right to an allotment on a neighbouring common, as happened at Orton (Whyte 2003). Earlier boundary perambulations might be a hindrance rather than a help. At Watermillock, a sloppy perambulation in 1740 was interpreted by the tenants of the neighbouring manor of Matterdale as giving some of them a right to use the western part of the Watermillock commons. The Matterdale tenants took the case to court and won, earning them an entitlement to allotments at enclosure (Figs. 23.6, 23.7, 23.8, 23.9, 23.10, 23.11).

The landscape that was laid out by the enclosure surveyors epitomised the ideals of the age of reason with networks of straight, wide roads providing access to square and rectangular allotments enclosed by drystone walls or sometimes hedges of a uniform appearance. In some enclosure processes, the highest lying, poorest areas of common were deemed too poor to be worth the cost of ring

Fig. 23.6 Field limekiln on enclosure allotment, Bowes Moor

Fig. 23.7 Sharp visual contrast between improved enclosed land and unimproved moorland, Stainmore

fencing and were left as stinted rough pasture whose boundaries were delimited simply by marker stones. In most enclosure processes, there was an increase in the extent of improved land, whether pasture or arable. Public enclosure roads were supported by infrastructure such as drains, culverts and bridges where necessary. Shares of turbary land were carefully demarcated and set out for peat cutting and areas were set aside for the quarrying of stone by allotment holders for making walls and roads, a right not previously enjoyed on many manors before enclosure. On the manorial and tithe-holders' allotments new farmsteads were sometimes built (Whyte 2007). The appearance of the landscape changed from open moorland to fields enclosed by drystone wall or, on some lower commons, hawthorn hedges surrounding improved pasture or sometimes arable. The surveys carried out as part of the enclosure process represented in many cases the first detailed maps of many upland areas and may have encouraged the land to be visualised and represented in cartographic form. Parliamentary enclosure was probably one of the most significant processes in altering spatial and environmental perceptions of land in Cumbria, even when it only impacted indirectly on a community through the enclosure of a neighbouring common. It sharpened the definition of boundaries and clarified who had, and did not have, customary rights. Before enclosure commons had been seen as so many units of resource and their actual areas were not accurately known. Parliamentary enclosure brought the practical mathematics of land surveying as well as the scientific rationality of agricultural improvement to land previously seen as marginal. Improvement of the land by the removal of stones, the application of lime and drainage could substantially alter its character as is sometimes evident where improved parliamentary allotments adjoined unimproved common with identical geology and soils (Whyte 2003).

The social impacts of parliamentary enclosure are less clear but were nonetheless important. It is often forgotten that the enclosure of open-field arable in lowland England could result in landowners getting back less land than they started with due to the need to deduct a proportion as the manorial and tithe

Fig. 23.8 limekiln parliamentary enclosure wall, Stainmore

owners' shares. In North West England, where most of the land to be enclosed was common pasture there was usually a substantial net addition to the original holding. The loss of former common rights by the customary tenants was balanced—often probably more than balanced—by an increase in the size of their holdings of 50 % or even 100 % in many cases. This may have improved the viability of many of the holdings worked by customary tenants, ensuring that a class of small owner-occupiers remained entrenched on the land through the nineteenth century and into the twentieth as holding enlargement is likely to have encouraged the commercial orientation of farmers. The enclosure awards show that efforts were made to be fair to the smaller commoners Commissioners usually allocated as manorial and tithe shares higher lying, poorer-quality land so that the

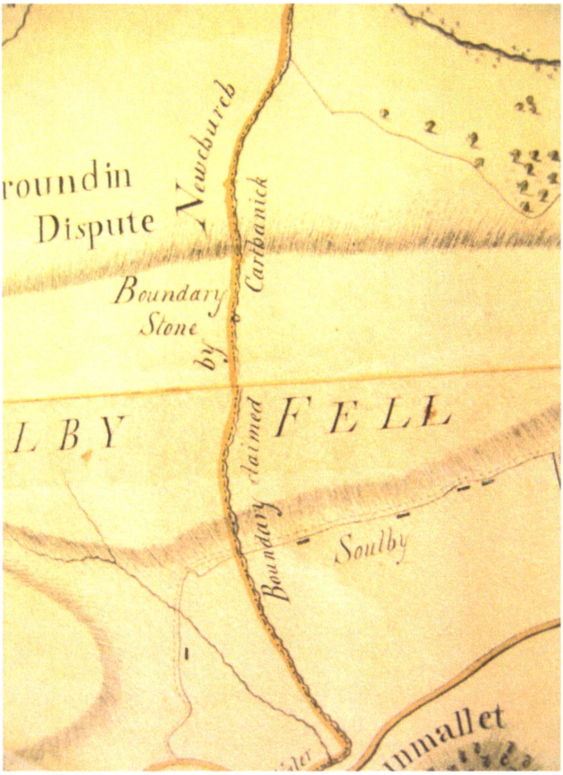

Fig. 23.9 Estate plan, 1806, marking disputed boundary between commons belonging to dacre and newchurch (watermillock) prior to the dacre enclosure

smaller customary tenants received better-quality land as close to their original holdings as possible and often contiguous with them. In the nineteenth century manorial lords often opted for such poor quality land because of its sporting potential (Done and Muir 2000). In addition holders of very small allotments could opt to have them amalgamated with those of other small farmers into areas of shared, stinted pasture.

For lowland England, it has often been claimed that parliamentary enclosure forced small landowners to sell out because of the cost of ring fencing and improving their allotments (Mingay 1997). In the North West some changes in landownership did occur at or around the time of enclosure. A study of landownership changes associated with parliamentary enclosure in Westmorland shows that under a fifth of small owners sold their allotments in the period between the passing of the enclosure act and the completion of the final award (Whyte 2003). In many cases, however, it can be demonstrated that those selling their allotments were tradesmen engaging in part-time farming who may well have taken advantage of the high prices which could be obtained for newly-created allotments to raise capital for their businesses. There was not, however, a mass sell-out by small farmers,

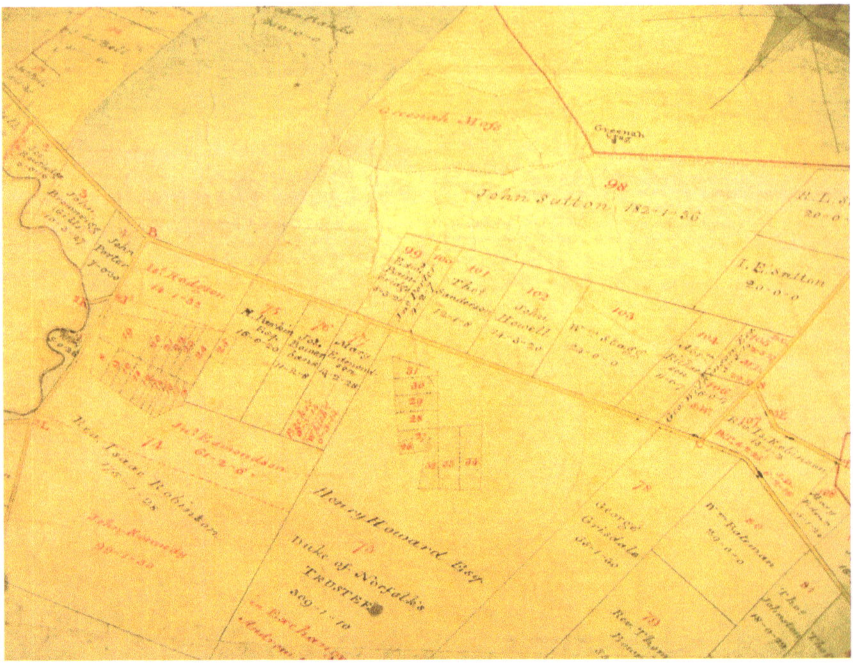

Fig. 23.10 Parliamentary enclosure survey, Greystoke, showing allotments, access roads, peat mosses and shares of turbary

Changes in landownership also occurred due to the widespread practice of meeting the public costs of enclosure by selling off portions of the commons at auction. In some areas, as on the east side of Windermere, allotments were bought up for villa construction. On the other hand ring fencing an allotment, providing additional infrastructure such as private roads and drains, and improving the land by liming and removing stones could be expensive (Turner 1973; Whyte 2006). Evidence presented in 1833 by William Blamire, MP for East Cumberland, to a royal commission on agriculture indicated that during the Napoleonic wars some farmers borrowed money to improve newly-acquired allotments only to experience difficulties when food prices fell after 1815, in some cases being forced to sell out (Chapman 2004). Contemporary newspaper advertisements do not, however, suggest a marked rise in farm sales.

A social impact of parliamentary enclosure which is harder to evaluate was the loss of the use of commons by labourers, cottagers and sub-tenants who did not have legal right of access to them. The extent to which they had been allowed *de facto* rather than *de jure* access to the commons before enclosure is far from clear though; it is likely to have varied from one common to another and to have become increasingly restricted in the course of the eighteenth century (Winchester 2000b). In some cases, such as Cartmel, loss of access to peat by labourers and the need to buy in coal instead may have had serious effects on household budgets (Stockdale

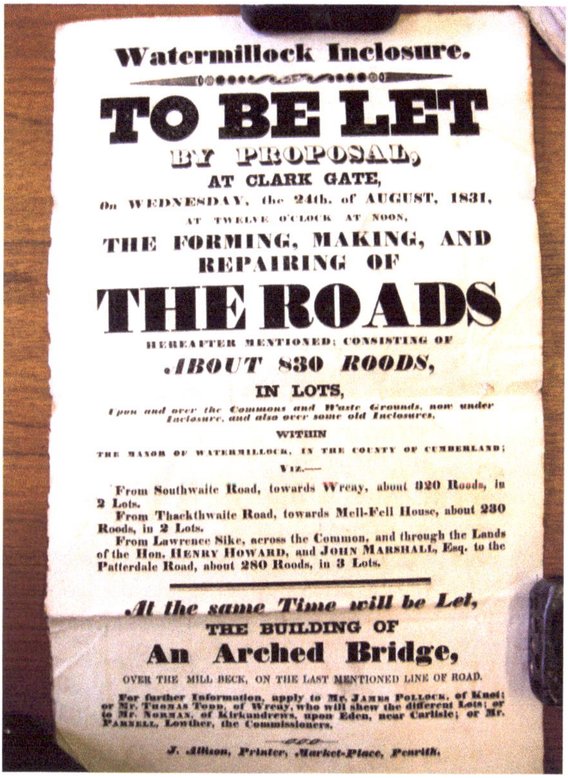

Fig. 23.11 Handbill advertising letting of contracts for building roads and bridges on the watermillock enclosure 1831

1872). On the other hand this may have been balanced by the extra work which was available in the short term in creating the new landscape of improvement, and in the longer term in farming it. The extent to which the infrastructure of the new landscape was created by largely unpaid family labour on small customary tenancies, by local craftsmen, or by itinerant gangs of wallers requires further research. For Watermillock, the parish register records a big surge of previously unrecorded married labourers having children baptised in the parish in the later 1820s and early 1830s when the new enclosure landscape was being laid out. A number of contracts for building the new enclosure roads survive which show that the work was undertaken by local people (Whyte 2003).

One less tangible loss resulting from parliamentary enclosure was that of familiar landmarks and rights of access across the commons. The sharp contrast between the traditional and new landscapes is graphically demonstrated by a comparison of Jeffreys' map of Westmorland dating from 1770, right at the start of parliamentary enclosure, and the modern Ordnance Survey map (Hindle 2001). In areas such as the country between Appleby, Kirkby Stephen and Orton, or along the steep face of the Cross Fell escarpment, the landscape has been transformed by enclosure. These changes involved a loss of freedom of movement using the networks of trackways shown by Jeffreys and their replacement by limited access via the public enclosure roads. There is often a marked contrast on the modern Ordnance Survey 1:25,000 map between the dense networks of rights of way through the 'ancient enclosures' surrounding settlements and the very few rights of way through the later allotments. Associated with this was the loss of access to recreational space, something which was acknowledged in the 1845 General Enclosure Act by its provision for setting aside small allotments for the recreational use of the poorer inhabitants of the community. The areas which were provided, however, were often in remote locations which, as with the recreation allotment on the Strickland Roger enclosure near Kendal, were too wet and boggy to be of much use; compliance with the 1845 General Enclosure Act was clearly only token (Whyte 2003) and represented a symbolic exclusion of ordinary people from the countryside.

Contemporaries who commented on the process of parliamentary enclosure generally saw the fencing and improvement of upland commons in North West England as beneficial to the economy and society in general though some did opine that the enclosure of high-lying fell-sides was hardly likely to produce much of a return. An example of the pro-enclosure view of a larger customary tenant is shown by Thomas Rumney of Mellfell (Rumney 1936). In 1835, he celebrated the completion of the Watermillock enclosure award and the increase of his estate from *c.*110 acres (44 ha) to 350 acres (141 ha) due to the award of allotments in lieu of his rights on the former common plus additional land purchased from the commissioners. This, effectively, secured his elevation to gentry status). He held a party on the summit of Little Mell Fell immediately above his home on land which had formerly been common and was now his freehold property. There was racing and Cumberland wrestling. Drink was taken and the event may have been the end of Rumney for he died suddenly a day or two later.

23.2 Conclusions

It would be going too far to suggest that in North West England, because of the nature of the land involved and the strength of customary tenures, parliamentary enclosure was, overall, a 'good thing'. It was certainly not always good for the land. A greedy, rapacious attitude towards newly enclosed land during the Napoleonic Wars led to the over-cropping of marginal land and the ruin of some of it for a generation or more. William Blamire commented that he would not accept such land as a gift: it was worth nothing but was still liable for tax (Whyte 2003). Equally we know very little about the people in the lower ranks of Cumbrian society at this time, the labourers, husbandmen and sub-tenants who were not awarded enclosure allotments, making it impossible to assess how they were affected by the economic and social changes associated with parliamentary enclosure. The differences between individual communities within the region were complex, reflecting numerous nuances in social structure which are hard to identify today. More detailed community studies for the eighteenth and nineteenth centuries are needed to assess the nature of social change with and without enclosure. As we have seen many communities avoided the need for enclosure at all and their commons survived to be run today by active commoners associations rather than moribund manorial courts (Straughton 2004). Yet, where parliamentary enclosure occurred, by enlarging customary holdings it may have helped to entrench a class of relatively small farmers on the land in Cumbria throughout the nineteenth and into the twentieth century (Shepherd 2003).

References

Appleby AB (1978) Famine in tudor and stuart England. Liverpool University Press, Liverpool

Chambers JD, Mingay GE (1966) The agricultural revolution 1750–1880. Batsford, London

Chapman J (2004) Parliamentary enclosure in the uplands. In: Whyte ID, Winchester AJL (eds) Society landscape and environment in upland Britain. Society for landscape studies supplementary Series 2, 79–88

Done A, Muir R (2000) The landscape of grouse shooting in the Yorkshire Dales. Rural Hist. 12:195–210

Hindle BP (2001) Thomas Jefferys' historic map of westmorland 1770. Manchester Geographical Society, Manchester

Hoyle RA (1987) An ancient right and laudable custom. The definition of tenant right in north western England in the sixteenth century. Past & Present 111:22–55

Humphries A (2008) Trends in pastoral communing in England. Pastoral Commoning Partnership, Carlisle

Mingay G (1997) Parliamentary enclosure in England. An introduction to its causes, incidence and impacts. Longman, London

Neeson J (1993) Commoners, common rights, enclosure and socialchange in England 1700–1870. Cambridge University Press, Cambridge

Rumney AW (ed) (1936) Tom Rumney of mellfell (1764–1835) by himself as set our in his letters and diary. Titus Wilson, Kendal

Searle C (1984) The 'odd corner' of England, a study of a rural social formation in transition: Cumbria *c.*1700–*c.*1914. University of Essex, Colchester (Unpublished PhD thesis)

Searle C (1986) Custom, class conflict and agrarian capitalism: the Cumbrian customary economy in the eighteenth century. Past & Present 110:106–133

Searle C (1993) Customary tenants and the enclosure of the cumbrian commons. North Hist 29:126–153

Searle C (1995) Cumbria's parliamentary enclosure movement: a case study in rural quiescence. Trans Cumberl Westmorl Antiq Archaeol Soc 95(2):247–269

Shepherd ME (2003) From hell gill to bridge end: aspects of economic and social change in the upper eden valley 1840–95. University of Hertfordshire Press, Hatfield, Hertfordshire

Snell K (1985) Annals of the labouring poor. Cambridge University Press, Cambridge

Stockdale J (1872) The annals of cartmel, 1978th edn. Michael Moon, Beckermet

Straughton EA (2004) Beyond enclosure: upland common land in England and Wales since 1800. In: Whyte ID and Winchester AJL (eds) Society landscape and environment in upland Britain. Society for Landscape Studies Supplementary Series 2, pp 89–98

Turner ME (1973) The cost of parliamentary enclosure in Buckinghamshire. Agric Hist Rev 21:35–46

Turner ME (1980) English parliamentary enclosure: its historical geography and economic history. Dawson, Folkestone

Whyte ID (2003) Transforming fell and valley: landscape and parliamentary enclosure in north west England. Lancaster, Centre for North West Regional Studies
Whyte ID (2006) The costs of parliamentary enclosure in an upland setting: south and east Cumbria *c*.1760–1860. North Hist 43:97–115
Whyte ID (2007) Parliamentary enclosure and the creation of new farms in north west England. In: Lawson J (ed) Aspects of lancashire history. Essays in memory of mary higham. Lancashire Local History Federation, Preston, pp 75–84
Whyte ID (2009) Political spaces and parliamentary enclosure in an upland context: Cumbria c. 1760–1840. In: Kumin B (ed) Political space in pre-industrial England. Ashgate, Folkestone, pp 95–119
Williams M (1970) The enclosure and reclamation of waste land in England and Wales in the eighteenth and nineteenth centuries. Trans Inst British Geogr 51:55–69
Winchester AJL (1987) Landscape and society in medieval cumbria. John Donald, Edinburgh
Winchester AJL (2000a) Dividing lines in a moorland landscape: territorial boundaries in upland England. Landscapes 1:16–32
Winchester AJL (2000b) The harvest of the hills. rural life in northern England and the scottish borders 1400–1700. Edinburgh University Press, Edinburgh

Part IV
Issues and Approaches for Future Commons and Cultural Landscapes

Chapter 24
Biodiversity Conservation and the Traditional Management of Common Land: The Case of the New Forest

Adrian C. Newton

24.1 Introduction

In the UK, the term 'common land' refers to land in private ownership, where traditional rights exist for people other than the landowner ('commoners') to use the land in specific ways. Such rights include the grazing of stock (common of pasture), digging of peat for fuel (turbary), collecting timber (estovers) and the taking of fish (piscary) (Aitchison et al. 2000; Short 2008). The use and management of common land is supported by national government legislation such as *The Commons Registration Act* of 1965 and *The Commons Act* of 2006.

In recent years, common land has become the focus of increasing interest and concern. Specific issues relate to the decline of their economic functions, which could potentially threaten their existence; the development of multiple use patterns; their resilience to socio-economic and environmental change; and the policy responses required to sustain them in future (Short 2008). It has also been suggested that commons are playing an increasingly important role in the provision of a range of functions, including nature conservation, heritage, landscape and ecosystem services, as well as supporting local livelihoods (Short 2008).

The high value of common land for biodiversity conservation is widely acknowledged (Aitchison et al. 2000). This value is illustrated by the fact that around 20 % of all Sites of Special Scientific Interest (SSSIs) in England include common land, and that 55 % of all commons contain SSSIs. Given their long history of human use, there is a widespread belief that the biodiversity value of common land is dependent on maintenance of traditional management approaches (Hindmarch and Pienkowski 2000). A recent review of pastoral commoning (i.e.

A. C. Newton (✉)
School of Applied Sciences, Christchurch House, Talbot Campus, Bournemouth University, Fern Barrow BH12 5BB Poole, Dorset
e-mail: anewton@bournemouth.ac.uk

I. D. Rotherham (ed.), *Cultural Severance and the Environment*, Environmental History 2, DOI: 10.1007/978-94-007-6159-9_24,

the grazing of common land with livestock) in England suggested that there has been a significant reduction in the numbers of grazing livestock on commons over the last 20 years, and especially during the past decade. As reported by commoners, this has been associated with a significant increase in scrub and bracken cover over the same period (Pastoral Commoning Partnership 2009), suggesting a decline in habitat condition. However, the precise impacts of this trend on different elements of biodiversity remain unclear.

It is now widely believed that traditional management approaches can make a positive contribution to biodiversity conservation in many different contexts (Berkes et al. 2000). This has led researchers to highlight the importance of social capital in biodiversity conservation and management, and the need to integrate the biological and social elements of conservation (Pretty and Smith 2004). In the UK, a belief in the efficacy of traditional land use approaches has led to the widespread reintroduction of livestock grazing on common land to support their conservation management. However, few attempts have been made to assess whether traditional management approaches are indeed effective in terms of delivering conservation benefits; available evidence suggests that this is not always the case (Sutherland 2000). This highlights the need for critically evaluating the impact of traditional management approaches on the biodiversity associated with common land.

This can usefully be illustrated by the example of lowland heathland, a habitat of high conservation value that is an important constituent of many commons. Heathland has become the focus of increasing conservation concern throughout northern Europe as a result of high rates of loss and degradation. A decline in the traditional uses of heathlands is a significant factor responsible for the widespread decline in habitat condition. Such uses typically included livestock grazing, controlled burning and cutting of vegetation for use as fuel and animal fodder, and the cutting of turf and peat (Webb 1998). As a result of end of this tradition, many heathlands have reverted to scrub or woodland through a process of natural succession, which now represents one of the main threats to heathland biodiversity (Rose et al. 2000). In response, livestock grazing has been widely reintroduced to lowland heaths to support their conservation management; Newton et al. (2009) identified 46 heaths where this has recently taken place. However, introduction of livestock to heathlands has been controversial, often attracting significant local opposition.

To evaluate the evidence for management of lowland heaths, Newton et al. (2009) conducted a systematic review of the scientific literature. Meta-analysis indicated that grazing can result in an increase in the ratio of grass to ericoid shrub cover, highlighting a potential negative effect of introducing grazing to heathland; no other statistically significant effects were recorded. This highlights the lack of a substantive evidence base to inform conservation management decisions, a problem that characterizes many conservation problems (Sutherland et al. 2004). Newton et al. (2009) also demonstrated most heathland managers believe that grazing is an effective approach to heathland management, contradicting the findings in the scientific literature. This example highlights the potential risks of adopting traditional management approaches uncritically (Sutherland 2000).

This paper explores the impacts of traditional management approaches on biodiversity, with specific reference to the common land of the New Forest. This area is situated on the south coast of England in the counties of Hampshire and Wiltshire (Fig. 24.1). The ecological characteristics of the New Forest have been greatly influenced by its history as a medieval hunting forest, and the survival of a traditional commoning system that became formalised in late medieval times (Tubbs 2001). This has led to the maintenance of populations of large, free-ranging herbivores, including deer as well as livestock. The 'perambulation' of the Forest, encompassing some 37,907 ha, refers to the area within which forest bye-laws apply, relating to the pasturage of livestock on common land. Around three-quarters of this area is referred to as the 'Crown lands', reflecting their status as Royal Forest. The unenclosed Forest is referred to by Tubbs (2001) as the largest area of semi-natural vegetation in lowland Britain, and includes large tracts of heathland, valley mire and ancient pasture woodland, three habitats that are now fragmented and rare throughout lowland Western Europe.

The New Forest is widely considered to be one of the most important areas for wildlife in the UK (Newton 2010a). This is reflected in the many conservation designations in the area. In 2005, the area was designated as the New Forest National Park, which extends over 57,100 ha, a substantially larger area than that included within the perambulation (Fig. 24.1). Some twenty SSSIs, six Natura 2000 sites, and two Ramsar Convention sites included at least partly within the

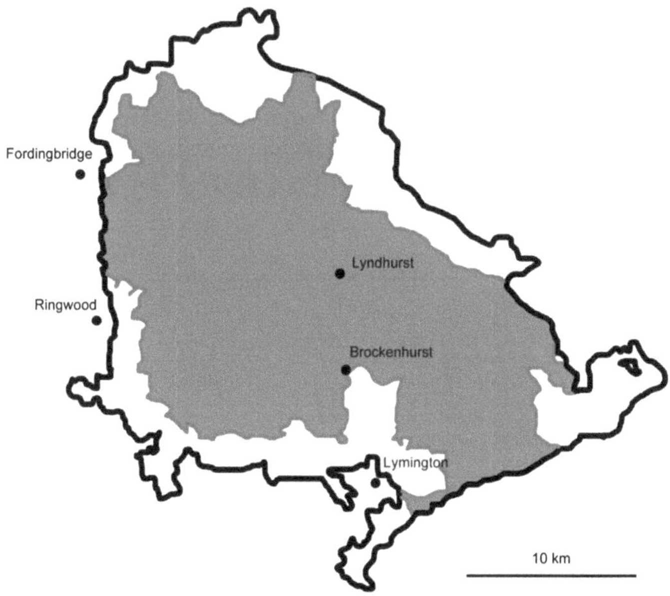

Fig. 24.1 Map of the New Forest National Park, with the area of the Perambulation shaded in grey (after Newton 2010a)

Park boundaries (Chatters 2006). In recent years, some 6,000–7,400 ponies, cattle, donkeys, pigs and sheep have been de-pastured on the Open Forest (i.e. the unenclosed common), by about 550 commoners (NPA 2008).

24.2 The New Forest Commons

Commoning in the New Forest was recently reviewed by the Pastoral Commoning Partnership (2009), on which the following account is partly based. The livestock de-pastured on the Forest are principally ponies and cattle, with smaller numbers of pigs and donkeys. During the past 50 years, pony numbers have normally been in the range 2,000–4,000 and cattle 1,500–3,000. For pigs, numbers have generally been around 200, although with occasional higher peaks. The number of ponies has tended to increase over time, whereas numbers of cattle have tended to remain within the same range, despite fluctuations (Fig. 24.2).

The New Forest is unusual in that grazing numbers are officially unrestricted, but are limited to those individuals whose properties have the right to pasture attached. In practice, overall limits to numbers of livestock can be set through management of payment schemes. Currently (2003–2013) there is a Countryside Stewardship Scheme in place, which is a 10-year agreement, under which the UK Government makes payments totalling some £460,000 each year. In return for de-pasturing their animals for laid down minimum periods, commoners receive an annual headage payment, which is currently set at £55 each for cattle, ponies and donkeys. The Scheme defines limits on the numbers of livestock on which payment can be claimed.

The economics of commoning, and its future prospects, have been the subject of a recent review (The New Forest Commoning Review Group 2007). This highlighted the poor economic returns from commoning, and suggested that this is undermining its long term sustainability. Despite payments through the

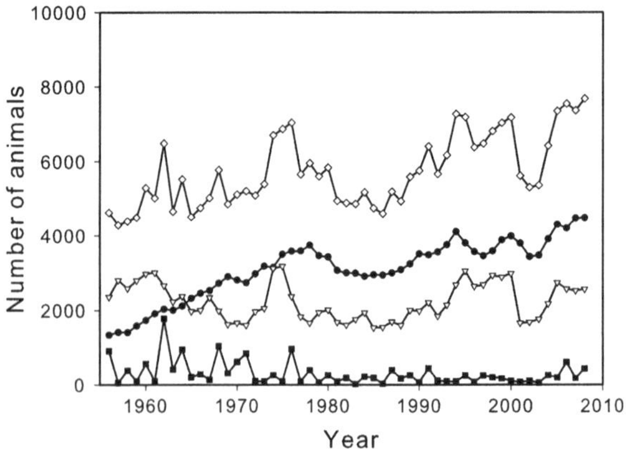

Fig. 24.2 Numbers of stock depastured in the New Forest. Data from the New Forest Verderers (http://www.verderers.org.uk/stock_depastured.pdf). Symbols: pigs, filled squares; cattle, open triangles; ponies, filled circles; total, open diamonds. After Newton (2010b)

Countryside Stewardship and Single Payment Schemes, the review concluded that *'all commoners are losing substantial amounts of money as a result of commoning'*.

There is general consensus that tradition plays a major role in maintaining the use of the New Forest common (Pastoral Commoning Partnership 2009). There is also a general consensus that commoning activity, or more specifically livestock grazing, is central to maintaining the high conservation value of the New Forest (Verderers of the New Forest 2005). This is reflected in all current management plans, including those relating to the Special Area of Conservation (SAC) (Wright and Westerhoff 2001), the Crown Lands (Forestry Commission 2008) and the National Park (New Forest National Park Authority 2008). But is this true? The following section evaluates this contention, with reference to current information regarding the status and trends of biodiversity in the New Forest.

24.3 Biodiversity in the New Forest

24.3.1 Biodiversity Importance

Many authors have suggested that the New Forest is of exceptional importance for biodiversity (Tubbs 2001). However, what evidence is available to support such claims? The biodiversity of the New Forest has recently been profiled in detail, with the objective of answering this question. Newton (2010a) presents information on the current status and trends of New Forest habitats and different groups of species. In the light of this evidence presented, can the claims regarding the conservation importance of the New Forest be sustained? The answer is a resounding 'yes'.

For all of the species groups considered by Newton (2010a), the New Forest is of national importance, and for many, it is also of international importance (Table 24.1). The species richness of many groups is high, sometimes exceptionally so. For example, more than two thirds of the British species of reptiles and amphibians, butterflies and moths, fish, bats, dragonflies and damselflies are found in the New Forest (Table 24.1). Even for those groups that are less well represented, at least one sixth of all British species have been recorded in the area. In every group considered, the New Forest is home to species of national conservation concern, and in some groups, the numbers of such species is very substantial; for example 155 vascular plants, 264 butterflies and moths, and 142 lichens (Table 24.1).

What accounts for the exceptional importance of the New Forest for biodiversity? High species richness can be attributed to a wide range of factors, which differ in their relative importance between different species groups. Its geographic location on the south coast of England is certainly a factor, with relatively mild winters and warm summers being conducive to thermophilic species. However, the

Table 24.1 Importance of the New Forest for different groups of species (based on information presented by Newton (2010b), except Snook (1998))

Species group	Significance of the new forest at national scale	No. species of conservation concern	Estimated total no. of species	Approx. percentage of total number of species in Britain (%)
Birds	Outstanding; particularly important for breeding waders, raptors and heathland; exceptionally rich in woodland birds	37	302[a]	17
Mammals other than bats	Small mammals generally scarce. Species present of conservation importance include dormouse, otter and water vole	3	19	35
Bats	Outstandingly rich; possibly the most important area in Britain. High species richness; may contain significant populations of Bechstein's and barbastelle bat, two of the rarest bats in Europe	13	13	81
Reptiles and amphibians	One of the most important areas in the UK. High species richness; particularly notable species include smooth snake, sand lizard and great crested newt.	12	12/13	92

(continued)

Table 24.1 (continued)

Species group	Significance of the new forest at national scale	No. species of conservation concern	Estimated total no. of species	Approx. percentage of total number of species in Britain (%)
Fish	Fairly high species richness, possibly of national importance	>2	22	88
Invertebrates	Nationally significant	544	5000–10,000	17–33
Dragonflies and damselflies	A national hotspot for diversity	9	31	69
Saproxylic beetles	One of the richest parts of Britain, and of European significance	53	326	55
Butterflies and moths	Outstanding national importance	72 RDB, and 192 NN	1488 (of which 33 are butterflies)	66
Other invertebrates	Exceptionally rich invertebrate fauna, at least in woodlands. Largest known British assemblage of Diptera	403 including Coleoptera, Hymenoptera, Diptera, Orthoptera, Hemiptera, Crustacea	1539 Coleoptera, 22 Orthoptera, 296 taxa of macro-invertebrate recorded from Forest streams	
Vascular plants	Nationally and internationally important, but perhaps not of exceptional importance at the international scale	72 RDB, 43 nationally rare or scarce	Approx. 540	36
Lichens	Outstanding international importance	64 RDB, plus 78 other species of conservation interest	421	18

(continued)

Table 24.1 (continued)

Species group	Significance of the new forest at national scale	No. species of conservation concern	Estimated total no. of species	Approx. percentage of total number of species in Britain (%)
Fungi	Of the highest importance nationally, and of high international importance, at least for some fungal groups (e.g. beechwood saprotrophs)	89	2600	22
Bryophytes	One of the best areas in lowland England for bryophytes	33	326	32

[a] Snook (1998)

RDB red data book; *NN* nationally notable

New Forest is not characterized by especially high endemicity. Rather, the New Forest can perhaps best be viewed as a refuge for species that were formerly more widespread and abundant, but have declined elsewhere (Rand and Chatters 2010).

A key feature of the New Forest is the maintenance of low-input pastoral patterns of land use that have declined both in Britain and throughout much of mainland Europe. It is this pattern of land use, relatively free from agricultural improvement and intensification, which accounts for the extensive areas of semi-natural habitats that characterize the New Forest today. As is made clear by Tubbs (2001) and by Wright and Westerhoff (2001), it is not just the presence of such habitats that is important, but their occurrence in an intimate mosaic, and on a scale that is now unique in lowland England. These characteristics can be attributed to the maintenance of commoning activity over a period of centuries.

24.3.2 Status and Trends in Biodiversity

Tubbs (2001, p. 365) stated that '*the biodiversity of the New Forest is now diminishing rapidly*'. To evaluate the evidence for this contention, Newton (2010b) compiled available information regarding the status and trends in different species groups, and the habitats with which they are associated.

Cantarello et al. (2010) presented an overview of the current condition of habitats in the New Forest, based on the Common Standards Monitoring (CSM) approach conducted by Natural England. Current results indicate that 463 assessment units (out of 576) are in unfavourable condition, representing 80 % of units, or 68 % expressed as a percentage of the total area. For 114 of the 463 units in unfavourable condition, the reasons for the condition being unfavourable remain unclear. For those units for which data are available, dry heathland and grassland habitats are principally threatened by overgrazing, although inappropriate scrub control is also a significant factor (Table 24.2). In wet heathland, wet grassland

Table 24.2 Assessment of threats to habitats in the New Forest, based on results of Common Standards Monitoring (CSM) assessments (Cantarello et al. 2010). Values presented are percentages of the total area classified as in 'unfavourable condition', attributed to each threat

Threat	Habitat type			
	Dry heathland and dry grassland	Wet heath, wet grassland and mire	Pasture, riverine and bog woodland	Inclosure woodland
Forestry and woodland management	3.17	0.73	35.3	45.4
Overgrazing	39.7	0.02	1.79	–
Inappropriate scrub control	34.2	11.5	10.5	–
Drainage	0.19	43.6	17.3	30.2
Public access/disturbance	0.72	–	–	0.42

and mire habitats, the principal threat is drainage. In woodland habitats, inappropriate forestry or woodland management practices are the principal threat, although drainage is also a significant factor accounting for unfavourable condition.

Available evidence indicates that at least 170 species have been lost from the New Forest in recent decades. This estimate is necessarily uncertain; many species are difficult either to locate or to identify, and might be rediscovered by future survey work. On the other hand, this estimate might be conservative, as information on many species groups (particularly the most specious) is lacking. The number of species that have been extirpated varies between different groups; losses of butterflies and moths are particularly high, but significant losses also appear to have occurred in lichens, saproxylic beetles and fungi (Table 24.3). A number of other species appear to be declining, although again, the lack of robust monitoring data limits the conclusions that can be drawn.

These data highlight a range of different causes of the decline or loss of species, which vary among different groups (Table 24.3). The widespread damage to ancient woodland habitats caused by forestry operations in the twentieth century appears to have had a significant negative impact on groups such as vascular plants, fungi and some invertebrates. Another key issue has been the increase in grazing and browsing pressure in recent decades, particularly in the Silvicultural Inclosures, which accounts for the losses of many invertebrates, especially the Lepidoptera. As for the assessment of habitat condition (see above), inappropriate habitat management interventions are widely cited, including scrub control, tree felling and burning of heathland (Table 24.3). The loss or decline of some species may be the result of processes occurring in the wider countryside, including agricultural intensification and land use change in areas adjacent to the New Forest (Table 24.3).

24.4 The Role of Traditional Commons Management in Biodiversity Conservation

The case of the New Forest presents an intriguing paradox. On one hand, there is general agreement that the New Forest is of exceptional importance for biodiversity, and that this importance can largely be attributed to the continuous maintenance of commoning activity over a prolonged period. On the other hand, there is clear evidence of biodiversity loss in recent decades, some of which can be attributed to commoning activity. Negative impacts of commoning include overgrazing, and associated management activities that are undertaken to increase provision of food resources for livestock, including scrub clearance, heathland burning and mire drainage.

Is the New Forest overgrazed? This controversial issue is considered by Newton (2010b). It is widely recognised that many of the distinctive characteristics of the

Table 24.3 Declines and losses of different species groups in the New Forest, and associated causes (threats). Based on information presented in Newton (2010a, b), synthesized from a number of sources

Species group	Trends	Threats
Birds	At least three species lost during the last century. While some species (such as nightjar and woodlark) are stable or increasing, others (such as Dartford warbler, snipe, curlew and redshank) are declining.	Species losses attributable to habitat loss and possibly climate change. Causes for declines in species often unclear, but may include inappropriate habitat management (e.g. Dartford warbler, sparrowhawk), disturbance from human recreation (e.g. ground-nesting birds), climate change, nest predation (e.g. Montagu's harrier).
Bats	No evidence of species losses. Insufficient data to determine trends.	Some forest management interventions may be negative (e.g. tree felling and holly pollarding). Possible disturbance from recreation.
Reptiles and amphibians	One extinction of a native species (Natterjack toad). Sand lizard lost but reintroduced.	Common toad declines may be caused by fungal disease. Inappropriate heathland management (burning) responsible for loss of sand lizard. Main threat to reptiles is inappropriate heathland management.
Fish	No evidence of losses. Insufficient data to determine trends.	History of catchment modification and drainage likely to have had negative impacts on fish populations, but evidence limited.
Invertebrates		
Dragonflies and damselflies	One extinction. Some evidence of historic declines in some species; others appear stable.	Drainage actions and scrub development responsible for species loss.
Saproxylic beetles	At least five species believed to be extinct; 27 further species not reported in past 25 years. Insufficient data to determine trends, although some species appear to have declined.	Extinctions caused by scrub clearance, and forestry/commoning activities involving the felling of large, old trees.

(continued)

Table 24.3 (continued)

Species group	Trends	Threats
Butterflies and moths	General decline of many species in recent decades; 124 species believed to have been lost.	Increased levels of herbivore grazing and browsing, particularly in the Inclosures, leading to a loss of structural diversity and food sources. Greater intensity of management for grazing (burning, reseeding, scrub clearance). Direct destruction of habitat caused by forestry operations (e.g. conifer planting, surfacing of rides in Inclosures). Economic pressures driving land use at the Forest margins (e.g. urban development, pony paddocks, lack of support for traditional woodland management).
Other invertebrates	Insufficient data to determine trends. Some extinctions are likely to have occurred as many rare species have not been recorded for a long time, e.g. New Forest cicada may now be extinct. Groups such as Orthoptera appear to have undergone significant declines.	Changes to the grazing regime and management of the heaths and woodlands are likely to have had a detrimental affect on many insect species and their habitats. Increase in grazing intensity since the 1960s is a particular issue, especially in Inclosures. The intensification of farmsteads within the Forest and the loss of small rotationally managed fields must also have been negative in the Forest, as throughout the wider countryside. Inappropriate ride management and widespread scrub clearance likely to have negative impacts.

(continued)

Table 24.3 (continued)

Species group	Trends	Threats
Vascular plants	One species known to have gone extinct in the middle of the 20th century: summer lady's-tresses (*Spiranthes aestivalis*), which was exterminated by over-collecting and drainage damage. Little evidence of declines in species, although few monitoring data available and impacts of human activity uncertain.	Invasion by exotic water plants (e.g. *Crassula helmsii*) is probably a major threat to flora associated with ponds. Other invasive species such as *Rhododendron* similarly pose a threat to terrestrial vegetation. In the 20th century, forestry practices involving creation of new plantations and planting up of ancient woodland undoubtedly caused enormous damage. Management practices and laissez-faire attitudes to grazing within the Inclosures during the second half of the twentieth century led to negative impacts on flowering plants.
Lichens	Few monitoring data available. Most uncommon species appear to be stable. However, some are clearly declining and some extinctions appear to have occurred. A total of 13 species were recorded from New Forest woods in the nineteenth century and have not yet been refound, and may therefore be extinct. In addition, four leafy species recorded since 1967 appear to have been lost and a further four are declining and rare.	The spread of holly, and hence increased shade, in the past 150 years is the most significant issue. Pollution is another significant factor, especially of sulphur and nitrogen. This may be responsible for difficulties in colonising rather than direct poisoning of the mature thalli. Death of trees has also caused loss of colonies.

(continued)

Table 24.3 (continued)

Species group	Trends	Threats
Fungi	Few monitoring data available. Little evidence of declines. Extinctions hard to evaluate although 18 species of conservation concern have not been seen in the past 50 years and may be extinct.	Substantial losses of semi-natural woodland through felling and establishment of exotic conifers in the twentieth century must have had a major deleterious impact on fungi. Other threats include deadwood removal, and possibly also commercial collecting and climate change.
Bryophytes	Four species of liverwort have apparently become extinct. Most species generally stable.	Some species threatened by scrub invasion.

New Forest, and the survival of many of its species, depends directly on the maintenance of large herbivore populations. It is for this reason that recent management plans (Wright and Westerhoff 2001, New Forest National Park Authority 2008) have placed the maintenance of the pastoral economy, and the tradition of commoning, as a principal objective. On the other hand, it is clear from the substantial losses of insect diversity that many entomologists believe that grazing pressure is too high. Some forest ecologists believe the same, reflecting the high browsing pressure and low density of tree regeneration throughout the New Forest woodlands (Newton et al. 2010).

Despite the importance of grazing to the maintenance of many species, overgrazing has clearly contributed to biodiversity loss. The high losses of invertebrate species, especially Lepidoptera, have largely been attributed to an increase in grazing pressure in recent decades, particularly within the Inclosures. Such losses are an indicator of a decline in the condition of the New Forest as an ecological system, and must have had a negative impact on the other species that depend on them, such as insectivorous birds and bats (Newton 2010b). The New Forest is characterized by a lack of small mammals and the birds and mammals that predate them, which is the result of an impoverished insect fauna caused by high grazing pressure (Putman 1986). As noted by Putman (1986), "*at least in some habitats and in some areas, grazing really is excessive – by whatever criteria*". This suggestion is borne out by the data on habitat condition presented above.

The SAC Management Plan (Wright and Westerhoff 2001) implies that management approaches should be based on tradition. This avoids the fact that some habitats are not currently in favourable condition (as noted above), and fails to address the fact that livestock numbers are currently higher than ever before and are continuing to increase.

As noted earlier, overall limits to numbers of livestock can be limited by the current Countryside Stewardship Scheme, which states that the number of animals (ponies, donkeys and cattle) is not to fall below 3,500 or exceed 7,000 (Verderers of the New Forest 2005). Yet, as indicated on Fig. 24.2, numbers have exceeded this total for each of the years 2005–2007. Although such totals do not necessarily provide an accurate indication of the livestock actually de-pastured on the Forest, this does highlight an apparent difficulty in regulating livestock numbers.

24.5 Conclusions: The Role of Traditional Management of Common Land in the New Forest

The New Forest presents a paradox: traditional management of common land appears to provide both an explanation of its high biodiversity value, and a cause of recent biodiversity loss. The solution to this paradox lies in recognition of the fact that the pattern of land use in the New Forest is dynamic, and continues to evolve.

Based on the analysis presented by Tubbs (2001, p. 161), grazing and browsing pressure in the New Forest is currently at a very high level, with livestock numbers now among the highest on record (Fig. 24.2). In the historic past, deer densities would have been much higher than at present; for example, around 8000 fallow and red deer were recorded in the Crown lands in 1670 (Putman 1986). Deer densities within the Crown lands are currently maintained at around 2,000 animals through a programme of culling (Forestry Commission 2007). Over the past two centuries, there has therefore been a shift from deer to livestock (ponies and cattle) in terms of the main contribution to grazing and browsing pressure. Taking account of the higher forage requirement of ponies than that of either cattle or deer the grazing and browsing pressures in the New Forest may currently be higher than at any time in the past, at least on those areas favoured by ponies.

Tubbs (2001) describes in detail how commoning activities have changed over time. In Medieval times, the commons were used intensively, including the widespread harvesting of bracken and gorse, turf cutting and collection of fuel-wood. Such 'traditional' commoning declined during the early twentieth century as a result of the changing socio-economic and political environment (Tubbs 2001). In contrast to the historic situation, few commoners are now full-time farmers for whom the use of common land is a central part of the farm economy (Tubbs 2001). For many commoners today, de-pasturing livestock on the Forest is primarily undertaken to continue family traditions and as a social habit, rather than to generate significant revenue. Participation in the social occasions associated with commoning now outweighs profit as a motive to engage in commoning activities (Tubbs 2001). This also accounts for the continual increase in the number of ponies de-pastured in recent decades (Fig. 24.2) (Tubbs 2001).

The social and cultural value of commoning therefore accounts for another New Forest paradox: its increasing popularity despite its low economic returns. This is illustrated by the fact that the number of people de-pasturing animals in the Forest increased by 50 % from 1987 to 2007 (Pastoral Commoning Partnership 2009), despite the lack of a significant economic incentive. Analyses of the economics of commoning consistently conclude that despite payments through the Countryside Stewardship and Single Payment Schemes, commoning generates little if any profit (New Forest Commoning Review Group 2007; Verderers of the New Forest 2005). As noted by Tubbs (2001), todays commoners provide a marked contrast to the subsistence farmers of the historic past. Most subsidise their commoning activities from other sources of income; some are recent incomers 'indulging an understandable whim' (Tubbs 2001).

Such conclusions have implications for understanding the resilience of the New Forest as an integrated socio-ecological system (Newton 2011). It is striking that commoning has persisted, and even expanded in recent decades, despite the oft-cited economic constraints affecting it (New Forest Commoning Review Group 2007; Tubbs 2001). It appears that the high social and cultural value of commoning confers a degree of resilience to an adverse economic climate. This raises questions regarding the impact of the financial incentives provided for commoning. On 1st March 2010, a new Higher Level Stewardship Scheme came into effect for the unenclosed Crown Lands of the New Forest, which will provide in excess of £1.6 million per year of public money for the New Forest over the next 10 years. Much of this will support commoning activities. This represents one of the largest agri-environment agreements undertaken to date in Europe.

As demonstrated here, however, the impacts of livestock on biodiversity in the New Forest are not universally positive. There is a risk that current trends could lead to further intensification of grazing, and further biodiversity loss. Importantly, the changing nature of commoning activity has not been reflected in management plans relating to the New Forest, which emphasize the conservation benefits of grazing. However, as noted by Newton et al. (2009) for the specific case of heathlands, little robust scientific evidence is available for such benefits. One of the principal benefits of livestock grazing often cited by the commoners is the control of scrub (Verderers of the New Forest 2005), yet little evidence is available concerning the ability of ponies and cattle to effectively control scrub colonisation (Newton et al. 2009). The fact that the Forestry Commission places such emphasis on manual scrub control highlights the fact that grazing is not effective in this regard. Furthermore, scrub is of high habitat value, and plays a crucial role in the ecological dynamics of the New Forest (Tubbs 2001; Newton 2010b). There is therefore a strong argument for increasing, rather than decreasing, the amount of scrub cover in the Forest (Tubbs 2001; Newton 2001).

Whereas it has been widely suggested that traditional management approaches can make a positive contribution to biodiversity conservation (Berkes et al. 2000), the case of the New Forest commons highlights the risks of simplistic assumptions about the conservation benefits of such approaches. Current commoning approaches are not necessarily equivalent to historical land use patterns, and may be poorly suited to maintaining habitat characteristics that have evolved under the latter. The New Forest has witnessed a decline of turf and vegetation cutting, and an increase in grazing pressure over the past century, and this has contributed to biodiversity loss. As we enter what may be an unprecedented era of environmental change, the maintenance of biodiversity in future will depend on a critical awareness of both the strengths and weaknesses of traditional approaches to the use of common land, and an acceptance that the past may be a poor guide to the future.

References

Aitchison J, Crowther K, Ashby M, Redgrave L (2000) The common lands of England. A biological survey. Rural Surveys Research Unit, University of Wales, Aberystwyth. Department of the Environment, Transport and the Regions, London

Berkes F, Colding J, Folke C (2000) Rediscovery of traditional ecological knowledge as adaptive management. Ecol Appl 10(5):1251–1262

Cantarello E, Green R, Westerhoff D (2010) The condition of New Forest habitats: an overview. In: Newton AC (ed) Biodiversity in the New Forest. Pisces Publications, Newbury, pp 124–131

Chatters C (2006) The New Forest—National Park status for a medieval survivor. British Wildlife (December 2006), 110–119
Commission Forestry (2007) New Forest District deer management plan 2005–2015. Forestry Commission, Lyndhurst
Commission Forestry (2008) The crown lands management plan 2008–2013. The Forestry Commission, Lyndhurst
Hindmarch C, Pienkowski M (eds) (2000) Land management: the hidden costs. Blackwell Science, Oxford, for the British Ecological Society, London
New Forest National Park Authority (NPA) (2008) New Forest National Park Plan. Consultation Draft (2008) New Forest National Park Authority. Lymington, Hampshire
Newton AC (ed) (2010a) Biodiversity in the New Forest. Pisces Publications, Newbury
Newton AC (2010b) Synthesis: status and trends of biodiversity in the New Forest. In: Newton AC (ed) Biodiversity in the New Forest. Pisces Publications, Newbury, pp 218–228
Newton AC (2011) Social-ecological resilience and biodiversity conservation in a 900-year-old protected area. Ecology and Society 16(4):13
Newton AC, Cantarello E, Myers G, Douglas S, Tejedor N (2010) The condition and dynamics of New Forest woodlands. In: Newton AC (ed) Biodiversity in the New Forest. Pisces Publications, Newbury, pp 132–147
Newton AC, Stewart GB, Myers G, Diaz A, Lake S, Bullock JM, Pullin AS (2009) Impacts of grazing on lowland heathland: a systematic review of the evidence. Biol Conserv 142:935–947
Pretty J, Smith D (2004) Social capital in biodiversity conservation and management. Conserv Biol 18(3):631–638
Putman RJ (1986) Grazing in temperate ecosystems. Large herbivores and the ecology of the New Forest, Croom Helm
Rand M, Chatters C (2010) Vascular plants. In: Newton AC (ed) Biodiversity in the New Forest. Pisces Publications, Newbury, pp 84–111
Rose RJ, Webb NR, Clarke RT, Traynor CH (2000) Changes on the heathlands in Dorset, England between 1987 and 1996. Biol Conserv 93:117–125
Short C (2008) The traditional commons of England and Wales in the twenty-first century: meeting new and old challenges. Int J Commons 2, 2. URL: http://www.thecommonsjournal.org/index.php/ijc/article/viewArticle/47/41. Accessed 17 June 2010
Snook AM (1998) Birds of the New Forest. A visitor's guide. Centurion Books, Fordingbridge
Sutherland WJ (2000) The conservation handbook: research. Management and Policy, Blackwell
Sutherland WJ, Pullin AS, Dolman PM, Knight TM (2004) The need for evidence-based conservation. Trends Ecol Evol 19(6):305–308
The New Forest Commoning Review Group (2007) New Forest Commoning Review 2007. A commoner led review of current issues relating to the practice of commoning within the New Forest National Park. New Forest National Park Authority, Lymington
The Pastoral Commoning Partnership (2009) Trends in pastoral commoning in England. Natural England Commissioned Report NECR001. http://www.naturalengland.org.uk. Downloaded 17 June 10
Tubbs CR (2001) The New Forest. History, ecology and conservation. New Forest Ninth Centenary Trust, Lyndhurst
Verderers of the New Forest (2005) Verderers' Countryside stewardship scheme grazing management plan. Verderers of the New Forest Lyndhurst, Hampshire
Webb NR (1998) The traditional management of European heathlands. J Appl Ecol 35(6):987–990
Wright RN, Westerhoff DV (2001) New Forest SAC management plan. English Nature, Lyndhurst

Chapter 25
Looking Back to the Future: Ancient, Working Pollards and Europe's Silvo-Pastoral Systems

Jill Butler

25.1 Introduction: A Common European Heritage

A few years ago, a strange piece of oak was dug up from gravel workings along the Trent near Nottingham. This was one of a number of sub-fossil pollards that have now been carbon dated at 3,400 years old. Similar cut trees, pollards and shreds, have been excavated from gravels in the River Meuse in the Netherlands and carbon dated at 1,800 years old. There is much evidence of pollards and cutting trees in works of art (Haeggström 2006). The oldest works of art are frescos from Akrotiri, Greece and two gold cups from a Tholos grave in Vaphio near Sparta, which are now in the National Museum in Athens dated from the fifteenth millennium B.C. Perhaps closer to home, there are the illustrations of pollards being cut in the Bayeux Tapestry and in many other major European works of art. Although it has been a widespread common practice, there is surprisingly little written record of this ancient cultural tradition.

In this chapter, it is not possible to fully explore what we currently know about the history of pollarding—and what we know is likely to only scratch the surface. We do however know that for many millennia, across Europe and into Asia, trees outside woodlands have been cut as pollards, providing rural populations with the essentials of everyday, sustainable living such as fuel, timber (for building and transport), tools, cork, rope, fodder, clogs, food, fruits, drinks and chemicals such as tannin and the constituents of gunpowder (Butler 2010). Depending on tree species and the product(s) required, the trees were established in different configurations. They were cut to a different form but often using the basic natural

Conservation Adviser, Woodland Trust.

J. Butler (✉)
Woodland Trust, Kempton Way, Grantham, Lincs NG31 6LL, UK
e-mail: jillbutler@woodlandtrust.org.uk

I. D. Rotherham (ed.), *Cultural Severance and the Environment*, Environmental History 2, DOI: 10.1007/978-94-007-6159-9_25,

frame of the tree, and were called by different names. These trees are still usually found in traditional agricultural landscapes—wood pastures, common land, orchards, hedgerows and in parkland. Many are being lost due to changes in land management practices and lack of understanding about the origins of these trees and how to look after them.

25.2 Contemporary Benefits of Pollards and Sylvo-Pastoral Systems

25.2.1 Productive and Sustainable Agroforestry Systems

Pollards or 'working' trees (Green 2001) were cut to allow arable cultivation or stock to graze under the trees without affecting the tree's long-term usefulness. The trees were cut usually at 2 m or slightly higher, so that the products were accessible. The trees were spaced sufficiently widely apart to allow light to reach the herb layer for pasture or crop and to maximise productivity of the tree (Fig. 25.1).

Clearly, a balance has to been maintained between too many and too few trees depending on community or owners' requirements. Too many trees and the productivity of the pasture or arable crop is reduced, too few trees and the product

Fig. 25.1 Archaeologists and tree specialists believe this sub-fossil is a pollard oak that was cut by axe. Picture credit: Ted Green

from the trees is reduced. Compare the density of trees in an orchard where the priority is for fruit with a wood pasture system where the priority might be for grazing.

Research undertaken in Norway (Austad et al. 2003) provides us with some interesting evidence of the productivity of these systems. The study investigated the outputs from a field of pollards (in cycle) without fertiliser inputs and compared it with other fields—all without trees, some with or without artificial fertiliser inputs. The results showed that the fields with pollards yield more crops than the artificially fertilised fields and in addition are more productive owing to additional products from the trees e.g. the fodder or wood fuel obtained on a cyclical basis. They concluded that the pollards were naturally fertilising the soils through leaf and twig fall but were widely enough spaced and had small enough crowns not to compromise the light reaching the ground. These multi-tier agricultural systems allow maximum productivity from a given piece of land without any need for external inputs on a sustainable basis. Trees are especially beneficial in fertilising soils; they have more permanent and often deeper root systems than crop root systems. The associated mycorrhizal fungi enable the tree to efficiently gather nutrients, especially minerals such as phosphorus that are unavailable to plants in any other way and these then can be shed and recycled through annual leaf fall or biannual twig drop or longer-term branch drop.

Mycorrhizal fungi are especially vulnerable to inorganic fertilisers and without them trees and meadow plants, will be less productive and more vulnerable to disease. The trees provide many other benefits in farms including shelter for the crop or the stock—a factor recognised in modern alley agro-forestry systems (Wolfe 2006). A synopsis of the research literature about agroforestry is available from the Organic Research Centre (Smith 2010). Trees also provide other ecosystem benefits such as regulation of water runoff and quality, improved air quality, pest and disease control and climate change mitigation and adaption. The Woodland Trust is therefore keen to see twice as much native tree cover in the UK as an essential part of a healthy environment.

25.2.2 Biodiversity and Ecosystem Services

As the old growth forests[1] of Europe were cleared and affected by human interference, the decaying wood and old tree habitats of ancient and veteran trees were lost outside of silvo-pastoral (royal hunting forests, deer parks or wooded commons) or silvo-arable systems. It is only in these ancient and traditional pollard landscapes, many of them dating back before mediaeval times, that the presence of

[1] Old growth: stands with more than 200 years' growth with a continuity of old trees reaching back into the past. *Space for Nature*. The Woodland Trust (2002).

Fig. 25.2 Recently cut pollards. Photo credit: Austad (2006)

old trees has provided continuity of habitat for many hundreds of rare and special species especially fungi, invertebrates and lichens.

However, even within these systems the trees are very vulnerable today due to the intensification pressures from agriculture, forestry and concerns for public safety. The ancient and pollard trees themselves, as well as the fungi, plants, animals and soils that they support are of great significance to European biodiversity. In recent years, new research into areas of pollards in Spain and in Turkey are discovering many species new to science associated with the trees (Fig. 25.2).

Many of the species, especially fungi, have been used to provide chemicals for medicinal or agricultural purposes. The importance of natural recycling systems is only latterly being understood and thought about in the context of future farming alongside all the other plentiful benefits of establishing trees on farms to improve sustainable, welfare-based farming www.woodlandtrust.org.uk.

25.3 Landscape

Pollards help create landscapes of great cultural character, distinctiveness, beauty and historic value—nowhere more so than where ancient and very old pollards remain. The pollard shape is an integral part of the rural environment of hedges, trees and small fields that it is easy to overlook or notice its disappearance. Pollards are however anchored in our imagination, sometimes even subconsciously, as a quintessential tree shape. Pollards have been represented in artists' landscapes from the paintings of early Flemish masters, to today's comic strip illustrations and the works of contemporary photographers. They can be rural trees, trees that have been swallowed up into urban areas or trees that are cut specially for our city and village designed landscapes. They are as important to us in terms of our heritage as our buildings and archaeological artefacts. As a practice it should be maintained where ever possibly to provide continuity of thousands of year old examples of archaeological ecosystem services.

25.4 Perpetuating Lessons from Tree Archaeology into the Future

Astonishingly, there is little study in Europe and further afield of this fascinating tree archaeology and cultural history which extends from Portugal to Turkey, Crete to Scandinavia revealing that pollarding transcends international borders. The loss of this cultural and biodiversity rich heritage is possible because it is the story of the common man and not a part of our heritage relating to the rich and powerful and the lure of mechanisation. The systems are disappearing across Europe. Apart from a few places, the tradition is largely now just a memory of the oldest people in rural communities. In the UK, there is evidence that it largely died out with the coming of coal and other sources of heating and power or changes from leaf fodder to root crops. It has carried on only where a different purpose such as charcoal making, fodder production, household heating supplies or fruit and production are still important. In many parts of Europe, the trees of two (or multi-) tier sylvo-pastoral or agro-forestry systems are removed as part of the intensification to single tier agricultural or forestry systems. However, in Portugal with the rise in costs of electricity and other power sources there is a resurgence in the traditional system in rural areas. The tradition of pollarding does not appear to have been written down and few people have attempted to record how it was done from their memories. The founders of the *Ancient Tree Forum* such as Ted Green and people like Read (2008) at Burnham Beeches, City of London and other European projects are now trying to redress the situation. Despite their contribution to European biodiversity and cultural history, the trees, the habitats, and the landscapes, are not properly recognised at European Union level. It is time to discover their value and give them the care and recognition they deserve. This is really a case of '*Looking Back to the Future*' in order to secure the conservation and sustainability of one of Europe's most precious natural environmental resources: the ancient, working pollards and Europe's associated silvo-pastoral systems. These are long-term, historical, cultural landscapes and the imminent threat to their future is that of 'cultural severance' (Fig. 25.3).

Fig. 25.3 Pollard black poplars near Calamocha, Spain. The wood provided light, strong roof timbers for buildings. Photo credit: Ted Green

References

Austad I, Hauge L (2006) Pollarding in Western Norway. 1er colloque europeen sur les trognes, Vendome 26,27 et 28 octubre. www.maisonbotanique.com
Austad I, Hamre LN, Rydgren K, Norderhaug A (2003) Production in wooded hay meadows. In: Tiezzi E, Brebbia CA, Uso JL (eds) Ecosystems and sustainable development, vol 2. Wit Press, Ashurst Lodge, Ashurst, Southampton, pp 1091–1101
Butler J (2010) Ancient, working pollards and Europe's silvo-pastoral systems—back to the future. Landsc Archaeol Ecol End Tradit 8(2):54–57
Green EE (2001) Pollarding, origins and practical advice. British Wildl 8:100–105
Haeggström C-A (2006) Pollards in art. 1er Colloque europeen sur les trognes, Vendome 26, 27 et 28 octubre. www.maisonbotanique.com
Read HJ (2008) Pollards and Pollarding in Europe. British Wildlife 19(4):250–259
Smith J (2010) Agroforestry: reconciling production with protection of the environment. A synopsis of research literature. Organic research centre, Elm Farm
Wolfe M (2006) Seeing the wood, the treesand the Catch 22. ERFC, www.organicresearchcentre.com

Chapter 26
Promoting Stewardship of *New Commons*: Lessons from WakeNature Partnership

Gary B. Blank and George R. Hess

26.1 Introduction

Protecting natural areas of landscapes at varying geographic scales of interest has preoccupied activists in the United States for a number of decades. Protected areas typically reside in the public sector of ownership and become a resource common to all members of the populace. They most often have not fit easily the notion of *commons* as the term is used in European, especially British, contexts. Prior individual ownership of property most typically would preclude utilization of resources and even access by neighbours. Indeed, limitation of use and access may be the factor that has made these properties most desirable for acquisition and protection. Thus, we might call such acquired and protected lands our "*new commons*".

WakeNature Preserves Partnership (WakeNature) arose to promote the stewardship of high quality natural areas amid burgeoning metropolitan development fuelled by economic growth in Wake County, North Carolina, USA (Fig. 26.1). With 38 % population growth from 2000 to 2008, Wake County experienced intensive infrastructure, residential, and commercial changes. This period of change followed an even longer transformation of the region dating back to the 1970s. Wake County's Town of Cary, North Carolina, for instance, was home to around 4,000 people in the mid-1970s but had nearly 100,000 inhabitants at the 2000 census. Responding to such growth, local land conservancies and governments have increased funds to protect open spaces by purchasing land or

G. B. Blank (✉) · G. R. Hess
Department of Forestry and Environmental Resources, North Carolina State University, Jordan Hall 5229, NCSU Campus, Box 8008 Raleigh, NC 27695, USA
e-mail: gblank@ncsu.edu

G. R. Hess
e-mail: george_hess@ncsu.edu

I. D. Rotherham (ed.), *Cultural Severance and the Environment*, Environmental History 2, DOI: 10.1007/978-94-007-6159-9_26,

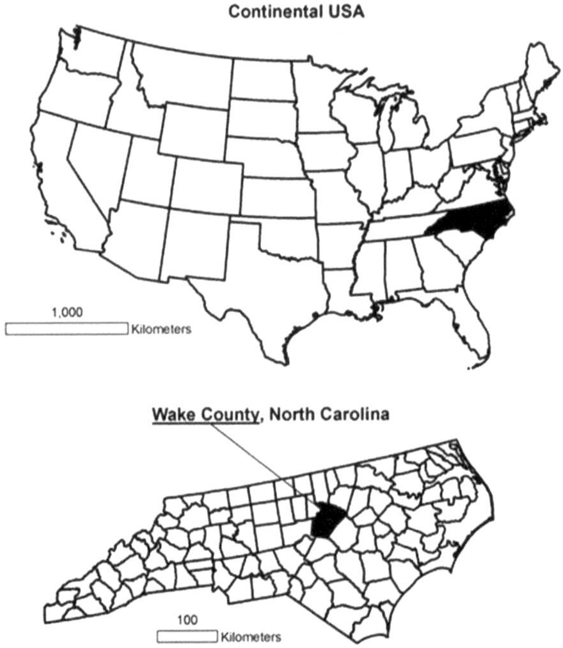

Fig. 26.1 Wake Country, North Carolina, USA has experienced significant population growth and urban development during the last four decades, resulting in declining amounts of green, open space

conservation easements that limit building. Wake County's Division of Parks, Recreation, and Open Space, for example, purchased 1,834 ha (4,531 ac) of open space between 2000 and 2008, at a cost of some 86 million USD (52 million GBP).

These open spaces are becoming the *new commons*, accessible to the public for a variety of uses, in a landscape that is almost entirely parcelled out to private owners who hold exclusive rights to the properties. Their creation is motivated by a sense of a vanishing natural and cultural heritage as more of the landscape is built, and justified for public benefits including protecting water quality, reducing flood damage, providing places for recreation and education, conserving wildlife habitat, maintaining scenic amenities, and preserving cultural heritage and a sense of place.

The *new commons* are being created primarily through two strategies: purchase of land by conservancies and governments, and government regulatory controls on land development. Purchase usually occurs when land in a desirable urbanizing location becomes available; cursory examination of its features suggests potential value to the public for recreational, environmental, or cultural benefits; and the owner is willing to donate or sell the land or development rights to a conservancy or government agency. As parcels destined to be built upon are sold and development plans are filed in municipal jurisdictions, regulatory controls usually come into effect through government requirements that some portion remain "natural" to provide environmental protection or recreational amenities; these lands frequently come under control of homeowners associations created when residential subdivisions are built. Natural heritage and unique qualities possessed by many of these sites often are only partially appreciated as they are acquired. Further, management requirements for maintaining suitable conditions to protect and enhance those qualities are only generally understood.

Thus, several things brought WakeNature to being: citizen activism, recognition among government and other agencies of the need to manage ecologically valuable natural areas appropriately, and action by a group of faculty, students, and other professionals interested in conserving natural resources in suburban environments. The group aims to '*organize and provide resources to identify ecologically valuable, publicly owned open spaces within Wake County [North Carolina] and to build capacity for appropriate management and long-term stewardship of those areas*' (WakeNature 2009). WakeNature hopes to develop organizational capacity among county and municipal governments to carry out these activities, and to establish itself as the organization that accredits high-quality nature preserves in the county. The long-term vision is of a collection of WakeNature Preserves—*new commons*, if you will—representing the finest natural resources in Wake County and available for the enjoyment and education of the public.

This chapter is about what we have learned in the process of pressing for enlightened treatment of emerging *new commons* in a dynamic landscape. To demonstrate its intent and provide examples for its client-partners WakeNature has, over several years, labored to develop a stewardship plan for Turnipseed Preserve, the first WakeNature Preserve in Wake County. By mapping and inventorying its features and developing a stewardship plan, the group is using the site as a proving ground while attempting to develop a replicable process others can follow. In the process, WakeNature has melded assets and needs of a diverse set of partners, who the relationship mutually benefits. Moreover, it becomes increasingly obvious that WakeNature also serves as a platform for community-engaged research and learning.

26.2 About the Lessons

As WakeNature participants, we identify lessons deriving from two tightly related realms of our experience: (1) building the organization and (2) accomplishing the organization's target tasks. Our observations pertain to substantially different aspects of the experience, and the values of the lessons diverge substantively in terms of who might find them important. Co-authorship of a book chapter about our group's collective experience creating WakeNature forged our thinking about organization-building (Hess et al. 2010). Reflecting upon participation and achievements of many contributors to the technical process and resulting management plan for Turnipseed Preserve provides the basis for our lessons about what enlightened treatment of the new commons requires.

To this date in 2012, WakeNature has existed 4.5 years, initiated around a local coffee shop table. From inauspicious beginnings as a conversation among academic types and agency practitioners a coalition began to take shape around the idea that shared expertise and perceptions of a problem could unite us. The circle of participants and their scope of activity grew through a web of contacts whose appearance at twice-monthly meetings may be regular or sporadic depending on other professional commitments. One point of contact anchors us all—the WakeNature wiki site at http://wakenature.wikispaces.com/. Created in January 2008 and expanded continually since, our wiki states who we are; declares our goals, vision, and aspirations; chronicles what we do; archives what we produce; and helps us build our agenda for moving forward. In some ways it symbolizes the loosely structured organization as it encompasses the diverse range of contributions partners make to the effort. Partners missing meetings can go there to catch updates, see the agenda and its outcomes, and they can move ahead in their own areas of initiative.

During the last two and a half years many of those initiatives have focused on the Turnipseed Preserve along Marks Creek, in eastern Wake County. There we have learned a number of important lessons from considerable team effort that has created the Turnipseed Preserve management plan. Chosen as a demonstration of what the partnership hoped to accomplish and could offer potential partners around Wake County, the Turnipseed Preserve offers a broad range of challenges for WakeNature.

Creating the management plan has drawn upon our diverse talents, perspectives, and areas of expertise. It has enabled teachers to engage classes in service-learning projects. It has prompted discussions of priorities among the partners. Its creation became a lynchpin in defining the process by which WakeNature Preserves would be designated. Only when everyone recognized that the management plan would never be completely done—that it would always be a work in progress that needed to establish certain standards of expectation, but not necessarily nail down every detail of action into the future—could we proceed to an important next step. That step was taken when Wake County's Board of Commissioners authorized submittal of the application to designate the Turnipseed Preserve as a WakeNature Preserve (Snow 2010). The designation was subsequently awarded in December 2010. Since that designation further work at Turnipseed has engaged students, county staff, and volunteers in a variety of activities called for in the management plan, thus implementing the stewardship focus that prompted creation of the partnership.

In the rest of this chapter we present eight lessons with explanatory comments where appropriate to illustrate those lessons in ways we hope are useful to others engaged in cooperative conservation efforts across diverse constituencies.

26.3 The Lessons

1) Most people—even ecologically minded types—do not really understand the relationship between conservation and management. The belief that something must be managed if it is to be conserved, as opposed to the notion that you can just buy the land and declare victory, is not prevalent among people who want to save places from development. In fact, *management* is often seen as anathema to

preservationists. However, set aside does not mean left alone, and taking the time to explain the need for active management of open spaces to decision makers unfamiliar with conservation is critical to the emergence of a new commons.

Manipulating vegetation and site conditions to assure persistence of desired species may require burning, thinning trees, removing exotic plant species, or a variety of other intentional actions. Such manipulation implies managing the site beyond merely acquiring it and letting it sit unattended in perpetuity. This is where idealism and preservation theory meet pragmatism and conservation practice. Conservation—particularly in an urbanizing environment—is a practical and necessarily hands-on activity that engages the manager of a resource in continual decision-making about utilization limits, tolerances, and reinvigoration of the resource.

In our local area examples abound, and members of WakeNature engage at differing levels of effort with these examples. For instance, a university tract proposed for consideration as a WakeNature Preserve, known as Lakc Raleigh Woods, contains some of the oldest intact deciduous forest in the city and an interesting complex of plant community types. Examining the Woods to develop a management plan, Dr Blank's graduate class determined that this woodland is seriously infested with invasive plants such as *Elaeagnus umbellata* Thunb. and *Microstegium vimineum.* Additionally, it is overrun by whitetail deer (*Odocoileus virginianus*) and the deer are browsing on the very botanical specimens for which university faculty sought to have the Woods protected from development. Lacking management in past decades, this protected area is at severe risk of rapidly declining in botanical value and losing the very ecological values for which it was set aside. Protection from development does not equal protection in this case.

2) Developing a management plan depends upon the property-owner's vision, knowledge of inventoried resources, and the natural communities' needs for attention. Not knowing specific details of what they had at Marks Creek, Wake County open space managers were understandably slow to establish directions for management. Developing a vision depends in part on knowing what is available and possible. In this case, the vision for the land emerged from a cyclical process of learning what was present—the raw material, understanding how that raw material might be transformed, and developing increasingly specific goals. The County managers, in fact, had to work through the process of learning with us to determine what was actionable and affordable in both short and long terms. Their vision is evolving and being guided by the combined expertise that WakeNature brings to bear: gathering baseline data through inventory procedures, organizing diverse resource information in a logical framework, and analysing the resulting material to identify priorities and project potential changes. Especially in an era of constrained budgets, bringing volunteer effort forward to do these things has enabled Wake County staff to accomplish this detailed planning task. The case of the lupines, cited in the next lesson, underscores the kinds of opportunities county land managers might well miss without being part of a group like WakeNature.

Even organizations dedicated principally to conservation and land protection find themselves inadequately staffed and at times in need of new perspectives on the questions they face. Triangle Land Conservancy (TLC), for example, called on WakeNature to assist with preliminary work on a management plan for its Swift Creek Bluffs site. The project, undertaken by another graduate class at NC State University, was expanded to incorporate two other adjacent properties in what we referred to as an assemblage. Three different tracts in various states of forest succession and owned by three entities (private non-profit, county government, and state agency) spanned a creek and a major thoroughfare, posing logistical questions. Resource inventories were found to be mostly complete on the TLC and county properties but only rudimentary on the state property. Municipal planning for greenways in the vicinity complicated the management situation and suggested the need for better definition of priorities on the county and state properties. A survey of neighbours, to determine interest and willingness to volunteer for maintenance efforts on the properties, and concerns over their management, provided TLC with better information than it was likely to obtain otherwise.

3) Making progress in new territory depends upon working with partners' interests and abilities. When we began our work on the Turnipseed Management Plan, we started with boundaries on a map outlining a geographic area purchased by Wake County because of generally recognized attributes and a few key natural features of interest. We had almost no details upon which to base plans for action, but details began to accumulate with our increasing experience and activities. Early on a group of student reptile and amphibian enthusiasts volunteered time and effort to visit the site at all hours to observe and to hear audible evidence of resident species. In some respects, the dynamic of service-learning and the desire to engage students at the site drove some of our information gathering. For instance, the forest inventory and global positioning system mapping of rock outcrops and boulders resulted directly from class or student engagement at Marks Creek. Individuals with expertise, curiosity, and time broadened the scope of our inventory as they had opportunity. For example, lichenologist Gary Perlmutter has spent untold hours examining the parcels and discovering species heretofore unknown to this region. John Connors, with the North Carolina Museum of Natural Science, championed propagating a plant that has become rare in Wake County, the blue lupins *(Lupinus perennis)*. This was because of their importance to butterfly species; and that interest pulled in Donna Wright from NC State University's Plant Biology Department to consult on the flower's reproduction cycle.

Ross Sullivan, a graduate student engaged in the class developing the management plan for NC State University's Lake Raleigh Woods, decided to map the *Elaeagnus umbellata* infestation mentioned above. His data were incorporated in the plan submitted to the university administration, and the problem was deemed a priority for attention. Subsequently, he proposed a study to experiment with control and possible eradication of this invasive species. Ultimately, his study promises to inform treatments that will be applied to this invasive species which is degrading the quality of the protected site. Moreover, his findings could be applied at other Wake County sites where this same invasive species is degrading their utility and functionality as exemplary *natural* areas.

In summary, partners have gone where our curiosity and talents have led us rather than followed a dogmatic prescription of how to get from our relative ignorance of the site to a final plan for its management.

4) A loose affiliation of individuals can evolve into an affiliation of their organizations. When WakeNature emerged in response to perceived need, the scope of the need was still undefined. Specific tasks to be achieved were not all evident, nor were specific persons needed to achieve the known tasks identified. Gradually the circle of participants enlarged, and people began to assume roles representing organizations in which they functioned outside of their affiliation with WakeNature. To varying extents persons brought the resources and perspectives of those organizations to bear on matters concerning WakeNature, such as developing a management plan for Turnipseed and defining the process for recognizing preserves. These two tasks converged when the demonstration site and its management plan became the test case for WakeNature Preserve recognition. A series of presentations to county officials was needed before permission to apply for preserve status could be granted. So as much as anything, it seems that WakeNature has earned legitimacy as an organization based on increasingly well-defined operational processes and the legitimacy of the organizations from which its members are drawn. Wake County's Board of Commissioners has recognized WakeNature officially and blessed the participation of one of its divisions in the partnership, and a county official has signed a Memorandum of Understanding between Wake County and WakeNature.
5) Partnerships will gain their own distinctive identities. Each partner comes to WakeNature with unique perspectives and experiences, from organizations with different operating systems and corporate cultures. As with any team effort, progress depends on shared information, negotiated responsibility, and a clear understanding of our identity as a team—who we are and what we are trying to do together. In the case of WakeNature, everyone is united in their concern for the ecological and human communities in Wake County and committed to working collaboratively to improve them. This shared concern allows individual members to identify with WakeNature's mission and work through differences in their personal and organizational cultures.

As we approached designation of the first WakeNature Preserve at the Turnipseed site, questions of partnership identity were highlighted through discussion of WakeNature signs and logos. What exactly does a WakeNature sign signify? Why should another organization (e.g. Wake County) allow a WakeNature sign to be posted on property it manages? The emerging answer to these questions was that the WakeNature imprimatur identifies a place as a high quality, well-managed natural area to people looking for such places.

Here, then, the *new commons* concept gains traction. We have to ask how the public will perceive and use the land in question, how they will benefit from our stewardship of the resources there, and what actions we must undertake to protect the values at risk. Do we need to burn the site to foster prairie conditions in a broader landscape transformed through agriculture to eliminate such features? Do we reintroduce lupines or Michaux's sumac in appropriate sites? If our identity is wedded to the perceptions of values encompassed on the properties where we work, then it is important to discover what we value in common among our partners and what the partnership values in common with the stakeholders among potential public users.

6) The process needed to create *new commons* takes patience and attention to people's needs for information in appropriately targeted forms and doses. Supporting Wake County Parks, Recreation, and Open Space in seeking permission to apply for WakeNature Preserve status for Turnipseed moved WakeNature into the realm of political action. We needed to present succinct, targeted messages to people attuned to sound bites and executive summaries. Because WakeNature comprises professionals from varied organizational cultures and disciplinary backgrounds, members have become attuned to creating mutual expectations while remaining sensitive to individual needs for explanation and more or less contextual detail.

The habit of targeting our attention and communicating efficiently evolved in our twice-monthly meetings as well. Creating the upcoming agenda on the wiki, soliciting agenda items and comments beforehand, then adhering to the schedule we collectively created has become routine. Occasionally last-minute additions to the agenda arise, but the 90-min time period on the first and third Thursdays of each month was a stable characteristic, no matter who led the meeting. Now, once a month meetings with committee meetings in between have changed the dynamic of our working relationships but not the routine of our meetings. Individuals arrive or depart as they need to, late or early, but the meeting starts and adjourns promptly.

Also instructive was the opportunity to work with a student communications team charged with developing a communications campaign for WakeNature as part of a service-learning course. The students had limited knowledge of natural resources and none of WakeNature when they began their effort. This put us in the position of having to explain ourselves in the most basic terms to an uninitiated group so that they could help us create messages that our target audiences—elected officials and the public at large—would understand. The students' naïve questions often gave us pause as we were forced to explain ourselves in plain language to the uninitiated.

7) Personal rewards are probably more important than the professional status attained through the partnership, particularly for academicians. There is another dynamic also, which is the dynamic between conservationists, in the academic sense, and the economic, social, and political systems in which conservation actually occurs. Publishing peer-reviewed papers may be necessary for advancing academic careers, but it's not sufficient for the creation and

management of *new commons*. It is difficult to attain professional status in academia with the kind of work we are doing in WakeNature. Although the academics among us have been writing book chapters (like this), engaging in service-learning, and directing graduate projects based on our efforts, this is not the type of work that typically leads to tenure. In contrast, agency and conservancy practitioners are more interested in taking direct action to solve problems. Within WakeNature we have all come to respect one another's professional needs and goals, and we work to create projects and approaches that move everyone forward. Nevertheless, a strong personal commitment is critical to continued involvement.

8) The value of 'fumbling together in the dark'. As several of us worked on a book chapter describing WakeNature's evolution as an example of participatory partnerships for social action (Hess et al. 2010), it occurred to us that we did not "act like academics" in our roles in the partnership. We worked with our non-academic partners to "make it up as we went along" using an approach that was inductive, reflective, and democratic rather than isolating ourselves in the ivory tower and emerging with "the answer". We are not trying to be anti-intellectual here, but merely noting that there may be something positive about fumbling together in the dark to combine well the theory-based approaches of academics and the experience-based approaches of agency and conservancy partners. We all have developed a great deal of respect for one another and our different perspectives. We are learning and acting together in a way that seems particularly well-suited to the task of building the *new commons* in today's social and political environment.

26.4 Conclusions

American urban parks are typically perceived to be tame, manicured, landscaped intentionally for aesthetic or utilitarian effect, or given over to recreation, meaning sport in many cases. Initiative to purchase and manage *open space* amid burgeoning sprawl of urbanizing America creates a new classification of land that falls outside the typical ideology of *park,* yet falls short of the full blown ideology of *wilderness*. In this context, we must redefine *Commons* in twenty-first-century America. WakeNature holds that not all of our *new commons* should be ball fields and playgrounds. Nevertheless, we also hold that simply cordoning off places that cannot be built upon will not guarantee protection of the important natural characteristics of the *new commons*. That action will not sustain elements that deserve observation and wonder by future generations.

Collective management of our landscape needs to reconcile with the fact that intensive attention to open spaces must replace custodial neglect in the name of preservation. Everyplace cannot devolve to the mythical pristine climax forest urged upon us by naïve beliefs about historical conditions. For one thing, alien and invasive species have too intrusive a head start at this point. Moreover, characteristics worthy of preservation may be artefacts resulting from and dependent upon styles of landscape management that have passed their economic viability and left behind floral and faunal legacies.

We come back, therefore, to part of our evolving organization's purpose statement: 'to build capacity for appropriate management and long-term stewardship of those areas' that we have identified as worth preserving. At this point in our American landscape history, young as our role in it may be compared to the millennia of European history, WakeNature partners agree that we cannot just let 'nature take its course'.

Acknowledgments Thanks to Michael Youth for creating the map in Fig. 26.1.

References

Hess GR, Blank G, Clayton P, Connors J, Holcombe K, Ramsey J, Reis K, Snow C, Steelman T, Wallace J (2010) Perspectives on partnership evolution: from passionate people to committed organizations. Chapter 18. In: Harter L, Hamel-Lambert J, Millesen J (eds) Case studies in community-based participatory research. Kendall-Hunt, Dubuque, pp 349–376

Snow C (2010) Application letter to WakeNature Preserve Partnership. http://wakenature.wikispaces.com/TurnipseedReview, Accessed 25 June 2010

WakeNature (2009) WakeNature Preserves Partnership, http://wakenature.wikispaces.com/ Accessed 25 June 2010

Chapter 27
End of Tradition, Reworking of Custom: Re-assembling *Satoyama* Woodlands on Tokyo's Urban Fringe

Jay Bolthouse

27.1 Introduction

Discontinuation of traditional management practices, and the related process of cultural severance, signals an end for many commons and commoning traditions. Examples of commoning traditions coming to an end can be identified around the world, but the causes of discontinuation vary according to local circumstance. Some commons are targets for abrupt severance from commoners through enclosure at the hands of powerful state and corporate interests (Vasudevan et al. 2008). Other commons and communities are more gradually cleaved apart through slow and silent economic and cultural transformations that render traditional management practices unnecessary. Yet wherever commoning traditions meet their demise, the result is the decoupling of socio-natural interactions that have co-evolved for generations, often since time immemorial. Discontinuation of traditional practices plainly results in radically altered ecological relationships, but it also extinguishes shared ways of being and doing that formerly defined the very character of places. Surveying this grave situation, one is certainly prone to wonder whether we are indeed witnessing the 'end of tradition'?

The question mark after the 'end of tradition' is imperative, however, for the end of tradition also marks a potential beginning. Indeed, while many common traditions are in the process of unravelling, there are numerous instances where new traditions are being rewoven from the loose threads. The regrettable end of older commoning traditions thus signals an opportunity for remaking commons through the reweaving of custom.

J. Bolthouse (✉)
University of Tokyo, Environmental Studies Bldg. #513, 5-1-5 Kashiwa-no-ha, Kashiwa, Chiba 277-8563, Japan
e-mail: jay@nenv.k.u-tokyo.ac.jp

I. D. Rotherham (ed.), *Cultural Severance and the Environment*, Environmental History 2, DOI: 10.1007/978-94-007-6159-9_27,

The remaking of custom and commons should not surprise us. The durability and socio-cultural significance of the commons ensures that efforts to reweave are both highly probable and nearly always possible. As Olwig (2003) suggests, "[T]hough the landscape of the commons often tends to be economically marginal, it is socially and symbolically central" (p. 19). The centrality of commons to social and cultural reproduction suggests that efforts to reconfigure older traditions to the shifting socio-political landscape are highly probable. Second, and interrelated, the locally determined and flexible nature of commons institutions enables them to pass through periods of great change. Indeed, common landscapes remaining today owe their continued existence not to fixed or immovable 'traditions', but to the constant adaptation of custom to shifting circumstance (Thompson 1993). The symbolic importance of the commons and their customary adaptability, suggests the current juncture is not a dead-end but an opportunity to find new paths forward.

This chapter examines the transformation of peri-urban *satoyama* woodlands in Japan to identify both the end of older commoning traditions, and the ways in which new commoning traditions are being reassembled. While *satoyama* woodlands currently cover expansive areas of mixed-use landscapes on the fringe of Japanese cities, these woodlands were historically essential components of rural village ecology and economy. As the name implies, villages (*sato*) communally owned and managed woodlands (*yama*) to regulate the use of essential fuelwood and green fertilizers (Totman 1989). However, common ownership was gradually extinguished along with modernization and the privatization of woodland ownership, and traditional management practices were abruptly discontinued as a result of the 'fuel revolution' of the early 1960s (Takeuchi et al. 2003). Yet the end of common ownership and traditional management practices has also fuelled a '*satoyama* renaissance' (Nakagawa 2004) and the emergence of new commoning traditions. The new common traditions forming around *satoyama* woodlands are driven by three key factors: a cultural reappraisal of the value of heritage woodlands, recognition of the importance of management to biodiversity conservation, and a socio-cultural desire to participate in woodland management. Certainly many of the specific factors shaping the reinvention of the *satoyama* commons are unique to Japan, but the general processes driving both the loss and the reconstitution of these commons are relevant more generally.

The paper begins by developing conceptual tools to understand the end and beginning of common traditions. After describing the deficiencies of Hardin's (1968) metaphor of tragic overuse, as well as much of the common pool resource literature that developed in response, I reiterate Olwig's (2003) call to integrate analyses of the institutional bases of commons with their symbolic importance and historical fluidity. Olwig's synthetic framework for analyzing the commons is then applied to a new strain of commons research termed the 'tragedy of the anti-commons'. First proposed by Heller (1998), the tragedy of the anti-commons identifies an inverse tragedy of resource underuse resulting from the fragmentation of ownership among multiple actors. The anti-commons concept is highly applicable to *satoyama* woodlands, and undoubtedly to other commons being radically transformed through processes of cultural severance. However, as per Olwig's

(2003) suggestion, the anti-commons concept must be expanded to include both historic and symbolic dimensions. In the second half of the paper, these conceptual tools are applied to a broad outline of *satoyama* woodlands in Japan, as well as a case study of a peri-urban area of Tokyo where shifts in forest policy have opened up an opportunity for building new commoning traditions. The concluding section discusses what the case of *satoyama* can tell us about the end and re-emergence of commoning traditions today.

27.2 Rethinking Tragedy

The present juncture—what I characterize as the end of commoning traditions and the potential for new ones—calls for conceptual tools that can illuminate how and why common traditions are being put back together. Although four decades have passed, Hardin's (1968) tragic metaphor of resource overuse maintains a firm grip on how we think about the commons. Hardin (in)famously used an analogy of a traditional common pasture and rational, self-interested herdsmen to assert that commons inevitably result in resource degradation. As intended, this deliberately simplified analogy has provided a narrative script through which to perform the privatization of diverse resources. Fortunately, in response to Hardin, a vast literature on common-pool-resources (CPR) has demonstrated that collective action is possible and that commons are by no means synonymous with degradation (e.g. Ostrom 1990). Yet despite the challenge to Hardin's narrative, the CPR literature unfortunately remains within the rational-institutional setting of his original argument. One key problem is that issues of cultural severance shaping commons today remain outside this framework. Hardin tells us that commons lead to inevitable tragedy, while the CPR literature suggests that commons can be successfully maintained so long as we can get the rules of the game right. This body of research tells us little about the values that society places on commons and how these values fuel the historical transformation of the material and symbolic commons. Understanding issues of cultural severance and the end of tradition and reworking of custom requires attention to these interrelated aspects of the commons.

Olwig (2003) provides a conceptual framework for synthesizing different approaches to the commons. He suggests that there are generally three approaches to the commons: (1) the rational institutional framework that the CPR literature uses to delineate the formal rules of collective action (2) the historical perspective that historians and geographers adopt to analyse the transformation of specific physical commons, and (3) the approach utilized in the humanities to identify representations of the commons. He asserts that 'our study of commons, institutions and landscape ought to take cognizance of the fact that the commons is not simply an institution, but also a symbol of the human ideals and values necessary to the maintenance of such institutions' (p 18). Accordingly, the narrow rational-institutional framework of the CPR literature must be synthesized with accounts of the fluid history and shifting symbolism of the commons: the 'study of the commons ought to be rooted in an understanding of the shifting relationship between the symbolism of the commons, and the evolving historical commons' (pp 18–19). This suggests that we should conceptualize the commons not as relics fixed in time and space, but as a set of common spaces and claims of rights to space that evolve and emerge at the dynamic interface between society and environment. The conceptual framework provided by Olwig offers an opportunity to make sense of the juncture of the end of common traditions and their replacement by new traditions by focusing on the interlinked historic and symbolic evolution.

27.3 Under-Use and Anti-Commons in Historic and Symbolic Perspective

Applying Olwig's synthetic conceptual framework to issues of underuse and cultural severance can help to illustrate both the *necessity* and *utility* of such an approach. While issues of underuse and cultural severance have remained largely outside the scope of the CPR literature, in the last decade scholars have begun to give attention to these issues, often using the term 'anti-commons'. Drawing upon Hardin's tragic metaphor, Heller (1998) coined the phrase 'tragedy of the anti-commons' to refer to an inverse tragedy of resource *underuse*. Heller suggests that fragmented property rights can lead to less-than-optimal resource use if no individual 'has an effective privilege of use' (p 624). The idea came to him while working in post-socialist Moscow where he noted that outdoor kiosks were booming with business while indoor shopping centres were empty. He reasoned that these anti-commons were created by officials determined to justly distribute property rights, rather than aiming for the efficiency of allocating property rights to one individual. Heller's analysis is eye opening, but problematic (cf. Mukhija 2005). While he alerts us to an overlooked potential for *underuse*, his preferred solutions point solely towards greasing the gears of capital accumulation through 'efficient' property rights. His failure to consider the environmental and social implications of anti-commons is a serious omission, but an omission that only glares back at us when placed into historic and symbolic context.

Placed into historical perspective, it becomes clear that the anti-commons concept is far from new. Rather, it is an old idea repackaged for present circumstance. Heller's anti-commons mirror the logic of enclosure. Drawing upon discourses of inefficiency and waste, historical enclosures targeted commons of that day as anti-commons ripe for assembly (Perelman 2000). The historical struggles waged over enclosure thus attest to the importance of 'positionality' to the anti-commons, since commons from one vantage point can be anti-commons from another. Yet despite the uncanny similarity between Heller's anti-commons and the logic of enclosure, his popular guide to the anti-commons (intentionally?) avoids the paradigmatic example of historical enclosures (Heller 2008). This is troubling because while the enclosures, or what Marx rather unfortunately termed 'primitive accumulation', are seen as historical, they are by no means behind us.

At present, the process of enclosure is alive, well, and advancing into new pastures. As part of what Harvey (2003) terms a fresh round of 'accumulation by dispossession', the enclosure of countless common resources—water, intellectual property, urban space etc.—has become a new frontier for growth (see Glassman 2006). It appears that version 2.1 of the anti-commons concept seems tailor-made to provide a discursive script necessary to rationalize future enclosures. Accordingly, placing this concept into historical perspective is an important intervention that also reveals the urgency of reclaiming its symbolic terrain.

Both historically and today, symbolic appeals to common rights to space and resources provide a potent means of resisting enclosure (Thompson 1993). To take a prominent example, in opposition to enclosures of the nineteenth century, the 'Commons, Forests and Footpaths Society' formed to secure access to remaining peri-urban commons through appeals to common law. From this anti-enclosure perspective, anti-commons were certainly not a result of inefficient property rights or land-use, but of the very process of enclosure that threatened to appropriate common space for private gain. Similarly, the many calls for 'reclaiming the commons' today can be seen as a response to present-day enclosures. In the language of Polanyi's (1944) 'double movement', we can conceptualize current appeals to the commons as a spontaneous reaction to protect the social from the neoliberal drive to create a nominally 'self-regulating market' (McCarthy 2005; Glassman 2006).

Yet while Polanyi's notion of a double movement has been used primarily to describe conditions of resource degradation, a similar tension can be identified between conditions of imposed underuse and responses aimed at sustaining a more balanced wiser-use. For example, as a result of global economic integration and free trade agreements, the EU has increasingly adopted a discourse of 'multi-functionality' to contest the imposition of conditions of underuse. Hollander (2004) argues that while a 'weak' version of multi-functionality—one that asserts that European landscapes are exceptional—is regressive, this response to issues of underuse is extremely important because a 'strong' version of the multi-functionality discourse—one that recognizes that landscapes beyond the EU are also multi-functional-contests the very logic of free trade. This logic asserts that anti-commons stem not from inefficient property rights or land use, but from an economic calculus that fails to account for the positive externalities of local production, and the necessarily place-based processes of social reproduction (Katz 2001).

Similar processes are shaping forests and forestry, and a brief look reveals that solutions often take the form of common regimes. For example, Schlueter (2008) calls for a new common property regime to solve the anti-commons tragedy plaguing small-scale forestry in Europe. In New England, Donahue (1999) documents the efforts of a suburban town to reclaim the commons through community agriculture and forestry on underused municipal land. In contrast to the logic of enclosure and efficiency, these authors suggest that underused anti-commons call for reworking custom to forge new commons. Furthermore, their work confirms that the issue is not only about property rights, but also about our capacity to recognize and build communal interest in common landscapes. In summary, the contrasting perspectives of the anti-commons outlined above constitute a struggle to define its ideological contours. To extend Olwig's (2003) argument, this struggle demands attention to not only the institutional, but also the historic and symbolic, dimensions of the commons.

27.3.1 Satoyama: From Commons to Anti-Commons and Back Again

The compound term *satoyama* combines *sato* [village] and *yama* [mountain/forest] and, although the term is rather ambiguous, it generally signifies a historically intimate and essential relationship between villages and nearby common woodlands (Takeuchi et al. 2003). These woodlands were historically managed as *iriai* [literally 'joint-entry'], a phrase referencing both common lands and the institution through which rights and responsibilities to these lands were specified and enacted (Totman 1989; McKean 1992). Although the management of common woodlands varies according to time and place, most were rotationally coppiced to produce fuelwood and forest litter was collected to produce organic 'green' fertilizers (Totman 1989) (Fig. 27.1). However, following rapid post-war growth, and subsequent devaluation because of the 'fuel revolution' of the 1960s, the relationship between human communities and woodlands was radically transformed. Woodlands near expanding urban centres were razed to produce space for urban growth, while more marginal woods were simply abandoned. These abandoned *satoyama* woodlands became sites for little else than potential future development, illegal waste dumping and the expansion of invasive bamboo grasses (Takeuchi et al. 2003). Thus, it appeared that the once vital relationship between *sato* and *yama* had been severed.

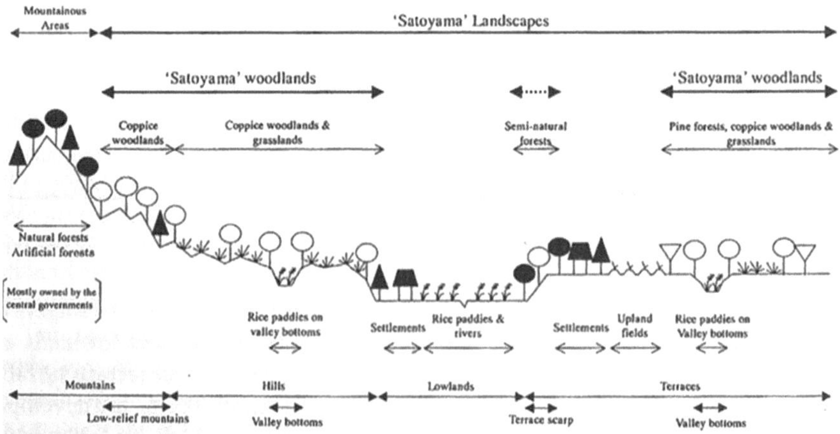

Fig. 27.1 Diagram of a typical *satoyama* landscape of the Kanto area (From Takeuchi et al. 2003)

However, at the same time as the relationship between *sato* and *yama* was being unceremoniously severed, the term *satoyama* was coined to fuse them back together. Although the term *satoyama* was used historically, it only came into widespread use after being coined by a forest ecologist named Shidei in the early 1960s. At that time, village woodlands were referred to in the scientific literature as *nouyourin* [literally agriculture-use-woodlands], a term that accurately described their connections with agricultural communities but offered little symbolic appeal. As Shidei (2006) recounts in his memoirs, he was dissatisfied with the scientific term *nouyourin* and coined *satoyama* because he believed it would have 'currency' with the public (Bolthouse 2007).

The currency of *satoyama*—its ability to circulate through representational practices—was ensured by the fact this it was citational of the nostalgic discourses of *furusato*, the symbolic home of the Japanese. These discourses gained prominence in rapidly changing postwar Japan. Ivy's (1995) conceptualization of "vanishing" illustrates how a cultural logic of nostalgia has shaped both *furusato* and *satoyama*. Ivy demonstrates that the 'traditional' is not swept entirely away by the 'modern' (i.e. it never 'vanishes'), but is instead suspended at a point of constantly *vanishing*. Suspended between presence and absence, vanishing practices, places and objects become the target of conservation and restoration. During the period of rapid economic growth following post-war recovery, *furusato* village landscapes were portrayed as vanishing—decaying and disappearing, yet remaining and possibly recoverable—through travel campaigns and other media. Resultantly, they became sites to rediscover and re-enact cultural identities (Ivy 1995). Shidei's selection of the term *satoyama* tapped into the nostalgia for the 'old-village' and served to situate *satoyama* woodlands within the vanishing cultural landscape of modern-day Japan. Accordingly, while the relationship between *sato* and *yama* was severed as a result of the fuel revolution of the 1960s, the appearance of the symbolic *satoyama* metaphor has served as a nexus for efforts to restore vanishing woodlands (Bolthouse 2007).

Amidst what Nakagawa (2004) terms a '*satoyama* renaissance', thousands of volunteer groups have formed throughout Japan to conserve and restore abandoned, overgrown woodlands. These groups have become particularly prominent since the 1990s, fuelled in part by the popular animated film *My Neighbor Totoro*, which brought the nostalgic *satoyama* landscape of the 1950s to the nation's attention (Yokohari and Bolthouse 2010). Although the volunteer groups managing *satoyama* woodlands are highly diverse, they are typically active near large urban population centres (Tokyo, Osaka, and Nagoya) and largely driven by the growing number of retirees in rapidly ageing Japan. These volunteer groups commonly aim to turn abandoned, impassable woodlands into park-like woodland spaces for recreation and community gatherings. Summarizing the outcomes of the *satoyama* renaissance, Takeuchi (2003) asserts that *satoyama* woodlands are becoming 'common' again.

27.4 The Evolution and Potential Resolution of an Anti-Commons Tragedy

Forest policy has had significant impact on the formation of an anti-commons tragedy. From its inception, forest policy in Japan has continually attempted to remedy the problem of small, fragmented ownership of woodlands such as *satoyama* (Tsutsui 2003). During the Edo period (1603–1867), most *satoyama* woodlands were owned and managed as commons. However, during the rapid drive to modernize during after the Meiji Restoration, a land tax reform assigned individual parcels to individual owners to facilitate the collection of tax revenues. The second Forest Law of 1907 tried to remedy the confusion prompted by this breakup of the commons by organizing woodland owners into cooperatives at the hamlet level. As the country moved into total mobilization for war, the Forest Law of 1939 mandated that all forest owners place their woodlands into Forest Working Plans at the level of the larger administrative district. However, this system of forced cooperatives was dissolved by the post-war American Occupation forces and replaced this with an optional system. Finally, the many Forest Laws and Revisions passed from the 1960s onwards increasingly focused on timber production in marginal mountain areas, opting not to support less profitable small-scale forestry in *satoyama* woodlands (Tsutsui 2003).

Increasingly, however, forest policy in Japan has shifted towards multi-functionality or, what is often termed, post-productive forestry. The revision of the Forest Law in 2001 exemplifies this shift. This law proposed that forests throughout Japan be divided into three zones; resource cycling forests, watershed protection forests, and co-existence forests. Co-existence forests are conceptualized as forests in or surrounding urban areas that will protect biodiversity and enhance the liveability of communities. This category represents a cautious, but nonetheless unprecedented, movement into the urban area by the Forest Agency. Most importantly, Co-existence Forest classification provides a 40 % reduction in inheritance taxes to owners who individually or collectively produce a woodland management plan for greater than thirty hectares of woodland. The landowners may then contract the work out to a management company or volunteer group or do the work themselves. In this way, the Forest and Forestry Law of 2001 has opened up a space for re-communalizing peri-urban woodlands, although one which problematically depends on volunteer labour to fulfil rather strenuous management plans.

27.5 Re-assembling *Satoyama* Woodlands on Tokyo's Urban Fringe

The following brief case study of Funabashi City, a densely populated suburb of the Tokyo metropolitan area (Fig. 27.2), situates the re-communalization of *satoyama* within broader shifts in forest policy in Japan. Similar to many of the

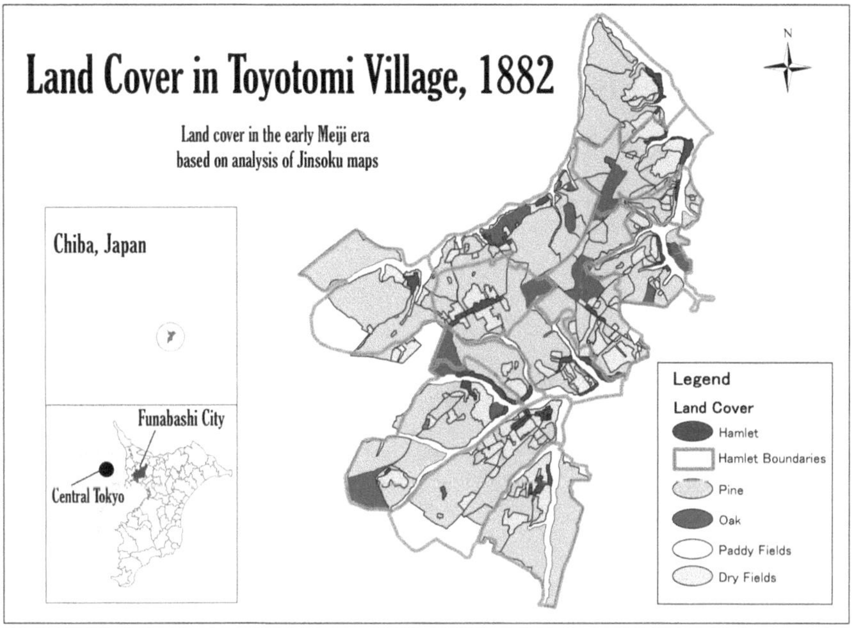

Fig. 27.2 Land cover in the former Toyotomi village in 1882

bedroom communities surrounding Japan's largest cities, Funabashi is partitioned into Urbanization Promotion (UPAs) and Urbanization Control Areas (UCAs) as a result of the New Urban Planning Law of 1968. Although this law has been criticized for failing to control development (Sorensen 2002), it has conserved large areas of farmland and woodland on the urban fringe. Following the New Urban Planning Law of 1968, the formerly rural and recently merged village of Toyotomi (an agglomeration of twelve smaller hamlets) was zoned as the UCA of Funabashi City. Figure 27.2 depicts the twelve hamlets of Toyotomi during the early Meiji era, illustrating that extensive paddy and dry fields as well as common woodlands surrounded each individual hamlet. Many of these woodlands have been developed or converted to farmland and orchards and remaining woodlands are highly fragmented, privately owned and under-managed. As a result of pine wilt disease, many have been invaded by thick bamboo grasses (*Pleioblastus chino*) and are targets for illegal waste dumping.

In Funabashi, a Forestry Extension officer of the sub-regional branch of the Forestry Agency served as the catalyst in turning neglected private forests into 'Co-existence Forests'. The Forestry Extension officer recruited volunteers from another volunteer forest management group who were looking to 'step-up' into a more challenging woodland management role. Secondly, he worked through a partially remaining network of a defunct local Forestry Cooperative to gather local woodland owners together, eventually convincing forty-three landowners to enrol ninety-six of woodland into this new scheme (Fig. 27.3).

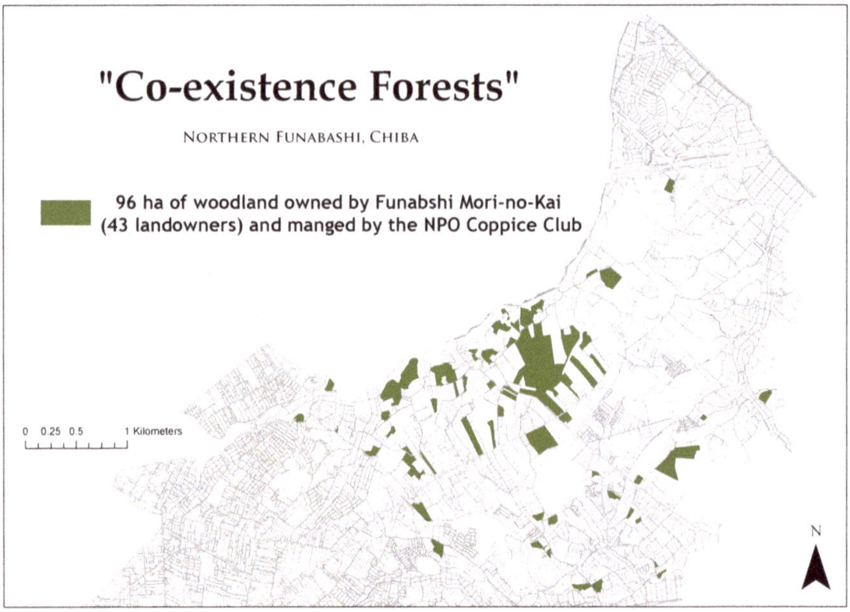

Fig. 27.3 'Coexistence Forests' in northern Funabashi city managed by the NPO Coppice Club

Fig. 27.4 At *left*, photograph of an abandoned woodland enclosed with mesh to prevent illegal dumping. At *right*, a bamboo fence installed by Coppice Club to discourage tipping

As a result of these efforts, the NPO Coppice Club was established in 2005. In contrast to more typical *satoyama* volunteer groups, Coppice Club takes a more active role in woodland management. The group meets eight times a month and each member has received certification in proper chainsaw utilization techniques. Their main activity is making woodlands passable and, in particular, clearing forests of invasive and aggressive bamboo grasses (Fig. 27.4). Woodland spaces cleared of bamboo grasses are then re-planted with deciduous oak species, which the group plans to harvest on twenty-year cycles to produce logs for shiitake cultivation. In addition, the group has also started to process and sell firewood; however, this enterprise is constrained by a weak local market for fuelwood.

27.6 Conclusions

Is the discontinuation of traditional management practices and cultural severance resulting in the end of long-existing commons and their traditions? Or is the passing of these older traditions only a precursor to the formation of new commons traditions?

The case study of *satoyama* woodlands in Japan outlined above provides evidence that older commons are indeed being reassembled in the present. It demonstrates that shifts in national forest policy have opened a potential space for local communities to claim common rights to woodlands. Although the activities of the NPO Coppice Club outlined above are not exemplary of the average volunteer group, they are illustrative of efforts to work through the anti-commons tragedy shaping socio-ecology of *satoyama* woodlands throughout Japan. Thus, it appears that Takeuchi's (2003) suggestion that *satoyama* woodlands are becoming 'common' again is accurate. However, it is important to pay careful attention to the historic and symbolic aspects of this phenomenon. *Satoyama* woodlands have become anti-commons through processes of privatization, devaluation, urbanization and the retreat of government support for small-scale and peri-urban forestry. It is the socio-cultural importance of *satoyama* woodlands that has allowed them to become an important site for reworking the relationship between woodlands and community in present-day Japan.

Hardin's well-rehearsed metaphor linking commons and resource degradation has little to tell us about the state of *satoyama* woodlands, or other underused cultural landscapes, today. Similarly, while the CPR literature provides resources for designing effective common property regimes, these designs risk being shelved without recognition of a communal interest in sustaining or reclaiming the commons. Fortunately, the socio-cultural and symbolic import of the historically shared spaces of *satoyama* woodlands has been recognized and there is a broad movement to sustain and reclaim them. Accordingly, this study also suggests that there is an important need for more research from the synthetic commons framework articulated by Olwig (2003).

Undoubtedly, the socio-cultural importance of commons, and the need to find solutions to the anti-commons, is not unique to Japan. The point made here is that part of the answer to this problem lies in, first, recognizing the common interest in the use of underused locally shared spaces and, second, discovering common schemes for sustaining these spaces. Perhaps the concept of the anti-commons provides a useful conceptual language for linking the many geographically diverse initiatives together.

References

Bolthouse G (2007) Reconstructing satoyama: metaphor and the reproduction of the vanishing Japanese rural landscape. Unpublished Master's Thesis, Department of Geography, Western Washington University, Bellingham, WA
Donahue B (1999) Reclaiming the commons: community farms and forests in a New England Town. Yale University Press, New Haven
Glassman J (2006) Primitive accumulation, accumulation by dispossession, accumulation by 'extra-economic' means. Prog Hum Geogr 30(5):608–625
Hardin G (1968) The tragedy of the commons. Science 162(859):1243–1248
Harvey D (2003) The new imperialism. Oxford University Press, Oxford
Heller M (1998) The tragedy of the anticommons: property in the transition from Marx to markets. Harv Law Rev 111(3):621–688
Heller M (2008) The gridlock economy. Basic Books, New York
Hollander G (2004) Agricultural trade liberalization, multifunctionality, and sugar in the south Florida landscape. Geoforum 35(3):299–312
Ivy M (1995) Discourses of the vanishing: modernity, phantasm Japan. University of Chicago Press, Chicago
Katz C (2001) Vagabond capitalism and the necessity of social reproduction. Antipode 33(4):709–728
Marx K (1976) Capital (Fowkes B, Trans), vol 1. Penguin, London
McCarthy J (2005) Commons as counter-hegemonic projects. Cap Nat, Social 16(1):9–24
McKean M (1992) Management of traditional common lands *(Iriaichi)* in Japan. In: Bromley D (ed) Making the commons work: theory, practice, and policy. Institute for Contemporary Studies Press, San Francisco
Mukhija V (2005) Collective action and property rights: a planner's look at the dogma of private property. Int J Urban Reg Res 29(4):972–983
Nakagawa S (2004) Mori-Tsukuri Tekisubukku. Yama to Keikokusha, Tokyo. (in Japanese)
Olwig K (2003) Commons & landscape. In: Proceedings from a workshop on new and old commons, Centre for Advanced Study, Oslo 11–13 Mar 2003
Ostrom E (1990) Governing the commons: the evolution of institutions for collective action. Cambridge University Press, Cambridge
Perelman M (2000) The invention of capitalism. Duke University Press, Durham
Polanyi K (1944) The great transformation. Beacon Press, Boston
Schlueter A (2008) Small scale European forestry, an anticommons? Int J Commons 2(2):248–268
Shidei T (2006) Shinrinwa Mori ya Hayashi Dewanai: Watashi no Shinrinron. Nakanishiya, Kyoto. (In Japanese)
Sorensen A (2002) The making of urban Japan: cities and planning from Edo to the twenty-first century. Routledge, New York
Takeuchi K, Brown RD, Washitani I, Tsunekawa A, Yokohari M (eds) (2003) Satoyama: the traditional rural landscape of Japan (9–16). Springer, Tokyo
Takeuchi K (2003) The nature of satoyama landscapes. In: Takeuchi K, Brown RD, Washitani I, Tsunekawa A, Yokohari M (eds) Satoyama: the traditional rural landscape of Japan. Springer, Tokyo, pp 9–16
Thompson EP (1993) Customs in common: studies in traditional popular culture. The New Press, New York
Totman C (1989) The green archipelago: forestry in preindustrial Japan. The University of California Press, Berkeley

Tsutsui M (2003) The formation of Forest Culture in the twenty-first century: towards a paradigm shift in forest policy. The Association of Forest Culture, 22–26 (in Japanese)
Vasudevan A, McFarlane C, Jeffrey A (2008) Spaces of enclosure. Geoforum 39(5):1641–1646
Yokohari M, Bolthouse J (2010) Don't freeze it, keep it alive: a conceptual perspective on the conservation of continuously evolving *satoyama* landscapes. Landsc Ecol Eng 7(2):207–216

Chapter 28
New Commons for Old: Inspiring New Cultural Traditions

Duncan Mackay

28.1 Inspiring New Traditions from the Old Commons

Just like the story of Aladdin's lamp, there has been a long history of commoners being double-crossed and commons being tricked away from their local communities. This often had severe impacts on self-sufficient peasant farmers who depended upon common land and rights of common for their livelihoods, as the following case-study of Otmoor shows. These rights of common, especially those without stint, were a unique historical golden thread to a time '*out of the memory of man*'. Such is the ancient tradition of common land. This chapter explores some of the reasons why the traditional elements of commons have been worth fighting for in the past and why this cultural tradition is still vitally important today to inspire future generations to care for, be part of, and, therefore, be proud of their local places. The paper suggests that the opportunity exists for everyone to become involved in their neighbourhoods as 'new commoners' and that new local commons could be established for the long term benefit of people, places and nature. These could be products of the UK Coalition Government policies[1] to create a 'Big Society', to instigate a tree-planting campaign and wildlife corridors, and to create a new designation, similar to Sites of Special Scientific Interest, to protect green areas of particular importance to local communities, as well as protecting the Green Belt and other designations.[2]

[1] Coalition Government '*Building the Big Society*' May 2010.
[2] Coalition Government '*The Coalition: our programme for government*' Cabinet Office May 2010.

D. Mackay (✉)
Natural England Estates Team, Block B, Whittington Road,
Worcester WR5 2LQ, Natural England
e-mail: Duncan.Mackay@naturalengland.org.uk

I. D. Rotherham (ed.), *Cultural Severance and the Environment*,
Environmental History 2, DOI: 10.1007/978-94-007-6159-9_28,

In the last two millennia common rights became an important democratic statement between state and individuals and so intriguingly bound up with the fundamentals of life that they became an early part of the essential legal fabric of English Law. The Statute of Merton of 1235 followed not long after the Great Charter (Magna Carta) of 1215 and the Forest Charter (Carta de Foresta) of 1217. In the six centuries following the Statute of Merton, common land, common rights and common fields were the metronome of rural agricultural and economic life ruled by weather, seasonal customs, courts leet and other regulatory mechanisms. These regular, everyday traditions formed the background rhythm that ticked on for centuries despite disruptions by wars, famines and plagues.

Common land has just about survived the enormous political, economic and social transitions of the past centuries. This survival contains important clues about what we intrinsically value about commons that shows how they could be reinvigorated and re-established for the benefit of current and future generations. This chapter proposes to create a new tradition out of the old.

In illustrating the powerful democratic attraction that commons hold for us this paper will describe two examples from the nineteenth century when local commons were threatened by an avalanche of Private Acts of Enclosure. These Acts of Parliament were dealt with wholly in Westminster and therefore agreed amongst like-minded, propertied 'gentlemen' with only one intention—for collective self-aggrandizement. Common land was simply added to their estates and emparked within walls or fences and commoners expelled; job done. In many places, there was little resistance by the dispossessed to this iniquitous land grab but gradually opposition hardened into riotous assembly.

28.2 The War for Otmoor 1814–1830

The common grazing land of Otmoor in Oxfordshire, was a naturally wet and marshy landscape that supported commoners' geese and cows as part of the local subsistence agriculture and '*had been used as a public common without stint…–from remote antiquity*'.[3] Yields per cow might have been low but that was not the point; the daily milk helped make the difference between family success and family failure with butter being sold at markets and milk fed to children. The geese were endlessly self-replicating and simply grazed the wet meadows where '*a single cottager sometimes cleared as much as £20 a year by geese*'.[4] The geese were the vital cornerstone of this cottage economy and supported many families in the surrounding hamlets, '*the greatest benefit was reaped by the cottagers, many of*

[3] Dunkin *History of Oxfordshire* (*The history and Antiquities of the Hundreds of Bullingdon and Ploughley*) 1823.

[4] Jackson *Oxford Journal* 11 September 1830.

whom turned out large numbers of geese, to which the coarse aquatic sward was well suited, and thereby brought up their families in comparative plenty'.[5]

The landed gentry and Lords of the Manor, however, saw Otmoor quite differently from the commoners, '*this dreary waste was surveyed with longing eyes by the surrounding landowners, most of whom wished to annex a portion of it to their estates, and in consequence spared no pains to recommend the enclosure as a measure beneficial to the country*'.[6] They were egged on by the prevailing, sneering journalism of writers like Arthur Young in his *Survey of Oxfordshire for the Board of Agriculture* in 1809 '*I made various inquiries into the present value of it by rights of commonage; but could ascertain no more than the general fact, of it being to a very beggarly amount…upon the whole, the present produce must be quite contemptible, when compared with the benefit which would result from enclosing it*'. The establishment looked down on commoners generally, '*in looking after a brood of goslings, a few rotten sheep, a skeleton of a cow or a mangy horse, they lost more than they might have gained by their day's work, and acquired habits of idleness and dissipation and a dislike of honest labour…God did not create the earth to lie waste for feeding a few geese, but to be cultivated by man, in the sweat of his brow*'.[7] However, even Young later came to recognise the rising social cost of impoverishment through enclosure and the consequential cost on landowners through parish relief rates. It mattered naught, even as Young grappled with the long term cost of ripping up a self-sufficient peasant economy and replacing it with something naggingly dependant on wealth creation and taxes.

Preparations for the Inclosure of Otmoor were made by those who owned land around it but the first Inclosure attempt failed. In November 1814, a new application was made after notices had been affixed to church doors in August although '*it was found impracticable to affix the Notices on the church doors of the other two Parishes on that day, owing to large Mobs, armed with every description of offensive weapons*'.[8] This display did not alter the outcome of Inclosure and despite a last ditch attempt to petition Parliament armed with an official report (from the Keeper of the Records in the Augmentation Office) that stated unequivocally that the soil of Otmoor was *not* owned by *any* Lords of *any* Manors the Bill was passed by Royal Assent on 12 July 1815. Four thousand acres of Otmoor were thus tricked away, divided amongst the main local landowners, Lord Abingdon, Sir Alexander Croake, Duke of Marlborough and others, fenced off and drained.

In 1830, after 15 years of contemptibly miserable crop yields, the drainage failed. The River Ray flowed onto adjacent farms and some twenty-two local farmers took the law into their own hands by breaking the enclosure dykes and flooding Otmoor once again. The defeated commoners took this as a sign that the Enclosure was illegal and sought to regain their lost rights of common. Fences

---

5 Dunkin ibid.

6 Dunkin ibid.

7 *Oxford University and City Herald* 25 September 1830.

8 *House of Commons Journal* 17 February 1815.

were cut, the new landowners remonstrated with loaded pistols and bones were broken. Local sympathy with the commoners was so strong that special constables refused to be sworn in; so the High Sheriff summoned the Oxfordshire Militia and Lord Churchill's Yeomanry Cavalry. The commoners, however, were determined to perambulate the moor on its seven mile boundary in the old tradition on Monday 6th September. Early in the day around 1,000 local men, women and children armed with billhooks, hatchets and reap-hooks marched to destroy any fences in their path and by noon the destruction was complete. At this point Lord Churchill's Yeomanry arrived and read the Riot Act but the moor-men refused to disperse and forty-four were forcibly arrested and sent to Oxford Gaol in wagons with a Yeomanry escort.[9]

The lamentations of the moor-men were palpably felt by all who had been touched by this injustice and, as chance would have it, Oxford was enjoying one of the town's great traditional feasts, St Giles' Fair, and was already bustling when the prisoners of Otmoor were dragged into town. Cries of "*Otmoor for ever*" soon carried to the fairgoers who besieged the wagon train, fought off the guards and released the prisoners. Many rioters were later tracked down by stronger military forces and tried at the assizes but the jury asked for clemency and the longest sentence was 4 months.[10] On Otmoor itself, the commoners turned out their cattle and geese once again through the broken fences and formed an Otmoor Association, which boldly declared, '*the Right of Common on Otmoor was always in the inhabitants, and that a non-resident proprietor had no right of Common thereon*'.[11] Today Otmoor is still a wet grazing marshland and RSPB reserve partly supported by EU grant aid for agricultural conservation and the spirit of the traditional land use lives on although in a highly diminished form.

28.3 The Battle of Epping Forest

The case that partly stimulated the founding of the National Trust, one of the world's foremost conservation bodies, took place in the late nineteenth century in Epping Forest on the Essex fringes of London. It involved a tradition to lop or pollard forest trees for winter fuel. It was an ancient customary common right and attended by much ceremony including, bonfires, beer drinking and, as, the church bells struck midnight, lopping commenced until two in the morning. The rights were current from the eve of St Martin's Day (11th November) until St George's Day (23 April). The first loppings, in heaps six feet high, were traditionally dragged from the forest on sledges by a team of white horses at dawn.[12] Through

[9] Hammond and Hammond *The village labourer 1760–1832* Longmans 1912.

[10] Jackson *Oxford Journal* 5 March 1831.

[11] *Oxford University and City Herald* 11 September 1831.

[12] Eversley, Lord, *Commons, Forests and Footpaths*, Cassell 1910.

tradition the right to lop had meaning and through meaning it was maintained by local people through time. However, by 1865, the year the American Civil War ended, the ancient, straggling, wood-pasture common that was Epping Forest came under threat of total enclosure and disafforestation from disreputable Lords of the Manor and the Crown both intent on making money. The aim of the enclosers was to take the timber, rip out the tree-roots and turn the land to arable production. All they needed to do was make the Crown an offer, buy out or threaten off any of the remaining commoners and proceed. In 1793, the Report of the Land Revenue Commissioners had stated that Epping Forest comprised 9,000 acres of open land and was already being visited by Londoners for recreation but by the time of the Report of the Committee on London Commons in 1865 this had been reduced to just 3,000 acres. Thus, most of the forest had gone but in 1865 there was one unforeseen obstacle to any further enclosure; the newly formed Commons Preservation Society.[13]

The Commons Preservation Society solicitor, Robert Hunter, recognised the weaknesses of the law for anyone without a legal interest in the specific common under threat to successfully oppose its Inclosure Act. However, when the Forest in Loughton, Essex, was enclosed by a farmer-parson, the Reverend Maitland, fate took a hand. A commoner called Thomas Willingale and his two sons broke through Maitland's fences, to exercise their right to lop on St Martin's Eve 1866. '*For this act, in vindication of their rights the three Willingales were summoned by the Lord of the Manor before the local justices…they were convicted of malicious trespass on private property and sent to prison for 2 months with hard labour…one of Willingale's sons was put into a damp cell…caught pneumonia, and resulted in his death…These high-handed proceedings caused great indignation in the district and in the East of London*'.[14]

The Commons Preservation Society quickly raised £1,000 to fight Maitland's enclosure and won an injunction to stop him felling the trees. When old Willingale came out of prison Maitland meanly saw to it that he could find neither work nor lodging in the manor and even allegedly offered £500 to bribe him against pursuing the lawsuit. The Society countered this by paying Willingale a pound a week so that he could enjoy residency and thereby maintain his claim as a resident of the manor. Sadly, in 1870, before the case came to court, Thomas Willingale died. Hunter now had a problem, without a common right holder he had no locus in law to act as a proxy to oppose the destruction of the Forest. Hunter redoubled his efforts and somewhat to his surprise found that not only did commoners have rights to their own manors but had customary rights to the *whole* of the Forest. Importantly he also discovered that the Corporation and City of London owned land in the forest that had rights attached to it. At first the Corporation was

---

[13] Eversley ibid. The Commons Preservation Society was a redoubtable group of social reformers, philanthropists, lawyers and liberal thinkers including WH Smith, Sir Charles Dilke, James Bryce, John Stuart Mill, the Duke of Westminster, Octavia Hill, Robert Hunter (later to be knighted) and John George Shaw-Lefevre (later to become Lord Eversley).

[14] Eversley ibid.

indifferent but eventually agreed to contest the case. The case lasted nearly 10 years of expensive legal jousting, of claim and counter-claim, of court-rolls and obscure principles of English law going back to Magna Carta, the Statute of Merton and even into Anglo-Saxon records of the origins of commons and forests. Hunter and the Corporation eventually won[15] but he must have silently vowed to never do this again. England's great commons could not be saved one by one, particularly if they involved 10 year legal battles on each occasion.

A few years later, in September 1884, Robert Hunter presented a paper[16] to the National Association for the Promotion of Social Science in Birmingham entitled '*A Suggestion for the Better Preservation of Open Spaces*'.[17] Hunter's principal suggestion was the need for a body to acquire land with commoners rights to establish a legal interest in commons and thereby to oppose any attempts at enclosure '*But the purchase of a Manor is as impossible for the Commons Preservation Society, or for any other association possessing no corporate existence, as the purchase of a farm or property entitled to Common rights. Some new agency must be called into existence in order to afford this species of protection to the public*'.[18] Hunter, ever practical, continued with the specific means by which commons could be saved for the nation as open spaces. '*The remedy which I suggest for these defects in the existing means of protecting open spaces will have occurred to those who have followed me so far. I advocate the formation of a Corporate Company…the main functions of this Company would consist in:*

(1) *The acquisition and holding of properties to which common rights are attached*
(2) *The acquisition of manors, manorial rights, commons and other wastes of manors, downs, moors and other open spaces*
(3) *The acquisition of square gardens, dis-used churchyards and burial grounds, and any other spaces suitable for town gardens, and of any buildings connected with such spots.*[19]

---

[15] Eversley ibid. It ended happily for Epping Forest after the Queen had put her seal to the Epping Forest Act. '*The Forest was thrown open to the public by Queen Victoria in person, at High Beech, in the presence of a great assemblage of people on May 6th 1862. Restitution was thus in a sense made by the Sovereign, of land which in very ancient times had probably been taken from the folkland for the purpose of a Royal Forest, and the Forest was dedicated for ever to the use and enjoyment of the public*'.

[16] Hunter '*A Suggestion for the Better Preservation of Open Spaces*' Annual Congress of the National Association for the Promotion of Social Science, Birmingham September 1884 (Reprinted by the Commons Preservation Society 1884).

[17] Hunter ibid. It was this paper that attracted the attention of Octavia Hill who was trying to acquire for the nation the garden at Sayes Court in Deptford, the London garden created by the diarist John Evelyn in the seventeenth century. In the ensuing correspondence between them Hill suggested that a '*Commons and Gardens Trust*' should be formed…over which Hunter prophetically wrote in pencil the alternative title '*National Trust?*' It was another 10 years before the National Trust actually came into existence and by then the owner of Sayes Court had sold up.

[18] Hunter ibid.

[19] Hunter ibid.

Hunter concluded his paper with the words '*I may add, however, that it is capable of being adopted locally as well as generally. There might be a Company for Birmingham and the neighbourhood, for a county, or for a district like the Potteries…the central idea is that of a Land Company, formed not for the promotion of thrift or the spread of political principles, and not primarily for profit, but with a view to the public interests in the open spaces of the country. An experience of nearly 20 years in defending open spaces has convinced me that an organisation such as I have suggested would cure a radical defect in the existing machinery for the purpose, and would supply a most valuable means by which to give effect to public opinion on the subject of open spaces*'.[20]

That idea for a common land-owning Company is now the National Trust.

28.4 New Commons for the Climate Change Generations

Today, we can still enjoy the fragmented remnants of this golden historical thread linking back to the time of the Normans, Anglo-Saxons and possibly even earlier. Seen in that sense, this immense historical thread *alone* might be a concept worthy of UNESCO World Heritage Site status for all common lands still in existence today. This is a global tradition that is now a precious rarity.

The *real* tragedy for commons,[21] as Hunter and Hill recognised, was the loss of land for nature-friendly, sustainable, local economies based on a democratically regulated formula that fed, furbished and spiritually nourished millions of people. For the climate change generations, this is a tradition that we might need to rediscover and quickly.

This chapter is not intended to create a pantomime out of a tragedy but to make a play for new cultural traditions and places that embody all the best principles and democratic practices of ancient common land systems and relights the flame of this sublime idea.

Encouraged by Elinor Ostrom's work on common pool resources,[22] this chapter simply re-thinks the quintessence of commons for the imminent climate-change generations and the survivors of the obesity decades. It suggests the need for the creation of 'new commons' around *all* communities as part of climate-change adaptation for people and wildlife and, in particular, through woodland planting, giving direct action to 'heat-island effect' mitigation measures for cities. It draws

---

[20] Hunter ibid.

[21] Hardin in *Science* 162 1968 pp. 1243–1248.

[22] Ostrom et al. '*Rules, Games and Common-Pool Resources*', 1994 University of Michigan Press.

on the principles of Natural England's work on inspiring and providing access for people[23] and Defra's policy shift towards urban fringe forestry.[24]

28.5 Community Orchard-Commons

In 1996, the author was awarded the Henry Ford European Conservation Award for Heritage. The proposal that achieved this success was for 'community orchards' with inherent common rights to take fruit. The concept was that such orchard-commons could be at the centre of community life as spatial social enterprises. Everyone in the ward or parish would enjoy common rights to take fruit, make apple juice and create other products. Planned to be at widely spaced traditional orchard planting ratios there would be lots of space underneath standard trees for all the simple but cohesive community activities that bind people and places together. Activities such as picnics, children's games, forest schools, foraging, Maypole dancing, harvest festivals, Apple Days and outdoor local celebrations for births, deaths and marriages could be encouraged.[25] Each child born would have a fruit tree planted for them to add to the connections between place and people through time and develop a true sense of personal meaning. With certified organic standards of care the orchard commons would become havens for disappearing wildlife such as stag beetles, epiphytes and woodpeckers.[26] Bee hives would thrive and honey would be a new local product. Each common right holder would have a real stake in their community thereby adding to local democratic engagement and understanding.

28.6 The National Trust, Local Land Companies and New Commons

In 2010, the National Trust launched its new strategy for the next decade.[27] It is a bold and inspirational document and begins with this sentiment

> '*Things that really matter become more obvious in tougher times. Instead of material wealth or status, we take comfort in family and community, places we love, the*

---

[23] Natural England *Natural England Position on Access*, May 2009; *Natural England Position on Inspiring People to Value and Conserve the Natural Environment*, October 2009 www.naturalengland.org.uk.

[24] DEFRA '*The Strategy for England's trees, woods and forests*' June 2007.

[25] Hall et al. '*Apples, Berkshire, Cider*' Two Rivers Press 1996.

[26] Mackay '*Apples and Pears: Stairways to an Urban Paradise?*' Urban Nature Vol 4 No 1 1998.

[27] National Trust '*Our Strategy for the Next Decade*' 2010.

> *appreciation of beauty, fresh air, and a sense of kinship, with each other, with the past and with the natural world*'.

Sitting alongside Sir Robert Hunter when the National Trust was founded was Octavia Hill; she was an astute and highly active social reformer but took as her cue some simple principles about democracy and the essential goodness of human beings and natural beauty. She was clear what special places like commons and town gardens meant to her and how they should be looked after for everyone, rich and poor, city and country dweller, young and old…'*for the everlasting delight of the people*' was the way she saw it. She was closely involved in the attempt to rescue Swiss Cottage Fields from development when that part of nineteenth century London was at the edge of the rapidly developing city. She failed, but the force of the sentiment still holds good today. We need new commons and we need them close to where people live.

The National Trust has recognised the new social reality that 80 % of the population live in towns and cities and are disconnected from the natural environment and its properties

> '*What we care for belongs to us all. It unites us. Yet over time, vital links have been frayed, for instance between our largely urban nation and its countryside…*'

And bravely it wants to do something about it

> '*Our charity has a role in helping us all to celebrate Britain's mosaic of fields and fells, villages, seashores and buildings great and small. We want to do so on a more personal basis, relishing the identities of the places we love the most.*'

This has begun to reveal itself in new allotments and community farms on Trust owned land.

The National Trust aptly describes itself as a '*rare thread of continuity in an unstable world*' but '*the durability of that thread depends upon our charity's capacity to meet people's changing needs*'. In its founder's words "*new occasions teach new duties*".

My understanding of Hill and Hunter's character suggests that if they were both alive today she would be campaigning for the new dispossessed and he would be devising mechanisms to create smart design solutions. In the introduction to Lord Eversley's book *Commons, Forests and Footpaths* in August 1910, Hunter wrote '*The {Open Space} movement…is animated by two ideas—one, that the people of this country should have some interest in the land of the country, the other, that the amenities of everyday life should be placed within reach of rich and poor alike…*'. The '*new occasion*' of modern times is that 40 million people live in towns and cities in England and the diseases of obesity, stress and inactivity are increasingly killing us off. Economic stringency is coinciding with the inevitability of climate change impacts and our built up areas will become potentially lethal heat-sinks before the end of the century if more trees are not planted within them and around them. These are not problems caused by the National Trust or solvable by them but I would like to think that these '*occasions*' might teach us '*new duties*'. One of the ways to address these changing needs and to answer Hunter's two animated ideas for an interest in land and access to its amenities is to establish local Land Companies to create 'new commons' in and around every population centre in the country.

The places that most of us love are the places nearest to us, the areas surrounding our homes, our streets, our schools, and the places where we work or seek relaxation. These places are where local identity and character are set and where we spend most of our time. However, very few of us are lucky enough to enjoy doorstep access to a National Trust property or a National Park. This is where Hill and Hunter would have pinpointed the problem and sought new solutions. '*For ever, for everyone*' is the Trust's strap-line and new commons in and around every community would certainly fulfil that clear ambition. Add in the triple-bottom line functions of trees for cooling, space for 'air and exercise' or local food, and, habitats for wildlife; and the land around our habitations could be transformed.

National Trust land is held inalienably which means it is virtually impossible to remove it as it requires both Houses of Parliament to agree to such a course of action (National Trust Act, 1909). This is probably the highest form of conservation property protection in the world. It is an excellent tool to secure permanence if local Land Companies *franchised* by the National Trust perhaps were to be established to create new commons. However, it would simply not be possible for the National Trust alone to manage such a potentially large estate in and around the fringes of every settlement in England. But, in combination with the UK Coalition Government's 'Big Society' initiatives, a partnership combination of *nominal* National Trust ownership, local benefactors offering land, local people and local management could actually address an enormous gap in provision *and* answer Hunter's two 'animated ideas' from 100 years ago. Indeed, Hunter foresaw the potential for *local* Land Companies and the potential for the expansion of new commoners.[28]

28.7 New Commons Created Through 'Big Society' and 'Localism'

Supported by the evidence from Elinor Ostrom's Nobel Prize winning economic work on common pool resources and the ability of communities of interest to successfully regulate themselves the opportunity to test the theory is overdue. Today this 'new commons' mechanism is not only feasible but with the new Coalition Government's policies for 'Big Society *not* Big Government' and

[28] Hunter ibid '*So long as sufficient land was kept to support beasts to turn onto the common, the property could be worked for a profit by the Company in the same way as by any private person. Farms and other properties might be let, and portions of them might be sold, with suitable covenants as to the exercise of Common rights. Thus an income could be derived, while at the same time a body of active Commoners would be by degrees created*'.

'localism' gaining greater foothold in thinking and action there is a new impetus to harness it. This could aid both the power of people (such as new Neighbourhood Groups) to act for their locality and to work in new partnerships with philanthropists, landowners, businesses, voluntary bodies, state agencies and local authorities. It does need a focus however, it needs land to kick start it with a real sense of community engagement and this may rely upon philanthropic giving. In the Victorian era, many philanthropists stepped forwards such as Sir Henry Peek to preserve Burnham Beeches or Augustus Smith to liberate Berkhamsted Common,[29] and something similar may be required today if this idea is to gain full expression during decades of austerity. The Coalition Government's programme for a tree planting campaign and the Conservative manifesto statement on the need to double the amount of tree-cover by 2050, require land to be realised. Added to the programme's intention to create a new designation to protect much loved local green space, these measures could coalesce neatly into the concept for 'new commons' and truly make a step change that would be of historical importance.

[29] Eversley ibid.

Chapter 29
Community Grassland Conservation on a Former Common in the Wye Valley, England

George Peterken

29.1 Introduction

For the last decade, two community-based grassland conservation projects have been operating in the Lower Wye Valley, the Parish Grassland Project in the communities of Brockweir, Hewelsfield and St Briavels (Glos.), and the Monmouthshire Meadows Group, which is active principally in eastern Monmouthshire (Peterken and Tyler 2006). The former started in 2001 as a response to problems caused and anticipated from the outbreaks of Bovine Spongiform Encephalopathy and Foot-and-Mouth disease; the latter in 2002 as an extension 'across the Wye' of the former.

The PGP has operated mainly in the Hudnalls, once a large wooded common forming an extra-parochial district of the Forest of Dean. Until the eighteenth century, this common had been a mosaic of bilberry-rich rough grassland and scrub woodland used informally by local people as grazing and a source of wood. However, in the late 1700s, the core of the Hudnalls was cleared and settled, leaving woodland only on the slopes overlooking the Wye. The woods are still subject to common rights—which since 1945 have rarely been exercised—but the majority of the land quickly became a distinctive tract of narrow, winding lanes separating tiny fields and large gardens, bounded by hedges, dry stone walls and numerous hedgerow trees, all free of common rights. The scattered stone-walled houses of the original colonisers have now been much extended and many more have been added, but today the Hudnalls is still a district of low-density housing, hidden amongst trees, overgrown hedges and irregular walls in various states of repair. Perhaps because of their shared history as an ancient common, the residents believe that they have a strong sense of location and community.

G. Peterken (✉)
Beechwood House, St Briavels Common, Lydney GL15 6SL, UK
e-mail: gfpeterken@tiscali.co.uk

I. D. Rotherham (ed.), *Cultural Severance and the Environment*,
Environmental History 2, DOI: 10.1007/978-94-007-6159-9_29,

Being so small (average field size of 0.45 ha), the fields largely escaped the modernising impacts of post-War farming, but were maintained as small holdings by long-term residents who also did other work, or they became horse-paddocks and large lawns. Many in-comers have bought houses and found that they own fields as well, and these fields have often been let to real farmers and small holders, usually by informal, word-of-mouth arrangements backed by small mutual benefits in kind.

Under these arrangements, perhaps 60 % of all the grassland remained semi-natural or only partly improved, supporting a somewhat acid version of MG5 grassland, diversified by springs and small streams. Several fields have locally uncommon plants, such as *Genista tinctoria*, *Orchis morio*, *Carex pallescens* and *Lathyrus liniifolius*, and many have quantities of *Rhinanthus minor* and *Dactylorhiza fuchsii*, but none has been judged to be worthy of SSSI status. If they are to survive as flower-rich grasslands, it will be because residents continue to make moderate use of the existing sward to mow hay, or graze horses, sheep or cattle, without the backing of any special designation.

When BSE and Foot & Mouth diseases struck with their attendant bureaucracy and restrictions, we faced breakdown of the informal arrangements that had kept the landscape intact. Fields, we feared, would be overgrown with bramble, bracken and blackthorn, our houses would increasingly be confined to clearings in developing woods, and our grand views of the Wye Valley would be obscured. Rather than leave these anticipated problems to develop, a group of residents called a general meeting to discuss setting up a parish organisation whose aims would be (i) to help residents to maintain their fields as flowery grassland, and (ii) to increase interest in, and knowledge of, their surroundings. In the event, over 60 people attended, and we agreed to establish the project.

29.2 What Have We Done?

Since 2001, membership has been maintained at around eighty households, each paying a £5 annual subscription. Indoor lecture/discussion meetings have attracted 20–60 people, which is good for an unconventional parish organisation. The best attended meetings have been on wildlife, particularly bats and butterflies. Meetings on local orchards and boundary trees, which we set up as mini-symposia of several 10 min contributions mainly from residents, were also popular. Field meetings—literally—have attracted up to 25 people to hear how particular residents use and enjoy their fields. Three open days attracted many interesting visitors, some from far-afield, but the weather was never ideal and the overall numbers only just made it worthwhile. The subsequent parties for helpers were far more successful! We have also run a wonderful flower-hunt and butterfly-chase for the local primary school.

With a graphic designer on the committee, we have been able to establish an attractive website (www.parishgrasslandsproject.co.uk) but this has attracted less interest and activity than we hoped. Far more successful was our 44-page illustrated booklet (Peterken 2005), *Flowers in the Fields*, and several local walks leaflets. All these have sold well at the village shop. Our *Newsletter*, published twice a year, has also been very well received (see website).

Initially, we helped several residents to apply successfully for entry to the Stewardship Scheme, but thereafter a major effort was put into acquiring machinery for grassland management. With the help of the HLF, Wye Valley AONB and Forest of Dean DC, we bought a mini-tractor and associated haymaking equipment suitable for work in our very narrow lanes, and small, irregular fields (where modern, large machinery sometimes cannot reach). The machinery has been operated by a local farmer, and the work programme was arranged by one of the committee. Anyone who wanted work done on their fields had to be a member of the project, and the annual operations had to be financially self-supporting after we had paid the farmer a reasonable rate for his time. We have also occasionally arranged applications of green hay, and tuition on looking after sheep, which are now the main grazers.

29.3 Has it been Worth the Effort?

The original *raison d'être* vanished surprisingly quickly: the diseases abated and stock returned, but the bureaucracy remained and sheep have largely replaced cattle. The PGP has helped to maintain flower-rich grassland on a few properties where it might otherwise have been neglected, but most land continues to be managed, or not managed, without any influence of the PGP. In fact, in the last decade, several more fields have been lost to total neglect and a new wave of mansion-building. No attempt has been made to tell anyone what they should do with their land—the PGP philosophy has been to help if asked—and we would have been told to get lost if we had tried. We can never measure what would have happened on the ground if we had never existed, but my impression is that we have had only a limited direct impact on the environment.

Our management service has been an education. It has depended on two individuals, Peter Chard, the motivator who organised the management programme, and John Childs, who operates and maintains the machinery, but who has been increasingly occupied by his local-breed cattle and pigs operation. It worked well enough for 3 years, but in 2007–2009, it was grossly inconvenienced by the prolonged wet summers. Whilst there have been many satisfied customers, some have been disappointed by what we regard as unreasonable expectations. These include that one can put haymaking in the diary irrespective of weather, that hayfields will be left as tidy as a lawn; that field owners can still sell their hay, even if they leave ragwort to proliferate; or that the time of a skilled farmer should be paid at less than the minimum wage. Ironically, one field owner failed to sell her hay after we cut it because the person who said he would buy it was convinced that the St John's Wort—a natural component of the sward—was actually ragwort.

He would not be persuaded otherwise by the botanists on the committee and so we were left with a dissatisfied field owner, a disappointed potential hay user; and a farmer who had wasted his time cutting the hay.

Peter Chard died last year, but fortunately Mike Topp joined the committee and has become the new organiser, so the service can continue. We roughly break even on annual operations, but we have been unable to charge enough to be able eventually to replace the machinery. Having tried, we now think it would be more practical, and inherently better, to direct work to local farmers and small-holders who already have their own machinery, even though this may mean 'writing off' the most awkward and least-accessible fields.

If our influence on the land has been marginal, considerable general interest has been developed in parts of the local community, and many individual residents have sought advice about their fields or about keeping stock. The environment in general and wildlife in particular have received beneficial publicity, and we hope that some of the local children will develop an interest in and enthusiasm for wildlife and farming. Several individuals and organisation with a specialist interest have paid visits, and perhaps we have had some influence elsewhere.

29.4 Community Spirit?

So, has the shared history as a community founded on an ancient common made any difference beyond the fact that the tiny fields amongst which we live owe their condition as semi-natural grassland to their unsuitability for modern agriculture? Socially, our modern society is leavened by several families that have lived on what was the common for many generations (and still may call themselves commoners), and some of these remain as part-time stock-keepers with other jobs. Otherwise, most residents are now incomers with an urban or sub-urban background and source of income, some of whom happily adopt a version of 'the good life', but others of whom have no interest in the land—or, indeed, the community.

The shared history as a common now has very little resonance. Even when the management of the remaining common woodland comes up for debate, it is hard to find representatives of the right-holders, and incomers only take interest when any felling is proposed. As for community spirit, there is almost no sign that residents turn up at meetings or help with activities for anything more than personal interest and a need for social contact—there is no residue of the community spirit and mutual obligations that once brought out neighbours to help with the haymaking. For example, in 2007, many people were interested in restoring old orchards, or establishing new 'traditional' orchards, and we had excellent discussions at two meetings on the topic, when forty or more residents shared their experiences of orchards, but no-one could be found to lead on establishing a 'parish orchard' when this was mooted. The main exception is the PGP committee itself, which has enjoyed a good balance between stability and turnover. Members have all made valuable contributions, partly out of a sense that they are contributing something worthwhile to local life, but also because each has a personal interest in some aspect of our aims and activities. Enlightened self-interest is the tenor of the times, and the residents of our parishes are no exception.

An interesting comparison can be made with the Monmouthshire Meadows Group, our 'sister' organisation across the Wye. This covers a much wider area and so there can be no parish-based sense of community. Instead, it draws more on a wider social community, made up of individuals, mostly with a professional background, who have moved into the countryside and take an interest in maintaining their surroundings: they are more likely to meet at a local concert or art exhibition than at the local pub. The MMG has maintained group cohesion by meetings that combine talks with good eating and pleasant chatter—a dining club with grassland overtones—but it has also helped many individuals to maintain and improve their herb-rich grasslands.

29.5 Outside Help

Although we portray ourselves as 'grass-roots' organisations, the only community-based grassland conservation organisations without a paid organiser, we have been supported by grants from the Heritage Lottery Fund, the Wye Valley AONB, and local authorities. Staff of the AONB office and of DEFRA (now Natural England), have been a great help in several ways. Gwent Wildlife Trust has collaborated closely with Monmouthshire Meadows to their mutual benefit. We have, however, had only limited contact with the Gloucestershire Wildlife Trust; and the Grassland Trust, despite paying us a visit, remains an enigma. Tantalisingly, a specialist in European grasslands, a consultant surveyor for agri-environment schemes, a leading scientist in a conservation 'quango', a former researcher at the Grassland Research Institute, and the former organiser of the Gwent Grassland Project all live within our parish, but all for various understandable reasons have not become involved.

29.6 Comment

We previously took the view that the two grassland projects would probably expire when the core group of enthusiasts and motivators could not be replaced, and this still seems the best prediction. The PGP committee lost its primary enthusiast and motivator with Peter Chard's death in 2009, but we have fortunately added two new enthusiasts and can continue, albeit with some changes in emphasis. The MMG will also have to replace core motivators shortly, and there is some prospect of them doing so. Realistically, all local organisations depend on a few individuals, and unconventional organisations such as ours will be fortunate to find an unbroken succession of core enthusiasts.

If we crease to 'trade', we would hope to find a 'legacy' organisation to continue some of our activities, but these are hard to envisage. Two parish councils cover the PGP's area, but their interest depends on whether an individual councillor is interested. The Wye Valley AONB Office shares our aspirations, and has helped in many ways. However, the staff have no capacity to act as back-up for organisations that run into difficulty. The County Wildlife Trusts share our aspirations, too, and in Monmouthshire at least the Trust will maintain interest through its own extended role in grassland conservation.

In short, though we thought we might be a special case due to our small-scale enclosure pattern, scatter of now moderately-affluent properties and history as a common, we have found that we depend on individuals just as much as other small, local enterprises. Indeed, perhaps individualism is more the spirit of the common than collective co-operation. The community does come together round the new village shop—staffed largely by volunteers and constructed with local timbers—but community involvement in grassland conservation has its limits.

References

Peterken GF (ed) (2005) Flowers in the fields: a natural history of the Hudnalls grasslands. Parish Grassland Project, St Briavels

Peterken G, Tyler SJ (2006) Flowers in the fields: community conservation in the Lower Wye Valley. Br Wildl 17(5):313–323

Chapter 30
Upland Wood Pastures

Peter Quelch

30.1 Definition and Key Features

Ancient wood pastures in the uplands are areas of grazed pasture, heath or open hill with a scattering of individual trees, or groups of trees in small clumps. Many trees in wood pastures are open-grown with full crowns, are often very old, and some of those old trees may have been cut and utilised by man in the past. These upland habitats are analogous to the more familiar lowland wood pastures and parklands in the UK, which also contain large numbers of veteran trees.

Upland wood pastures grow on a range of soil types, and so may be found in several ecological niches corresponding with the site types occupied by the normal range of semi-natural woodlands as described in the National Vegetation Classification (Rodwell 1991). One of the most conspicuous types of ancient wood pasture in the upland landscapes of Scotland and North England is dominated by old alders of a 'low-pollard' tree-form (Fig. 30.1), with occasional veteran ash and often with old single-stemmed hazels. This is on a wet but nutrient-rich site type, which would otherwise carry slope-alder woodland (NVC community W7). Ash/elm/hazel wood pastures occur on somewhat drier nutrient rich sites (which would normally grow woodland of NVC community W9), while oak/birch wood pastures occupy better-drained and more acidic soils (NVC W17 and W11).

There are equivalent upland wood pastures dominated by birch or by Scots pine, yet with some exceptions, and rather illogically, these types of open woodlands have generally not been considered as wood pastures. It seems to the author, that any woodland eco-type might have a wood pasture equivalent, since the term 'wood pasture' describes a structural type or phase of woodland, rather than referring to distinct woodland eco-types. However, the long adaptation of the

P. Quelch (✉)
6, Glengilp, Ardrishaig, Argyll, Scotland
e-mail: peter.quelch@btinternet.com

I. D. Rotherham (ed.), *Cultural Severance and the Environment*, Environmental History 2, DOI: 10.1007/978-94-007-6159-9_30,

Fig. 30.1 Holly air tree in old alder. Royal Cottage w/p

trees to grazing and to open conditions confers a set of unique features to today's wood pastures. The veteran trees, the strange tree forms, the associated epiphytes, and the open canopy structure all combine to make upland wood pasture a distinct and recognisable habitat in the landscape.

Thorny tree and shrub species such as hawthorn, blackthorn, holly, juniper and crab apple are favoured in the well-lit conditions of upland wood pastures as they can withstand heavy browsing by livestock. Aspen often survives in micro-habitats which cannot be reached by herbivores, typically on craggy refugia within wood pastures.

The key feature that helps define ancient wood pastures in either upland or lowland situations is the presence of open-grown veteran trees, especially if they have been worked in the past. A long history of grazing under the tree canopy is implied for that type of tree-form to have developed. The open-grown trees provide shelter for livestock and both wood and foliage from the trees has been utilised for a variety of products. Ancient wood pastures therefore have the same structure and benefits as silvo-pastoral types of agro-forestry, whereby tree canopies co-exist with actively grazed pasture at ground level. Indeed, there is some evidence that in historic times seasonal crops may have been taken from under open tree canopies in a form of silvo-arable agro-forestry, and this aspect of woodland history is under active research.

Ancient wood pastures, both upland and lowland, display a range of tree forms with characteristics defined by the longevity of the trees and by their open and well-lit situation compared to normal woodland trees. Since many veteran trees in wood pastures are hollow, with hollow branches as well as a hollow main stem, a wide range of niches is created for nesting birds and animals. The hollow stem can build up humus soil that supports epiphytic plants, ferns, shrubs, and other trees.

A remarkable feature of ancient wood pasture trees is the growth of epiphytic trees, which have been dubbed 'air trees'. These epiphytic trees send down roots inside the hollow veteran host tree, eventually reaching the ground and so drawing

Fig. 30.2 Alder growing over rock, with rowan air tree. Glengyle (North) w/p

Fig. 30.3 A tall birch air tree in alder. Glengyle (North) w/p

up moisture and nutrients independently of the host tree. By far the commonest type of 'air tree' combination is a rowan growing inside a hollow alder, but there are many other species of air trees for example holly and yew. Typically, air trees are species that have berries and are sown in bird droppings, and in Sweden, the air trees are known as 'Bird trees' (Professor C.A. Haeggstrom, pers. comm.). However, well-developed birch air trees have also been recorded (Figs. 30.2, 30.3, 30.4 and 30.5).

Veteran trees in wood pastures frequently develop huge burry swellings on the stem, which may be a response to the light conditions. Massive basal 'skirts' of burr also develop in response to prolonged livestock browsing. The grazing animals continually bite back annual shoots at the tree base, which then grow burry wood tissue around themselves. Over very long periods, this can cause spectacular

Fig. 30.4 View looking west of Glengyle w/p. Glengyle (North) w/p

Fig. 30.5 Alder with rowan air tree. Glengyle House w/p

burry skirts to develop, which forever afterwards testify to a period of open-growth of this tree in a grazed wood pasture situation (Fig. 30.6).

Another feature typical of open wood pastures is that when trees are blown over, they very often continue to grow, as long as some roots are still attached. The prostrate trees develop new upright stems from branches and become ‘phoenix trees’, which may persist for long periods. Within woodlands, such blown trees would tend to die in the shade of their neighbours. Branches of open-grown trees

Fig. 30.6 High altitude alders, nearest with a burry skirt. Glengyle House w/p

may also layer in the soil once they are old enough to reach the ground. Other trees on very steep sites may fall or slip during soil erosion and landslides, yet re-grow and thereby persist in the landscape. There is a range of in-built survival mechanisms in wood pasture trees and shrubs, which of course contribute to them often becoming long-lived veteran trees.

30.2 Location

Wood pastures can be found in the uplands usually on the mid and upper slopes, but also sometimes in valley bottoms. They can occur in any location that semi-natural woodland may once have occupied. Upland wood pastures are often adjacent to crag and gully refugia, whereby remnants of genuinely natural-origin woodlands have survived through protection from grazing, exploitation and fire in deep gullies, gorges and ravines, and around crags. Often the woodlands in these gullies can extend to a high elevation, perhaps near to the previous natural tree line. In many upland landscapes, for example in the Border hills, the Trossachs, and in other parts of the Scottish Highlands, these are almost the only surviving locations for native woodlands containing oak, elm, ash and hazel in natural situations. Rare tree and shrub species like aspen and rock whitebeam (*Sorbus rupicola*) also tend to survive mainly in these refugia.

Wood pastures often survive on the less rocky terrain between woodland gully refugia, for example in the slopes above Loch Katrine and Loch Arklet, Stirlingshire. Comparison of the current location and extent of remnant woodlands in gullies with the extent shown on old maps often reveals very little change in those wooded gullies since the first accurate surveys in the mid-nineteenth century. The way in which wooded gullies have survived in the landscape over the last 150 years demonstrates that they may indeed be survivors from a more heavily

Fig. 30.7 Alder with rowan air tree. Portnellan wp

wooded past. Thus, one can infer from isolated wooded gullies in otherwise bare landscapes that these landscapes may indeed have been more wooded in past centuries. Research with old maps often backs up this hypothesis.

It does appear from observation that many upland wood pastures on steep valley-side terrain have a similar origin to the gully refugia woodlands that often accompany them. Wood pastures tend to be very open woodland with many modifications and features developed in response to a long history of grazing (Fig. 30.7). This contrasts with the restricted narrow ribbons of the adjacent gully woodlands which tend to have a typical woodland species composition, though often richer in less common herbs, shrubs and tree species because of the lack of herbivore impact. Wood pasture on the other hand is composed of relatively few grazing-resistant tree species, together with some browse-resistant thorny shrubs.

Modification of those trees in the past by pollarding or high-cut coppicing has made them extraordinarily stable and long-lived, protected against wind-throw by their low stature. Since their pollard-like shoots are above stock browsing height, they are resistant to damage by livestock and deer. The alder/ash/hazel type of wood pasture also occupies a site type, which is not prone to damage by fire, which may also account for its longevity in upland landscapes, which are often burned.

Old map research and documentary evidence tend to reinforce the hypothesis that the unenclosed upland wood pastures are often the modified remnants of previously more dense pasture woodlands of natural origin. Most woodland in the Highlands has a very long history of either summer cattle grazing or winter sheltering of livestock, depending on the woodland location and elevation. Few unenclosed woodlands have escaped the influence of herbivores. Many upland wood pastures also survive on sites that were once the hunting forests of the royalty and nobility in medieval times, for example Glenfinglas Forest north of Brig o' Turk, and Menteith Forest on the southern shore of Loch Katrine.

30.3 Problems of Inventory

It is unfortunate that wood pastures are often omitted from modern OS maps. Clearly, cartographers and surveyors need workable conventions for recording woodland. Since many wood pastures fall below the standard, minimum canopy density for mapping the consequence is that ancient wood pastures may be omitted from ancient woodland inventories. The Native Woodland Survey of Scotland (NWSS) currently being undertaken by Forestry Commission Scotland does not cover woodlands below 20 % canopy cover, or those with mapped polygon areas below 0.5 ha. Neither does that survey recognise ancient wood pasture as a distinctive woodland type.

Scottish Natural Heritage have compiled an inventory of wood pastures in Scotland which meet the Habitat Action Plan description including upland examples, but this inventory is partial and not based on comprehensive survey, nor is it published. Rackham reports that there are similar problems with the inventory of low density wood pastures in many parts of the world (Rackham 1998) This can lead to the resource being under-recorded and officially un-noticed, which is probably true of British upland wood pastures until relatively recently. The situation changed in 2008 when upland wood pastures were added to the lowland wood pasture habitat in the official biodiversity process. This change in attitude to the habitat is reflected in the excellent descriptive booklets on wood pasture recently published by the government agencies (Stiven and Holl 2004; FCS 2009).

30.4 Upland Wood Pasture: Is It a Sustainable Habitat?

There is often an implication in the published guidance on wood pastures that wood pastures have been managed sustainably for a very long period by careful husbandry. However, the history of many upland sites show that this is often far from the case. Unenclosed natural origin woodlands have often declined, appearing on later maps as open wood pastures after long periods of heavy grazing and occasional grassfires. Sometimes the reverse process can happen, whereby wood pastures have regenerated to dense woodland following the removal of livestock. In this less common scenario, the changes to the character of the wood pasture habitat can happen very quickly, as discussed below.

Wood pastures can be lost completely, even sites, which were shown as well wooded on eighteenth and nineteenth century maps. Brian Choille, in Glenloy, west of Fort William, is shown as semi-natural woodland of normal density on the first edition OS in 1860, but that previously wooded area is now an open wood pasture. The adjacent wood pasture to its east as recorded on that survey of 1860 has now only a tiny scatter of veteran trees, and is all but lost.

Because of that vulnerability to overgrazing, deer and fire, wood pastures in the uplands must be seen as an inherently unsustainable land use. To survive they must have periodic episodes of regeneration, including regeneration of the principal tree species such as oak, ash, hazel and alder, but because of generally high deer levels on the hills in Scotland, this is rarely achieved. Ideally, the less common tree and shrub species like aspen, bird cherry, juniper, gean, crab apple and hawthorn should also regenerate periodically, but again the prevailing levels of deer usually preclude regeneration of these species. Fortunately, bird-sown species such as rowan and holly, as well as birch that self-seeds prolifically, usually do regenerate once livestock grazing is managed carefully and deer levels are reduced.

Foresters normally see upland wood pastures as a degraded form of native woodlands. Certainly no-one has queried this approach to native pinewoods where for many decades, since the publication in 1959 of Steven and Carlisle's milestone survey of the surviving natural pinewoods, the emphasis has been to enclose and regenerate any open pine wood pastures. So who can blame foresters for seeing the broadleaved equivalents in a similar way? The normal response by both foresters and conservation managers has been to enclose the main part of a wood pasture, thus excluding all livestock and deer. Sometimes this does indeed result in a period of regeneration (albeit artificially within the confines of the fence). However, in many cases rank vegetation growth particularly of bracken does not allow that to happen.

It is encouraging that many land managers, of both nature conservation and forestry backgrounds, are now experimenting with low-level seasonal grazing by livestock, backed up by deer control or exclusion. There are indications that this kind of management may be very successful in maintaining upland native woodlands of a wide range of types, including previous wood pastures. Some land managers with agricultural knowledge are paying attention to cattle breeding, health and condition, and are raising surplus good quality livestock for sale. In this way, grazing aimed specifically at habitat enhancement can be profitable, rather than be a net cost as is much of conservation land management.

There are many other advantages in managing upland native woodlands through this type of conservation grazing regime. Open glades are maintained, grassland and heathland species encouraged alongside typical woodland species, and a wider variety of biodiversity maintained than through woodland regeneration alone. The resulting open woodlands with grassy glades and connecting paths also reflects the historic appearance and structure of these woodlands.

It seems to the author that most upland wood pastures should be protected and regenerated with as much vigour as any other ancient native woodland. This is particularly so because ancient woodland is irreplaceable. However, there have been good legal reasons why both small owner-occupier and tenant farmers, now and in the past, have not wanted to allow their wood pastures to turn into dense woodland. For example, it can often mean permanent loss of grazing land from their holding. This points the way to a need to promote a form of sustainable grazing management for wood pastures on farms and estates today that is not seen as a threat by the estate owners or the farmers. Again, there are encouraging signs that this form of land stewardship is being accepted and is reflected in the grant aid programmes for land managers (FCS 2009). Agri-environment payments can help owners and farm tenants provide the public benefits of biodiversity in historic and attractive landscapes.

30.5 Wood Pasture Dynamics

An ideal habitat structure for upland wood pastures would be one in which regeneration takes place more or less continuously, perhaps among thorny shrubs or in small patches of regeneration. This would result in a succession of groups and individual trees covering a range of ages from sapling to veteran through an unenclosed landscape. Yet this ideal and sustaining structure is unusual to find in the uplands today. The reality is that most ancient wood pastures are overgrazed and composed more or less of only veteran oak, or ash, alder, birch or pine. All of these species can show extraordinary longevity in open wood pasture situations, especially if pollarded in the past. The characteristic associate tree and shrub species of the better soils occupied by ancient oak or ash pollards are generally not regenerating.

It is somewhat ironic that when wood pastures are enclosed and do start to regenerate, their characteristic open structure can be quickly lost. Even the character of the veteran trees changes, and the micro-habitats that they provide can alter. Thus, alder, ash, and single-stemmed hazels despite having maintained a pollard-like tree-form over the last century or more, in the absence of grazing can quickly send up masses of basal shoots. In the case of hazel, this new growth is likely to cause the ancient single stem to be deprived of nutrients and it then decays and dies. A similar response has been observed with alder.

These are not just changes of structure but also of appearance, even the ambience of the habitat, which are admittedly very subjective qualities. The open character of wood pastures is quickly lost, bracken and brambles spread, and the feeling at least initially, can be one of loss rather than renewal. However, this is more of a problem of how people perceive change in the landscape. Once one realises that wood pastures are a unique product of certain traditional land-use practices, and then one can accept that if those practices are no longer realistic in today's agriculture and the habitat is bound to change.

There is a lack of public understanding of what upland wood pastures are, of how they arose, and what their future should be. Through observation and historic woodland evaluations, the author now regards many wood pastures as characterful but in the long-term essentially temporary woody habitats.

To take an extreme view it may seem that wood pastures are the accidental product of a history of two centuries of intense sheep grazing in previous natural woodlands. However, there is increasing evidence emerging from historic woodland surveys and from dendrochronological study that at least some of today's historic wood pastures have an earlier management history. If the living trees do indeed reach back to the eighteenth and even the seventeenth centuries, then they can provide evidence about trees management in pre-improvement agricultural landscapes.

There is little obvious evidence of a deliberate and planned management system for upland wood pastures during the nineteenth century. There may well have been more careful husbandry before that, but we know little of the detail. Large numbers of alders of low pollard form were apparently cut in the past, but in a way and for reasons we do not yet fully understand. It is amazing that these historic trees have survived into the twenty-first century, but clearly, there are limited numbers of them and they will not last forever. There is a strong incentive to study and yes even to celebrate them today, for their unique history cannot be replicated in the future.

So the bulk of the habitat that we today call upland wood pasture is a form of bio-cultural heritage, which cannot realistically survive long into the future with the same structure and features. This example of eco-cultural heritage is increasingly recognised as unique in time and place. The main threats to its future are the lack of recording and a consequent lack of status, combined with gradual natural loss and little replacement.

The main management technique, now as before, would be a low intensity and seasonal livestock-grazing regime with reduced deer numbers, which would allow some young trees to establish in perpetuity. Planting of replacement trees in strong individual shelters is both expensive and artificial. Practical difficulties may force us to recognise that it may be a delusion that upland wood pasture can be a sustainable habitat under modern land-use conditions? No one would suggest for example that the old system of transhumance to upland shieling huts in the summer would be a viable practice to revive in the twenty-first century.

30.6 Loss of Tradition

If it is agreed that today's wood pastures are indeed unique in time, then we cannot expect them to continue with the same structure and features indefinitely into the future. There has indeed been a complete break with traditional land use practices in wood pastures. Yet the effects of the loss of traditional management on the habitat have been long drawn out, and for several reasons only partial. One is that the traditional land use of subsistence farming changed during the nineteenth century to other types of upland grazing regime, either of large hill sheep flocks or of red deer herds maintained for sporting, and very often both simultaneously. Therefore, the effect on the previous traditional pasture woodlands was slow and almost imperceptible. This is in stark contrast to the effects of these land use changes on the human population and their traditional culture, which for most of the Highlands was also a Gaelic culture. The human effects were often sudden, dramatic and far reaching—'the end of tradition' indeed!

Yet few things are absolute, and there is some evidence that some of the old ways persisted in the work and life of the new type of shepherd who, after the clearances, husbanded large areas of the Highlands, and who replaced the numerous subsistence farmers of the small townships. Many of those old townships were abandoned, and now form the subject of a popular form of archaeology, recording the township ruins and investigating their past under a community archaeology-training programme called Scotland's Rural Past (RCAHMS 2011).

The idea that traditional ways of working trees by farmers persisted into the nineteenth century is under active investigation. By using scientific tree-ageing techniques, we hope to find out until what date pollarding and coppicing was last carried out in various sites. Was it contemporary with the last of the pre-improvement farms or did it continue after abandonment of the old townships?

One such investigation (Mills et al. 2009) has demonstrated a 10–12 year pollarding cycle during the eighteenth century in an old ash pollard which survives today. Dendrochronological analysis shows the last pollarding event for that tree to be in 1790, at about the same time that Murlagan farm on South-East Loch Katrine was abandoned. The farm tack for Murlagan appears in Montrose estate rentals up to 1779. Further studies combining documentary and map history, settlement archaeology, tree form and dendrochronology are continuing to unravel the story of man's place in the ancient woods and wood pastures of the Highlands.

30.7 Future of Upland Wood Pastures

The future of upland wood pastures should be the subject of wide debate. There needs to be informed discussion about the future management of upland wood pastures between landowners, farmers, and deer managers, with advice from landscape historians, conservationists, and other specialists. Upland wood pastures are not simply a habitat, nor simply a cultural landscape worth preserving, even though they are indeed both these things. Wood pastures together with adjacent semi-natural woodlands occupy large areas of private land, as well as extensive areas in public or conservation body ownership.

It is essential to find new ways to manage these lands, and every reason to learn from traditional management in doing so. We have an opportunity now to be more imaginative and to develop new sustainable management techniques to reflect twenty-first century needs. It would be good to see upland wood pastures (and many other native woodlands too) maintained sustainably by seasonal cattle and sheep regimes, producing saleable products, and covering their maintenance costs as far as is feasible. This is in contrast to the way many nature reserves are currently managed, with costs being covered by subsidy or fundraising, which can only be sustained in periods of favourable economics.

Settlement archaeology informed by documentary history shows that far more people once lived on these lands. Traditional ways of life have been lost, yet not well recorded. We have little knowledge of how these wood pastures were used in the past. Why for example were the alders cut in the way they have been? Was leaf foliage used to feed livestock as in Norway and Cumbria, and if so which species of tree leaf and for what benefits? What was the role of goat husbandry in forming these habitats?

These and similar questions provide interesting research topics, and the results of that research may inform future land use practices. With the predicted rise in cost and reduced availability of oil derived machine fuel and also artificial fertilisers, then alternative forms of land husbandry which utilise natural resources in a sustainable way will surely become more important in the near future. Research into historic land use in Scotland demonstrates that over the centuries traditional pre-improvement subsistence farming had developed a subtle and holistic control of nutrient flow throughout the landscape. Similar analyses have been carried out throughout Europe, which demonstrate a rural 'folk wisdom' of greater depth than we tend to recognise (Emanuelsson 2009). We can learn from those sustainable low-input low-output processes of the past.

Landscape history studies can help inform the development of potential new ways of life in the hills based on the best of historic traditional techniques. Pre-improvement farming traditions like transhumance to summer shielings have long since disappeared. Yet woodland archaeologists are observing ancient trees, which survive on the hills today and seem linked to those old ways of life. The trees of upland wood pastures have a story to tell, and a history, which has yet to be fully unravelled.

References

Emanuelsson U (2009) The rural landscapes of Europe: how man has shaped European nature. The Swedish Research Council Formas, Uppsala

FCS (2009) Management of ancient wood pasture. Forestry Commission Scotland, Edinburgh

Mills C, Quelch P, Stewart M (2009) The evidence of tree forms, tree-rings and documented history around Bealach nam Bo, Loch Katrine. Unpublished report to Forestry Commission Scotland, Aberfoyle

Rackham O (1998) Savanna in Europe. In: Kirby KJ, Watkins C (eds) The ecological history of European forests. CAB International, Wallingford, pp 1–24

RCAHMS (2011) Scotland's rural past—community archaeology in action. Royal Commission on the Ancient and Historic Monuments of Scotland, Edinburgh

Rodwell JS (1991) British plant communities, woodlands and scrub, vol 1. CUP, Cambridge

Steven HM, Carlisle A (1959) The native pinewoods of Scotland. University of Aberdeen, Aberdeen

Stiven R, Holl K (2004) Wood pasture. Scottish Natural Heritage, Edinburgh

Part V
Conclusions and Overview: The Implications of Severance for Future Landscapes

Chapter 31
Concluding Thoughts on the Implications of Cultural Severance on Landscapes, Ecology and People

Ian D. Rotherham

31.1 Inevitable Change

Progress in the modern economy generally equates to socio-economic development with rural depopulation and urban growth, to technological provision of needs, and a separation of people from nature. This process is an on-going part of human cultural evolution, but it has major environmental consequences. In terms of the severance of people and landscape there has been a rapid de-coupling of communities from their local environment. There are consequences of the cessation of traditional land uses. For individual sites, these can be especially problematic:

1. Eutrophication due to non-removal of biomass (for fuel, animal bedding, fodder).
2. Lack of micro-disturbance from grazing or other working animals, and from subsistence activities (including transhumance use etc).
3. Lack of propagule dispersal, particularly seeds through grazing stock moving from site to site.
4. Successional change due to abandonment (the rate varying with the landscape and its location, so upland zones in the UK for example are more resilient than lowland ones).
5. Decreased value for local communities and abandonment or replacement by other uses (building development etc).
6. Fragmentation and isolation.
7. Displacement of native species by exotics.

I. D. Rotherham (✉)
Faculty of Development and Society, Sheffield Hallam University Pond Street, Sheffield S1 1WB, UK
e-mail: i.d.rotherham@shu.ac.uk

I. D. Rotherham (ed.), *Cultural Severance and the Environment*,
Environmental History 2, DOI: 10.1007/978-94-007-6159-9_31,

Key issues and context for the discussion are established by Ted Green, Kenneth Olwig, and Ian Rotherham (this volume), in the first part of the book. Many of the chapters which follow (e.g. Samojlik et al. and Graham Bathe), then establish the clear importance of understanding history in order to provide a context of past change to inform the present (Rotherham 2005, 2007, 2009c, 2011). The historic changes help to establish a context for ecology, economy, and management.

31.2 Loss of Heritage and the Challenge for Conservation

At a regional level, there is also the serious risk of the loss of a unique cultural heritage that not only drove the ecology of the former landscape, but might also be a vital link to a future heritage-tourism economy. A fundamental problem is that the gross changes are driven by economic 'progress' but the conservation responses are generally not. On this scale, most conservation initiatives are cosmetic rather than economic. There are many conservation and environmental initiatives across Britain and throughout Europe, and these are to be celebrated and encouraged. The work of bodies such as the National Trust in Cumbria for example, is closing the gap between nature and the local economy. However, this is against a backdrop of cultural landscape abandonment that is probably unprecedented in human history. Even the local cultural knowledge is often lost, so we no longer know how these areas were managed even fifty or so years ago. This is happening across Europe, especially around the Mediterranean, and in the former Eastern Bloc. It is happening in Britain too. Recent ethnological research with older farmers in the Peak National Park indicated a rapid abandonment of family farms and a loss of local insight into landscape management. Those now seeking to conserve these areas and their unique wildlife heritage often have little idea of how these systems were created, or even how they looked in the recent past. The same family has managed some of the farms in the area continuously for over six generations and the impact of such a loss of unique local knowledge is incalculable (Fig. 31.1).

The challenge for now is to record local cultural knowledge and insight, to rebuild and celebrate local connectivity with nature, to value local traditions and uses, and to apply the knowledge in a meaningful way (e.g. Rotherham 2006). It is neither possible nor desirable (socially and economically) to stop the clock, but we need to find long-term economically sustainable solutions to these problems. The approaches must be more ambitious and more radical than anything that we have achieved so far. Webb (1986, 1998) considered the issues and conservation management options for the European heathland component of this discussion, and his prognosis was less than positive.

Fig. 31.1 Abandoned coppice on commonland SSSI

31.3 Some British Examples

Of all the regions of Europe Britain demonstrates both a cultural depth in its landscapes (e.g. Bowden, Herring, Greenaway, all this volume), and with very strong regional flavour to its character and the associated ecology. The homogenisation described already and proceeding with increasing intensity throughout the nineteenth and twentieth centuries has led to a critically impoverished fauna and flora across much of the country. Furthermore, the process of separation of people from nature and natural resources is manifested at every level from the psychological to the practical, and from the individual to the community. In most urban areas the bulk of people have little direct connection with their local or regional landscape beyond a visual interaction, and perhaps one based on leisure or recreation. Direct economic and resource-based links have been severed, and even in many rural areas, the majority of the local people have nothing to do with land or resource management. Research also shows that even the memory of such direct attachments may be lost in between fifty and a hundred years (Fig. 31.2).

Rotherham (this volume) describes just a small example of the more obvious wildlife species affected by the changes. However, the reality is that the impacts are so severe that the vast majority of those species that we treasure today for conservation have been adversely affected. Furthermore, I suggest that the losses and declines in habitat are so drastic that many species have been lost altogether and we do not even know the true scale of this destruction. My research on

Fig. 31.2 Overgrown common heath SSSI

wetlands for example has demonstrated that in mediaeval times and earlier, extensive wetlands existed in areas where today it is almost inconceivable (see Rotherham 2010, 2013a, b, for example).

31.4 An International Context

This book provides a broad, international context and debate for these issues. Major chapters by for example Petr Mares et al., Agnoletti, Bürgi and Stuber, Johann, Chakravarty-Kaul, Szabo, and others set a genuinely international scene and confirm the changes and the issues. These are global trends, each acted out a local level, and the problems are growing accordingly. Furthermore, despite some

Fig. 31.3 Overgrown commonland SSSI with Rosebay

modest progress in recognition of the issues (e.g. Agnoletti 2006, 2007), our consensus is that these are generally overlooked by key decision-makers (Fig. 31.3).

31.5 Re-connecting with Nature

The evidence is clear that cultural severance, as manifested by the end of traditional management, and its consequences, will cause massive changes in landscape and ecology. Some of these have already happened and to a large degree are irretrievable. We can restore some aspects of these lost ecosystems but should not kid ourselves that the new will be the same as the old. However, the real risk now is that the pace of landscape transformation is so accelerated and the degree of ecosystem redundancy is so absolute, that combined with other factors like eutrophication and climate change, we are on the cusp of an unprecedented loss of distinctive landscape-types and huge resulting extinction of species. Remarkably though the scientific and practitioner literatures are quiet on these subjects almost to the point of absolute silence. This in itself is academically interesting but from a viewpoint of future sustainability most perplexing and worrying. The implications of the omission of this core understanding of the root causes of environmental change, which now sweeps across both Europe and the rest of the world, are huge. This short paper only hints at the real potential losses that have happened, that are occurring now, and which will take place in the future. The contributions in the book however, do provide a broad sweep of evidence to exemplify and to justify these assertions.

Some of these changes and losses are driven by the long-term trends in human politics, economics and social evolution (e.g. Rotherham and Jones 2000), and cannot be repaired and cannot be halted. However, it is necessary to find some solutions and some positive responses to these challenges, and there are examples of possible ways forward. It is obvious though, that even the most bold and imaginative of these initiatives can neither halt the fundamental processes at work, nor can they replace the human-nature interactions of subsistence communities operating over long time-periods in predictable ways and at local landscape levels. The best we can hope for is a long-term sustainable ecosystem, which, whilst much changed from the original, retains a significant proportion of the regional distinctiveness and some of the richness of fauna and flora.

There are now modest projects to both record and retain knowledge of traditional management. There are other attempts to try to reconnect our new, largely urban populations and communities with local nature. These are excellent initiatives and should be supported. However, I argue here that the long-term issues are not educational or even ecological, but political and economic. These are the long-term drivers of change and need to be addressed and if necessary harnessed to secure a more sustainable future. Furthermore, it is naïve to pretend or to imagine that resolving this crisis can be anything other than hard, and with hope only of partial success at best.

This volume also reports current and future attempts to re-engage communities and commons. Gary Blank and George Hess for example, discuss a community-based case study from the USA, and George Peterken presents an account of a community-based group in the Wye Valley in England. Again, in England, Duncan Mackay considers the potential of new commons for old and Jay Bolthouse describes a fascinating case study focused on woodlands in from Japan.

31.6 Future Nature and a New Economy for the Countryside

A solution to all the issues raised is not possible, since time moves on. The past is history and cannot be re-created even if that was desirable. However, to imagine blithely that history is not important and that possible future sustainability does not relate to the problems that we have presented is foolhardy in the extreme. The challenge now is to minimise the detrimental changes and effects of severance and to try to develop a new functioning rural economy to deliver the ecology and landscapes that we value and need. It is also clear that there are massive risks and costs associated with failure to do this. With climate change and increasingly extreme weather events for example, the cost in human suffering and in economic impacts, of the separation of people from nature are evermore clear (Fig. 31.4).

The end of tradition and the impacts then of cultural severance are as big a threat as climate change to biodiversity and ecology and the 2010 conference

Fig. 31.4 Re-grown pollard

looked at the threats to biodiversity from cultural change and the abandonment of traditional management. We have heard much about climate change and the threats that this may pose in the future but '***The End of Tradition***' is bigger and more current. The threats from global cultural change and abandonment of traditional landscape management increased in the last half of the twentieth century and ten years into the twenty-first century show no signs of slowing down. Their impacts on global biodiversity and on people disconnected from their traditional landscapes pose real and serious economic and social problems, which need to be addressed now. The conference addressed the fundamental issues of whether we can conserve the biodiversity of wonderful and iconic landscapes and reconnect people to their natural environment. In addition, if we can, how can we do so and make them relevant for the twenty-first century. We covered the lessons of archaeology, history and ecology and looked at the challenges for modern-day management. Examples at the conference and presented in this book were drawn from rural and urban commons, wooded landscapes, heaths, moors, coasts and wetlands. In particular, we wished to raise critically important issues of the loss and abandonment of tradition in terms of future sustainability, landscape quality and biodiversity. For long-term success, these need to be embedded at a local level both socially and economically.

31.7 We Need to Take Action

In order to address these issues we must recognise the '***Eco-Cultural***' nature of landscapes and their biodiversity and begin to re-establish links with nature. Immediate action is now being taken in limited conservation areas to mimic traditional management methods and their impacts in conservation sites. This is in nature reserves and similar areas, requires massive injections of money and resources, and relies on skills and knowledge which themselves a rapidly being lost. Yet this action is needed in order to prevent, or at least limit, extinctions; but much wider application is required at a landscape level. In themselves however, these interventions and initiatives do not provide long-term solutions. History tells us that long-term changes and management of the landscape are driven by economics and politics, and it is these, which need to be harnessed if there is to be any real, long-term progress. Presently, as wildfires rage across severed landscapes in many parts of the globe, and in Indonesia and Brazil, tropical forests continue to be cleared, there is little sign that this will be the case.

Importantly, if we are to address these challenges, then we must establish social and economic links between nature, landscape and ecology—though this will require a **political shift in thinking and planning** and a **paradigm shift in conservation and environmentalism.** Future visions for a sustainable environment must recognise the lessons of human history and the impacts of culture and of history. In the context of all these it is essential to maximise ***ECOSYSTEM*** function benefits to humanity not diminish them and as climate change happens, natural and human-induced, help nature to respond and to minimise the damage. So far, such actions are too limited to have the necessary impacts in response to environmental change and particularly climate change—whether human-induced or natural. Chapters by Butler and by Newton consider future visions and the need to understand the ecological dynamics in achieving these. Policies need to be formulated but based on sound knowledge of history, ecology, society and management.

31.8 Conclusions

Throughout the book, the importance of history to both understand and inform our views on the environment is established and demonstrated. In particular, Shaw and Whyte, Whyte, Waddell, Hooke, Ruano, and Estébanez et al., and Szabo, all show how history, community and environment are inter-woven through time. This is the unifying theme that runs throughout the volume and that supports the assertion that the issues raised are truly global in nature, and are critical to both ecology and humanity. That said, the problems are not easy to resolve, but a first step is to raise awareness that they exist.

The contributions to this book have raised fundamental and important issues about how we understand and then might management our environment. This is a significant debate and even more so given the paucity of current awareness and the severity of the threats. Some of the issues are to an extent insolvable and indeed those surrounding say indigenous peoples are too complex to be debated here. There are major questions of human rights, of the integrity and value of unique social groups and their cultures, and the accessibility for all people, of health care, education, and living standards that most of us take for granted. For all subsistence-based rural communities there are the challenges of reasonable aspirations to modern living and 'quality of life'. Yet, all these changes ultimately, it seems, spell the end for these landscapes, the community's and cultures, and for their ecologies. This applies from a remote, aboriginal tribe in the Amazon, to a traditional pastoralist in the Mediterranean. There is no easy answer, but we must at least be aware of, and prepared for the consequences of what will inevitably happen.

References

Agnoletti M (ed) (2006) The conservation of cultural landscapes. CAB International, Wallingford

Agnoletti M (ed) (2007) Guidelines for the implementation of social and cultural values in sustainable forest management. A scientific contribution to the implementation of MCPFE—Vienna resolution 3. IUFRO Occasional Paper No. 19 ISSN 1024-414X

Rotherham ID, Jones M (2000) The impact of economic, social and political factors on the ecology of small english woodlands: a case study of the ancient woods in south yorkshire, england. In: Agnoletti M, Anderson S (eds) Forest history: international studies in socioeconomic and forest ecosystem change CAB International, Oxford, Wallingford, pp 397–410

Rotherham ID (2005) Fuel and landscape—exploitation, environment, crisis and continuum. Landsc Archaeol Ecol 5:65–81

Rotherham ID (2006) Historic landscape restoration: case studies of site recovery in post-industrial south yorkshire, England. In: Agnoletti M (ed) The conservation of cultural landscapes, CAB International, Oxfordshire, Wallingford, pp 211–224

Rotherham ID (2007) The implications of perceptions and cultural knowledge loss for the management of wooded landscapes: a UK case-study. For Ecol Manage 249:100–115

Rotherham ID (2009c) Cultural severance in landscapes and the causes and consequences for lowland heaths. In: Rotherham ID, Bradley J (eds) Lowland heaths: ecology, history, restoration and management. Wildtrack Publishing, Sheffield, pp 130–143

Rotherham ID (2010) Cultural severance and the end of tradition. Landsc Archaeol Ecol 8:178–199

Rotherham ID (2011) The implications of landscape history and cultural severance in environmental restoration in England. In: Egan D, Hjerpe E, Abrams J (eds) Integrating nature and culture: the human dimensions of ecological restoration. Island Press, Washington, pp 277–287

Rotherham ID (ed) (2013a) Trees, man, and grazing animals—a European perspective on trees and grazed treescapes. Earthscan, London

Rotherham ID (2013b) Re-interpreting wooded landscapes, shadow woods and the impacts of grazing. In: Rotherham ID (ed) Trees, man, and grazing animals—a European perspective on trees and grazed treescapes. Earthscan, London

Webb NR (1986) Heathlands. Collins, London

Webb NR (1998) The traditional management of European heathlands. J Appl Ecol 35:987–990

Index

A
Abandoned, 290, 296, 302, 304–308
Abandonment, 62, 64, 77, 78, 80–83
Access to high quality of grasses, 98
Acid grassland, 258
Agri-environment payments, 426
Air trees, 420
Aleros, 151
Altitude, 97, 98
Amphibians, 357
Ancient wood pastures, 419
Anti-commons, 388, 390–392, 394, 397
Apple days, 408
Archaeology, 218, 375
Areas of outstanding natural beauty, 259
Arnedo, 149
Arthur young, 403
Ashampstead, 233
A Suggestion for the Better Preservation of Open Spaces, 406
Augustus Smith, 411

B
Banjar kadim (long fallow), 95
Bats, 357
Bere, 189
Berkhamsted common, 411
Bernardino Corvera, 156
Big society, 401
Bio-cultural heritage, 428
Biodiversity, 59, 61, 62, 65, 68, 81, 311–314, 317–319, 373
Biodiversity conservation, 353
Birds, 367
Bluebells, 239, 243
Bodmin Moor, 249–254, 257–259
Bow pointers, 152, 154, 156
Bridges, 329
Burnham Beeches, 411
Butterflies and moths, 357

C
Capitularies, 180
Carlos I, 151
Carolingian dynasty, 180
Carta de Foresta, 182
Catholic Monarchs, 150
Cattle, 356
Chamba, 88
Character areas
Charters, 231
Chippenham, 189
Chute, 189
Clarendon forest, 189
Classification, 129, 131
Climate change, 3, 5, 6, 8
Coaxial field systems, 251
Colonial codification, 88
Commissioner, 157
Commissioner Garmendia, 158
Common, 177, 217–222, 224, 264, 266, 270, 271, 273, 413, 416
  parliamentary, 276–279, 281, 282, 284, 286
Commoners, 177
Common fields, 323–325
Common land, 14, 91, 242
  wooded, 279, 284
Common property resources, 94

I. D. Rotherham (ed.), *Cultural Severance and the Environment*, Environmental History 2, DOI: 10.1007/978-94-007-6159-9,

Common rights, 108, 110, 116, 117, 240
Common rights to take fruit, 408
Commons, forests and footpaths, 409
Commons preservation society, 405
Common uses, 205
Common waste, 111, 113, 114
Community, 413, 416–418
Community farms, 409
Community orchards, 408
Complementarity of scales, 97
Conflict, 110, 116, 117
Conservation management, 354
Convertible husbandry, 257
Coppice, 209, 396, 397
Cornwall, 249–251
Corporation and City of London, 405
Countryside and Rights of Way (CROW) Act, 259
Coursing run, 235
Cricket, 243
Cricket team, 245
Cultural commons, 31
Cultural landscape, 4, 5, 13, 17, 18, 25, 63, 68, 70–72, 82, 87, 134, 136, 140, 141, 144
Cultural severance, 4–6, 12, 13, 18, 22, 26, 28, 437, 438
Cultural tradition, 401
Custom, 31
Customary, 3, 4, 8
Customary grazing runs, 98
Custumal, 182
Czechia, 291, 292, 294, 295, 297–302, 307
Czech republic, 205

D
Damselflies, 357
Dartmoor, 251, 253–255
Deer, 355
Deer grazing, 246
Deer levels, 426
Deer park, 233, 234
Deer park topography, 235
Defra's, 408
Dendrochronology, 429
Disafforestation, 181
Diseases of obesity, stress and inactivity, 409
Displaced persons, 246
Doctor Hernando Suárez de Toledo, 152
Dog's mercury, 239, 243
Domesday book, 178
Domesday book evidence, 107, 108
Domesday record, 232
Donkeys, 356
Dragonflies, 357
Drainage, 325, 329, 331
Drifts, 256
Drirkar, 99
Duke of Marlborough, 403

E
Eco-cultural, 8, 12, 13, 18, 21, 26–28
Eco-cultural landscape
Ecological evidence, 238
Ecosystem function
Ecosystems approach, 311
Ecosystem services, 373
Ecotourism, 18
Elinor Ostrom, 407
Enclosing forests for conservation, 101
Enclosure, 31, 36–38, 112–116, 324
  nature, 278, 283, 285, 286
End of tradition, 11
England, 217, 218, 225
Enlightenment, 211
Environmental risk, 101
Epping forest, 404
European bison, 195, 196, 200, 201
European landscape convention, 259
Exmoor, 253
Exotic trees, 243
Exploitation systems, 161
Extensive de-forestation, 103

F
Fences, 329, 330
Fens, 326
Firewood, 246
Fish, 357
Flint implements, 233
Flora, 243
Fodder, 371
Foraging, 408
Forest, 111, 112, 114
Forestation, 64
Forest charter (Carta de Foresta) of 1217, 402
Forest dynamics, 161
Forest eyre, 185
Forest law, 179, 235
Forest of dean, 189
Forest ownership, 134, 137, 142
Forestry management plans, 206
Forestry uses, 161
Forest schools, 408
For ever, for everyone, 410
For the everlasting delight of the people, 409

Freeboard, 236
Free riding, 102
Fungi, 50, 362
Future nature, 438

G
Gaelic culture, 428
Geese, 402
Geology, 229
GIAHS, 61, 68
Gillingham, 189
Gleaning, 324
Globalism, 31
Goonhilly, 249
Governance, 222
Governor Thouvenot, 156
Grand Duchy of Lithuania, 191, 197, 201
Grassland management, 415
Grazing, 356
Great Charter (Magna Carta), 402
Greens, 263–265, 269–271
Ground flora, 238, 239
Guildford, 189
Guipúzcoa provincial council, 156
Gully refugia, 423

H
Habitat condition, 353
Hardinian perception, 87
Havos, 256
Heat-island effect, 407
Hedges, 329, 330
Hendre, 256
Henry Ford European Conservation Award for Heritage, 408
Historical ecology
Historical uses, 161
Historic landscape characterisation, 249
History, 60, 61, 64, 65, 68, 70, 293, 307
History of the commons, 233
Hodonínská Dúbrava, 205
Horca, 150, 154, 158
Human utilisation, 17, 18
Hunting forests, 424

I
Importance of complementarity of private–public governance of resources, 103
Inclosure of Otmoo, 403
Infrastructure, 329
Institutional uncertainty, 101
Intakes, 258
Integrity, 69, 75–78
In the lowlands, the period post-1947 brought the curtain down, 103
In the uplands, post-1947, 104
Inventory, 425

J
Japan, 388, 389, 393, 394, 397
Javier Ignacio de Echeverria, 150
Jerónimo Tavern, 156
José Odriozola, 148
Juntas, 151
Juntas Generales, 151, 152, 154
Justice in Eyre, 182

L
Land company, 407
Landscape, 32, 265, 270, 273, 304–306, 374
Landscape archaeology, 249
Land use, 240, 290, 297, 298, 301, 306, 307
Land use change
Land-use patterns of rotating fallows, 89
Laund, 238
Laws of Wessex, 178
Ley, 257
Ley husbandry, 258
Lichens, 357
Livestock, 354
Lizard, 249, 250
Long fallows, 93
Long-term models, 161
Lord Abingdon, 403
Lord Churchill's Yeomanry Cavalry, 404
Lords of the Manor, 177
Loss of heritage, 434
Loss of tradition and custom, 3
Lowland heathland, 354

M
Magna Carta, 182
Management, 218, 219, 225, 226
Manor, 263, 268–272
Manor courts, 331
Manorial documents, 240
Manorial regulations, 323
Manorial system, 178
Manorial waste, 232
Maps, 207
Marquis of Rocaverde, 148
Marquis of San Millán, 157

Marshes, 326, 328
Meadows, 326, 327
Merton (Statute of), 181
Methodist camp meetings, 243
Mining and industry, 114
Mixed cultivations, 71, 73
Monarchical hunts, 197
Moorland, 263, 271, 273
Multiple woodland uses, 211

N
National Trust, 404
National Trust Act, 1909, 410
Natural England's, 408
Natural pinewoods, 426
Nature conservation
Navy, 158
Navy commissioners, 157
Neighbourhood groups, 411
New allotments, 409
New commoners, 401
New commons, 378, 379, 381, 383–385
New occasions teach new duties, 409
Norman conquest, 179
Norman kings, 178

O
Occupation, 229, 233
Octavia hill, 406
Old growth, 373
Open-grown, 47
Oral history, 124, 126, 128, 130, 131
Ordinatio Forestae, 182
Otmoor, 401
Otmoor for ever, 404
Ownership, 136–138, 140, 144
Oxford, 404
Oxfordshire Militia, 404

P
Palaeoecology, 313
Pamber, 189
Pannage, 208
Parish councils, 246
Parker's lodge, 236
Parks, 112, 114, 119
Partnership, 377, 380, 383–385
Pasture woodland, 231
Pastures, 328
Paths, 329
Pendolero, 151
Pendón, 150, 154, 158
People and wildlife, 407
Phoenix trees, 422
Picturesque tradition, 31
Pigs, 356
Pillow mound, 238
Place name evidence, 232
Playground, 243
Poaching, 237
Pollarding, 186, 424
Pollards, 371
Pollen analysis, 312, 313
Pond, 236
Ponies, 356
Pottery industry, 233
Pounds, 255, 256
Practices, 3, 4, 8
Pre-conquest charter evidence, 107, 108
Principle of complementarity, 96
Private Acts of Enclosure, 402
Public common without stint, 402
Public good, 330

R
Real Compañía Caracas, 155
Real Sociedad Bascongada de Amigos del País, 150, 157
Real tragedy for commons, 407
Reciprocity was thus weakened, 101
Regional Government of Guipúzcoa, 159
Re-naturalization, 80, 81
Report of the Committee on London Commons, 405
Reptiles, 357
Reserves
  public open, 283, 285
Resilience, 313
Ribblesdale, 313–315, 318, 319
Rights of common, 234, 240
Riot Act, 404
Roads, 329, 330
Robert Hunter, 405
Roughtor, 252
Royal Armada, 152, 155
Royal Army, 155
Royal forest, 177, 191, 192, 194–196, 201
Royal Navy, 156, 158
RSPB, 404
Rules in use, 88
Rural economy, 438
Rural landscape, 59–62, 64, 66, 67, 70, 75, 78, 80, 82

S
Salisbury Cathedral, 188
Saproxylic beetles, 362
Satoyama, 388, 392–394, 396, 397
Savernake forest, 177
Saxon estates, 231
Scale efficiency, 97
Scheduled monuments, 259
Scotland's rural past, 428
Scrub clearance, 362
Seasonal grazing, 426
Seasonal pasture, 108, 110, 113
Settlements, 242
Sheep, 356
Sheffield, 263–265, 268–273
Shredding, 186
Shrouding, 186
Sir Henry Peek, 411
Sites of special scientific interest, 259
Slovenia, 289–291, 294–298, 302–304, 308
Socio-ecological system, 368
Soils, 229, 233, 239
Spaces, 283, 285
Spain, 161
Stag beetles, epiphytes and woodpeckers, 408
Statute for the Preservation of Woods, 186
Statute of Merton of 1235, 232, 402
Stewardship, 379, 380, 384
St Giles' Fair, 404
Suárez de Toledo, 153
Survey, 218
Sustainability, 246, 311, 313
Sustainable forest management, 140
Swiss cottage fields, 409
Sylvo–Pastoral systems, 372

T
Tavern, 157
Terraces, 73, 74, 76
The cultural commons, 32
The historic capture of the waste, 100
The Pang valley, 231
The tradition innovated, 103
Thomas Willingale, 405
Town, 218–223, 225, 226
Traditional knowledge, 141, 144
Traditional management, 205, 311
Traditional management approaches, 354
Traditional orchard planting ratios, 408
Tragedy of the commons, 15, 16, 99, 185
Transhumance, 91, 255–257
Tre, 256
Tree generation, 52
Trees, 238
Trespassing, 255

U
UNESCO World Heritage Site, 407
Up-keep of roads, 246
Upland-lowland complementarity, 102
Uplands, 312–314, 319
Urbaria, 206

V
Vascular plants, 357
Veteran trees, 420
Village common lands, 94
Villarreal de Bérriz, 148
Volunteers, 247
Vulnerability, 69, 77–81

W
WakeNature, 377, 379–381, 383–385
War Council, 154
Wastelands, 100
Wastes, 328
Westminster (Statute of), 181
West Penwith, 249, 250, 254, 255
Wetland, 107, 112
Woodbank, 209
Woodland, 107, 108, 114, 115, 118, 328, 367, 388, 392–395, 397
Woodland history, 231
Woodland Management Central Europe, 205
Woodland meadows, 208
Wood-pasture, 47, 108, 111, 112, 114, 115, 117–119, 208, 234, 372
World War 2, 246
Written sources, 206

Y
Yorkshire, 313, 314

Z
Zomia, 88

MIX
Papier aus verantwortungsvollen Quellen
Paper from responsible sources
FSC® C105338

If you have any concerns about our products,
you can contact us on
ProductSafety@springernature.com

In case Publisher is established outside the EU,
the EU authorized representative is:
Springer Nature Customer Service Center GmbH
Europaplatz 3, 69115 Heidelberg, Germany

Printed by Libri Plureos GmbH
in Hamburg, Germany